Flame-Retardant Polymeric Materials

Volume 2

CONTRIBUTORS

Robert H. Barker
College of Industrial Management and
Textile Science
Department of Textiles
Clemson University
Clemson, South Carolina

Avraham Basch
Israel Fiber Institute
Ministry of Commerce and Industry
Jerusalem, Israel

J. Economy
IBM Research Laboratory
San Jose, California

Mendel Friedman
Western Regional Research Laboratory
Science Education Administration
U.S. Department of Agriculture
Berkeley, California

Christine W. Jarvis
College of Industrial Management and
Textile Science
Department of Textiles
Clemson University
Clemson, South Carolina

Y. P. Khanna
Polytechnic Institute of New York
Brooklyn, New York

Jack Kracklauer
Market Development
Arapahoe Chemicals
Boulder, Colorado

Menachem Lewin
Israel Fiber Institute
Ministry of Commerce and Industry
Jerusalem, Israel

J. Ronald Martin
Textile Research Institute
Princeton, New Jersey

Bernard Miller
Textile Research Institute
Princeton, New Jersey

Eli M. Pearce
Polytechnic Institute of New York
Brooklyn, New York

R. V. Petrella
Olefin Plastics Department
The Dow Chemical Company
Midland, Michigan

Edward D. Weil
Stauffer Chemical Company
Dobbs Ferry, New York

Flame-Retardant Polymeric Materials

Volume 2

Edited by

Menachem Lewin
Israel Fiber Institute

S. M. Atlas
Bronx Community College
of the City University of New York

and

Eli M. Pearce
Departments of Chemistry
and Chemical Engineering
Polytechnic Institute of New York

PLENUM PRESS · NEW YORK AND LONDON

Library of Congress Cataloging in Publication Data

Main entry under title:

Flame-retardant polymeric materials.

Includes bibliographical references and indexes.
1. Fire resistant polymers. I. Lewin, Menachem. II. Atlas, Sheldon M. III. Pearce, Eli M.

TH1074.F58   668   75-26781
ISBN-13: 978-1-4684-6975-2   e-ISBN-13: 978-1-4684-6973-8
DOI: 10.1007/978-1-4684-6973-8

A Division of Plenum Publishing Corporation
227 West 17th Street, New York, N.Y. 10011

Preface

Flammability has been recognized as an increasingly important social and scientific problem. Fire statistics in the United States (Report of the National Commission on Fire Prevention and Control, "America Burning," 1973) emphasized the vast devastation to life and property—12,000 lives lost annually due to fire, and these deaths are usually caused by inhaling smoke or toxic gases; 300,000 fire injuries; 11.4 billion dollars in fire cost at which 2.7 billion dollars is related to property loss; a billion dollars to burn injury treatment; and 3.3 billion dollars in productivity loss. It is obvious that much human and economic misery can be attributed to fire situations. In relation to this, polymer flammability has been recognized as an increasingly important social and scientific problem. The development of flame-retardant polymeric materials is a current example where the initiative for major scientific and technological developments is motivated by sociological pressure and legislation. This is part of the important trend toward a safer environment and sets a pattern for future example. Flame retardancy deals with our basic everyday life situations—housing, work areas, transportation, clothing and so forth—the "macroenvironment" capsule within which "homosapiens" live. As a result, flame-retardant polymers are now emerging as a specific class of materials leading to new and diversified scientific and technological ventures.

From the humble beginnings of flame-retardance treatments of existing polymers, the field is now developing into the design and engineering of new heat-resistant molecules, polymers, and commodities which are inherently flame-retardant. It is an interdisciplinary development and involves several scientific, engineering, legal, medical, and sociological consequences.

The objective of this series is varied. In many cases it is to give an up-to-date summary of the state of the art in flame-retarding polymeric materials so as to be an aid to those involved in solving these problems. Interpretation as to mechanism and conjecture about future approaches has been encouraged on the part of the authors. Since polymer degradation is the precursor to flammability, suitable importance has also been placed on this area. There will be occasional chapters which also deal with a specific test method, some of which may have historical importance. In the first volume, we have included a general article on the development of the oxygen index test because of its broad interest and significance. Future issues will also be concerned with similar subject areas.

We are hopeful that through these means, meaningful solutions to a number of the flammability problems and their subsequent positive social ramifications will be accomplished.

M. LEWIN
S. M. ATLAS
E. M. PEARCE

Contents

CHAPTER 5

Flammability of Cotton–Polyester Blend Fabrics

CHRISTINE W. JARVIS AND ROBERT H. BARKER

CHAPTER 6

Factors Affecting the Combustion of Polystyrene and Styrene

R. V. PETRELLA

CHAPTER 7

Phenolic Fibers

J. ECONOMY

CHAPTER 8

Flame-Resistant Wool and Wool Blends

MENDEL FRIEDMAN

CHAPTER 9

Smoke and Tenability: A Perspective on the Materials Approach to the Fire Problem

JACK KRACKLAUER

Structure, Pyrolysis, and Flammability of Cellulose

Menachem Lewin and Avraham Basch

1. Introduction

Pyrolytic studies of natural polymers are attracting attention from several viewpoints. Important advances have recently been made in the development of new and improved flame-retardant chemicals and processes as well as of new heat-resistant polymers. These depend, to a large extent, on the understanding of the processes occurring during thermal decomposition of various substrates and of the parameters influencing them. The fact that cellulose is the most abundant natural organic material and is, therefore, the main fuel in fires serves to focus attention on it. The possibilities of producing oils of high calorific value and charcoal from cellulosic substrates by suitably directed pyrolytic decomposition lends further interest to detailed pyrolytic studies.

2. The Fine Structure of Cellulose

The fine structure of cellulose has been extensively studied for decades in an effort to elucidate its molecular and supramolecular structure. A wide variety of methods and techniques was applied in these studies, which in-

Menachem Lewin and Avraham Basch • Israel Fiber Institute, Ministry of Commerce and Industry, Jerusalem, Israel.

cluded X-ray diffraction,[1] hydrolysis,[2] infrared absorption,[3] dichroism[4] and near-infrared spectroscopy,[5] deuteration,[6] chemical substitution,[7] dielectric properties,[8] electron microscopy,[9] and others. While most investigators have tended to accept the basic concepts of ordered (crystalline) and less-ordered regions (LOR's), others have disagreed. One school of thought[10] has claimed that all cellulosics are completely noncrystalline and merely possess different degrees of order in their supramolecular structure. The opposite viewpoint that at least cotton is 100% crystalline has also been advanced.[7,11] Recent publications have reviewed the various theories of cellulose structure and conformation[12–17] and have indicated that there are problems still unresolved.[18]

3. The Major Products of Cellulose Pyrolysis: Levoglucosan and Char

3.1. Levoglucosan Formation

Levoglucosan (1,6-anhydro-β-D-glucopyranose) was first shown by Pictet and Sarasin[19] to be a major product of cellulose pryolysis. Madorsky et al.[20] investigated the degree and rate of "tar" formation during cellulose pyrolysis and showed by infrared and mass spectroscopy that levoglucosan was the major tar component. Their data indicated different tar yields from different celluloses and showed that cellulose oxidized at the C-6 hydroxyl or substituted (cellulose triacetate) yielded little or no levoglucosan upon pyrolysis. Holmes et al.[21,22] and Schwenker and Pacsu[23,24] investigated the tar and levoglucosan yields of untreated, oxidized, and flame-retardant treated cottons and showed a direct proportionality between these yields and cotton flammability.

Levoglucosan formation during the pyrolysis of cellulose was systematically investigated by Golova et al., who demonstrated widely differing yields from various cellulosic substrates[25] (see Table 1) and ascribed them

TABLE 1
Levoglucosan Yields from Different Celluloses[a]

Cellulose description	D.P.	Levoglucosan yield (%)
Cotton fibers	1000	60–63
Cotton fibers, mercerized	1200	36–37
Cotton fibers, dissolved in cupraammonium and precipitated	1000	14–15
Viscose fibers, unoriented	380	4.0–4.5
Viscose fibers, oriented	400	4.8–5.0

[a] Taken from Ref. 25.

to "differences in packing density of the macromolecules." These studies also indicated that although the initial degree of polymerization (D.P.) of the cellulose does not influence the yield of levoglucosan,[26] the major portion of the levoglucosan is released only after the D.P. drops to 200.[27] Levoglucosan is not obtained from the pyrolysis of cellobiose.[26]

3.2. Other Products

A derivative of levoglucosan, levoglucosenone (1,6-anhydro-3,4-dideoxy-β-D-glycerohex-3-enopyranose-2-ulose), has recently been identified as a product of cellulose pyrolysis.[28] It has not been definitely established whether levoglucosenone is a primary cellulose decomposition product[29] or an oxidation product of levoglucosan.[30] 1,6-Anhydroglucofuranose[31] and trace amounts of other carbohydrates[32] have also been identified in the pyrolysis tar.

A large number of other materials have been identified as cellulose pyrolysis products. At low temperatures (160–250°C), water, CO, and CO_2 are evolved.[33] Schwenker and Beck[34] showed the presence of at least 34 compounds by pyrolysis gas chromatography and were able to identify many of them (Table 2); Lipska and McCasland[35] detected at least 59 compounds with molecular weights lower than 150. Glassner and Pierce[36] showed that

TABLE 2
Pyrolytic Degradation Products of Cellulose
Indicated by Gas–Liquid Chromatography[a]

Compound
Volatile gases: CO, CO_2
Formaldehyde
Acetaldehyde
Acrolein
Propionaldehyde
n-Butyraldehyde
Glyoxal
Furfural
5-Hydroxymethyl furfural
Acetone
Methyl ethyl ketone
Methanol
Formic acid
Acetic acid
Lactic acid
Water
Levoglucosan

[a] Taken from Ref. 35.

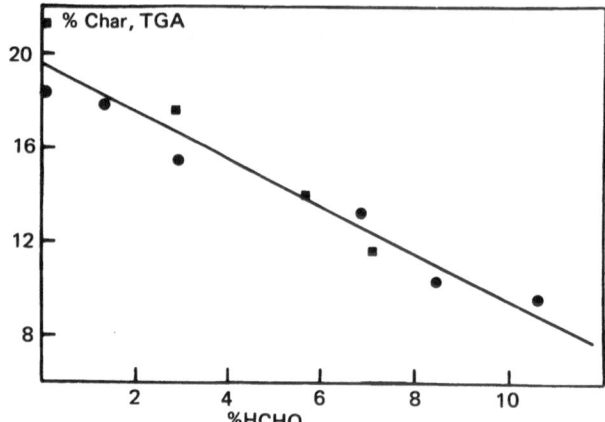

FIGURE 1. Char residue by dynamic TGA vs. HCHO content for cross-linked rayon (circles) and Vincel 28 (squares).[41]

the chromatograms obtained from pyrolyzed cellulose and pyrolyzed levoglucosan were identical, suggesting that the observed peaks from cellulose were actually secondary decomposition products of levoglucosan.

3.3. Char Formation

The pyrolysis of pure cellulose in vacuum yields from 6 to 23% char, depending on the nature of the cellulose.[37] Apart from cellulosic fine structural considerations (see Sec. 4.3) the char yields can be influenced by a variety of factors. Prolonged heating at low temperatures increases the final char yield,[38] while rapid temperature increase and removal of the volatile degradation products are not favorable to char formation[39] The effects of pyrolysis temperature and atmosphere on char properties have been studied spectrally.[40] It has been shown[41] that cross-linking with formaldehyde reduces char formation (Fig. 1).

4. Vacuum Pyrolysis

4.1. Mechanism, Rates, and Products

Theories concerning the mechanisms of cellulose pyrolysis have been concerned mainly with explaining the origins of the two main pyrolysis products, levoglucosan and char. Madorsky et al.[20] advanced the idea that the char arises from a dehydrative process and that this process competes with an alternate decomposition path which results in levoglucosan formation. This approach was reinforced by results obtained by Kilzer and Broido.[42] Using thermogravimetric analysis (TGA) and differential thermal analysis (DTA) techniques, they found that cellulose, when heated at temperatures below 250° C for extended periods of time, lost up to 10% of its weight as water and upon subsequent high-temperature decomposition yielded an increased proportion of char. They postulated that two decomposition processes are competitive and temperature controlled. During the first reaction, which occurs between 200 and 280°C, water is lost and "anhydrocellulose" is formed. This anhydrocellulose decomposes further at elevated temperatures to various volatile products, carbon dioxide, and char. The second reaction, involving thermal scission of glucoside linkages and levoglucosan formation, predominates at temperatures above 280° C. The postulated pyrolysis paths are depicted schematically in Fig. 2.

McKay,[43] who investigated cellulose pyrolysis by infrared, gas chromatographic, and DTA techniques, arrived at a similar conclusion. Arseneau,[44] on the basis of DTA and TGA, concurred and measured an energy of activation (E_{act}) of 36 kcal/mol for the first reaction and 45 kcal/mol for the second. Derminot and Rabourdin-Belin[45] investigated the products of pyrolysis of various celluloses by pyrolysis gas chromatography and found no significant differences among them. McCarter[46] examined the pyrolysis rates and energies of activation of different celluloses at high heating rates and discerned no differences. According to this approach, crystallinity or other fine structural parameters also do not play a significant role in pyrolysis. The processes depend, as Barns[47] summarized, mainly on temperature.

There are, nevertheless, findings which indicate a strong influence of fine structure on the rate and products of cellulose pyrolysis. Major[48] investigated the crystallinity and D.P. of cotton heated in air at 170° C and concluded that degradation occurred mainly in the less-ordered regions (LOR's). Shimazu and Sterling[49] studied the crystallinity and accessibility of cotton and rayon heated at 150° C. Deuteration and infrared spectroscopy indicated dehydration occurring mainly in the LOR's. Using TGA and gas chromatography, Kato and Komorita[50] investigated the heat stabilities and pyrolytic products of celluloses of varying degrees of crystallinity. They found that the crystalline regions of cellulose were more resistant to heat

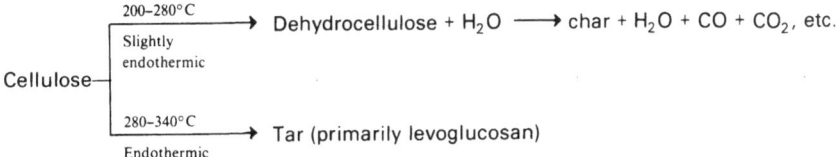

FIGURE 2. General scheme for the pyrolysis of cellulose.[42]

than were the LOR's. They also studied the yields of furfural and acetaldehyde obtained during pyrolysis and concluded that "the yields of these volatile compounds were dependent upon crystallinity of cellulose and pyrolysis temperature." Philipp *et al.*[51] compared the rates of pyrolysis of cotton and rayon under nitrogen and concluded that cellulose accessibility was the determining factor. Hurdoc and Schneider[52] found that the infrared absorption band at 900 cm^{-1}, which is attributed to bonds in the LOR's decreased in intensity during pyrolysis. Comparative pyrolytic studies on celluloses whose crystallinities had been decreased via ball milling were carried out by Chatterjee and Conrad[53] and Patai and Halpern[54] and on cellulose decrystallized with liquid ammonia by Weinstein and Broido.[55] These studies indicated that the fine structural changes accompanying decrystallization influenced the rates of pyrolysis.

Halpern and Patai[56] made a detailed study of rates of formation of levoglucosan at different pyrolysis temperatures (Table 3). They concluded that the bulk of the levoglucosan is formed from the crystalline regions via unzipping from reactive free ends after the LOR's have disappeared and the D.P. has dropped to 160.

TABLE 3

Thermal Decomposition of Cellulose *in Vacuo*
(1-g samples)[a]

Hour	Weight (mg)	Levoglucosan (mg)	D.P. of residue
		250°C	
1	8	0.13	—
2	15	1.3	211
3	26	3.1	212
4	40	7.0	237
5	62	13.6	—
6	80	21.3	190
		275°C	
1	51	11.0	223
2	177	65.0	162
3	307	96.0	156
4	408	134.0	157
5	493	171.0	158
6	586	200.0	—
		300°C	
1	620	252.0	157
2	786	274.0	—
4	811	286.0	—

[a]Taken from Ref. 56.

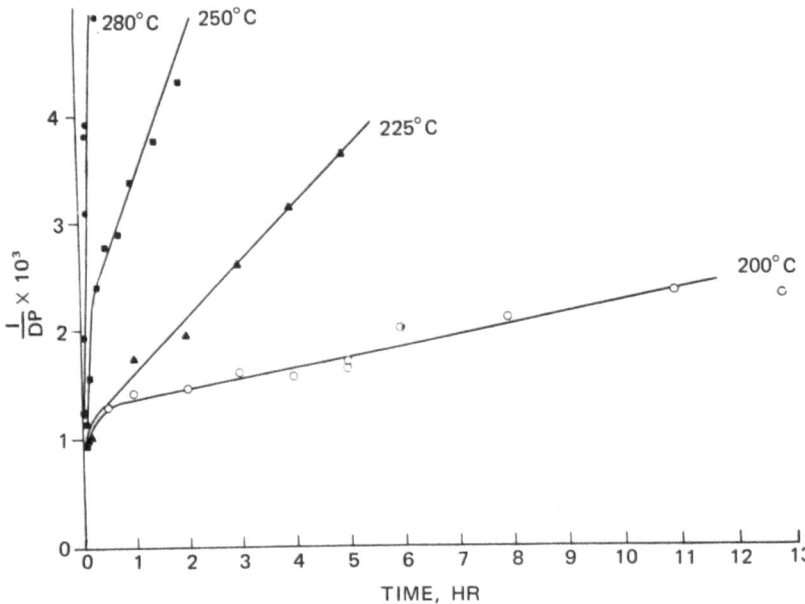

FIGURE 3. Relationship between reciprocal D.P. and time of treating of cellulose at different temperatures (*in vacuo*).[64]

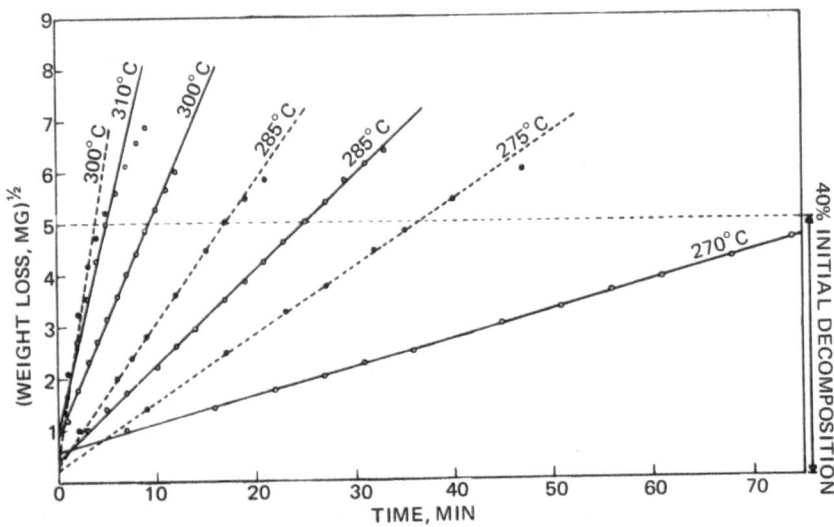

FIGURE 4. Square-root relationship between weight lost during the initial stage of pyrolysis and time; open circles, absorbent cotton; closed circles, ball-milled cellulose.[53]

4.2. Kinetics

The kinetics of cellulose pyrolysis have been examined by a number of investigators with widely differing results. They have been variously reported as being zero order,[57,58] first order,[59,60,61] and sequential zero and first order.[62,63] An important step in the pyrolysis must be depolymerization, and its kinetics were measured over the temperature range 200–280°C by Fung[64] (see Fig. 3), who found a consistent first-order reaction and calculated an E_{act} of 35.4 kcal/mole.

Chatterjee and Conrad[53] derived a theoretical expression for the kinetics of cellulose pyrolysis based on the assumption of a two-step reaction sequence: (1) Initiation, i.e., random scission of glycosidic bonds producing reactive molecules, and (2) propagation, i.e., decomposition with weight loss of the reactive molecules formed by the scission step. Their derivation pre-

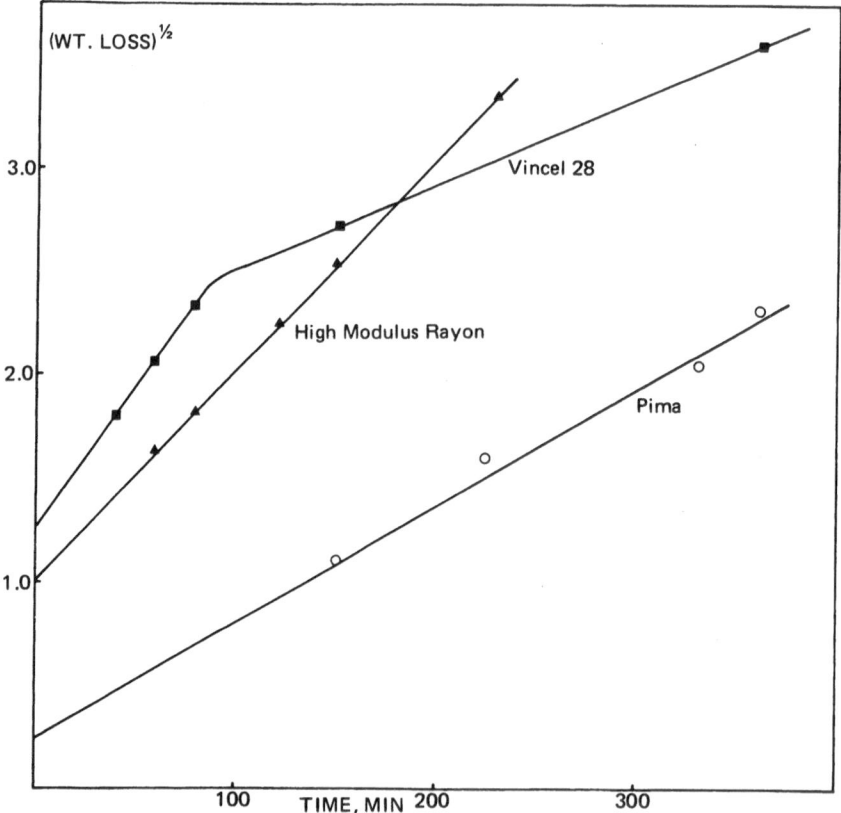

FIGURE 5. Kinetics of vacuum pyrolysis at 251°C: (weight loss)$^{1/2}$ vs. time.[65]

dicted a linear (weight loss)$^{1/2}$ vs. time relationship and was experimentally realized (Fig. 4) for both native cotton and decrystallized, ball-milled cotton.

This square-root relationship was confirmed by Basch and Lewin[37,65] on 17 different purified celluloses which differed widely in their fine structural parameters (Fig. 5). They showed that the pyrolysis rate increases with decreasing crystallinity and with increasing orientation and is inversely proportional to the square root of the D.P. (Figs. 6 and 7). These experimentally derived correlations are in line with the reaction sequence cited above. The initiation step consists of random chain scissions, and therefore the rate of the bulk decomposition reaction should be controlled by the number of free chain ends initially present, i.e., the initial D.P., and by the rate of formation of new chain ends. This in turn is governed by the % LOR's in the cellulose and by the specific rate at which the chains in the LOR's will cleave.

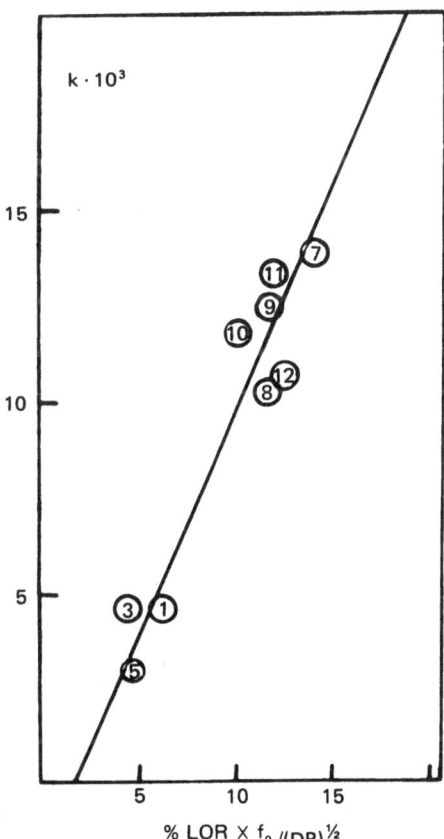

FIGURE 6. Vacuum pyrolysis rate constant vs. (% LOR's $\times f_o$)/(D.P.)$^{1/2}$. $r = 0.940$.[65]

% LOR $\times$ f$_o$ /(DP)$^{1/2}$

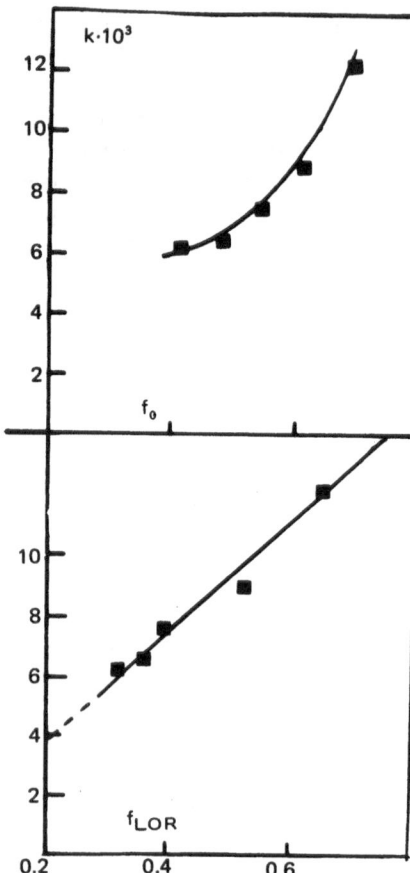

FIGURE 7. Effect of overall orientation (f_o) or LOR orientation (f_{LOR}) ($r = 0.982$) on the vacuum pyrolysis rate.[37]

The decrease in hydrolytic stability with increasing orientation can be explained as a result of straining the hydrogen bonds in the LOR's[66,67] and has been demonstrated also in the pyrolysis of stressed hydrocelluloses.[68]

4.3. Structural Changes

4.3.1. General

The effects of heat on cellulose manifest themselves in a broad range of structural and property changes. The rapid reduction of D.P. has already been noted (see Fig. 3), and it has been suggested[67] that the chain breaks occur at strain points at the crystallite-amorphous boundaries. The number of hydroxyl groups decreases,[69] while carboxyl groups increase.[70] Even

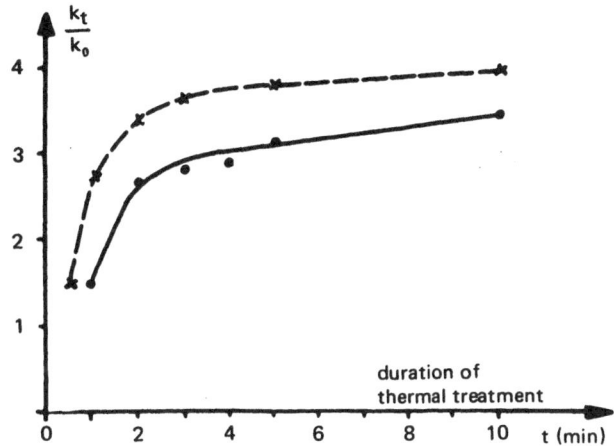

FIGURE 8. Increase in relative rate constant of cellulose hydrolysis due to thermal pretreatment. Cotton (•———•), viscose (×−−−×).[71]

very short heating times decrease the accessibility of cellulose to water and dye molecules and strongly increase its rate of acid hydrolysis[71]; see Fig. 8. An increase in the crystalline lattice spacing has been shown,[72] while the chain orientation is reported to be unaffected.[73] The melting of cellulose during brief laser irradiation has also been reported.[74]

4.3.2. Annealing

The degradative pyrolysis of cellulose may also involve significant changes in both the degree and nature of its crystallinity. Although amorphous ball-milled cellulose showed no crystallinity upon heating at 250° C,[54] liquid-ammonia-treated cellulose recrystallized quickly at the same temperature[55]; see Table 4.

The recrystallization kinetics of amorphous cellulose have been studied by differential scanning calorimetry (DSC) as a function of temperature and relative humidity.[75] Atalla and Nagel[76] have shown that cellulose can be precipitated from phosphoric acid solution into glycerine as either cellulose I or cellulose IV, depending on the temperature of precipitation, and that mercerized cellulose increases in cellulose II crystallinity upon heating at 150° C in glycerine.[77]

The source of the crystal growth upon heating is in dispute. While Philipp *et al.*[51] maintained that cellulose IV crystallinity arises from nuclei of this crystalline form present in the parent cellulose, it has also been stated[78] that other crystalline modifications are converted to cellulose IV upon heating.

TABLE 4
Crystallinity Index of NH₃-Swelled Cellulose Heated
Isothermally at 250°C[a]

Time of heating (hr)	Weight loss (%)		Crystallinity index
	Pure	Swelled	
Control	—	—	[b]
0[c]	1	2	68
2	3	12	70
4	5	20	61
25	29	59	51

[a]Taken from Ref. 55.
[b]Pattern similar to ball-milled cellulose.
[c]Samples withdrawn as soon as the reaction temperature was attained (about 40 min from start-up).

Basch and Lewin[65] have demonstrated that under comparable pyrolysis conditions, samples with initially high crystallinity show no significant crystallinity change, while samples with an initial low crystallinity show an appreciable rise in crystallinity (Fig. 9). A marked exception was hydrolyzed cotton, which, despite its initially high crystallinity (Table 5), showed a marked increase in crystallinity upon pyrolysis. This was interpreted as due to small crystallites formed during hydrolysis, which grow to measurable proportions (by X-ray diffraction) during pyrolysis. A sharp contract to the

TABLE 5
Vacuum Pyrolysis at 251°C[a]

No.	Sample	X-ray crystallinity (%)		Change (% crystallinity)
		Original	After 150 min	
1.	Cotton Deltapine	61.4	63.7	+2.3
2.	Cotton Deltapine, hydrolyzed	64.9	72.5	+7.6
3.	Cotton Pima	71.5	70.8	−0.7
4.	Cotton Pima, mercerized	65.0	65.3	+0.3
5.	Ramie	70.4	70.3	−0.1
6.	Ramie, hydrolyzed	70.9	69.5	−1.4
7.	Tire yarn	50.0	57.1	+7.1
8.	High-modulus yarn	67.9	67.5	−0.4
9.	Vincel 64	51.1	54.9	+3.8
10.	Vincel 28	63.2	66.0	+2.8
11.	Textile rayon	48.0	51.2	+3.2
12.	Evlan	46.9	49.8	+2.9

[a]Taken from Ref. 65.

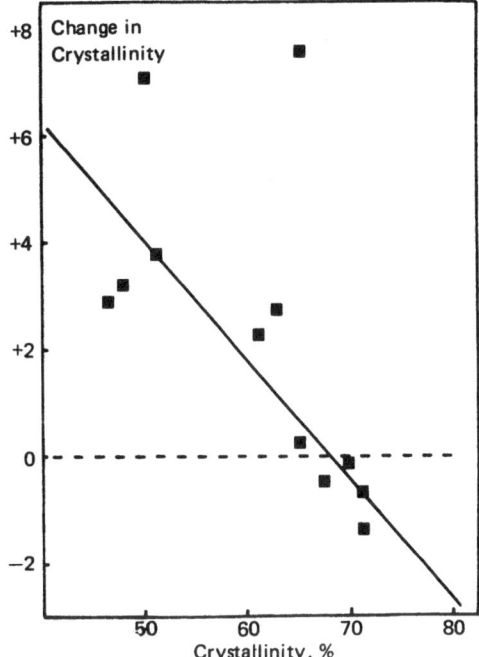

FIGURE 9. Change in X-ray crystallinity after vacuum pyrolysis (150 min at 251°C) vs. original X-ray crystallinity.[65]

behavior of hydrolyzed cotton was provided by pyrolyzed hydrolyzed ramie, which showed no crystallinity increase. This phenomenon was considered to be the result of the high orientation of ramie, which prevents or strongly limits crystalline growth.

TABLE 6
Properties of Rayon Fibers[a]

| Sample (% stretch) | X-ray, original | Crystallinity ratio | | f_o | D.P. |
		After heating	Change from original[b]		
20	0.405	0.747	0.325	0.41	326
30	0.448	0.672	0.250	0.48	319
40	0.417	0.663	0.241	0.54	322
50	0.414	0.689	0.267	0.61	305
60	0.425	0.626	0.204	0.69	310
Av.	0.422				

[a] Taken from Ref. 79.
[b] Original crystallinity ratio taken as 0.422 for all samples.

The role of orientation in limiting crystalline growth was demon-strated[79] by pyrolyzing a series of rayon fibers which differed only in orientation. It was shown (Table 6) that as the orientation of the cellulose increases, the change in crystallinity upon pyrolysis decreases. The large increase in crystallinity achieved under these conditions (vacuum, 200° C for 6 hr) can be seen from the X-ray diffraction diagrams obtained before and after pyrolysis (Figs. 10 and 11).

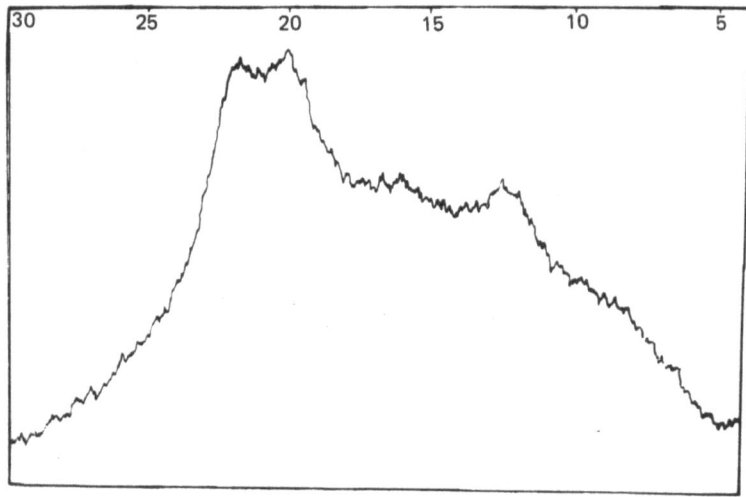

FIGURE 10. Rayon—20% stretch—before annealing.[79]

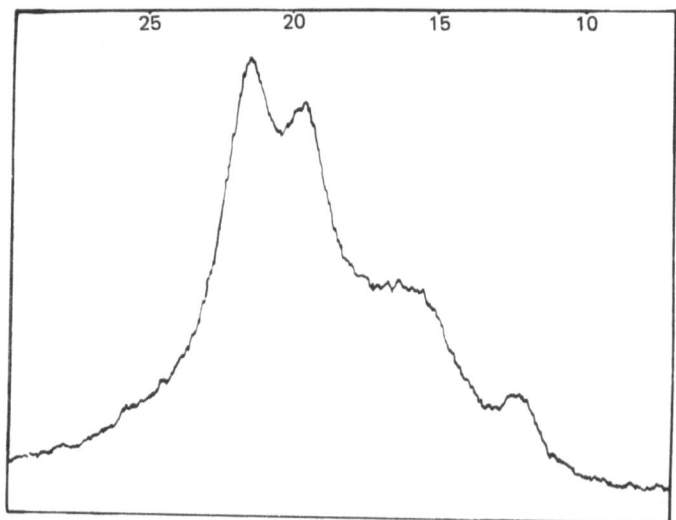

FIGURE 11. Rayon—20% stretch—after annealing.[79]

4.3.3. Thermal Autocross-Linking

The formation of cross-links in cellulose upon heating was first demonstrated by Back and Klinga,[80] who measured energies of activation for this process of 20–28 kcal/mol,[81] and later by Philipp and Stohr.[82] This cross-linking process is the reaction of neighboring hydroxyl groups in the LOR's with loss of water and results in a rapid "initial weight loss." It manifests itself in kinetic studies as an intercept on the time ordinate above zero (see Figs. 4 and 5) and was shown by Basch and Lewin[65] to be proportional to the % LOR's in the cellulose.

In further work[37] involving a series of rayon fibers differing only in orientation (see Fig. 12), it was found possible to separate the initial rapid reaction responsible for the initial weight loss from the bulk decomposition reaction in the following way: The extrapolated linear portion of curve 1 was displaced along the ordinate by an amount equal to its intercept on that axis. This yielded (curve 2, Fig. 12) the (weight loss)$^{1/2}$ vs. time plot of the slow bulk decomposition reaction. Graphical subtraction at suitable intervals of curve 2 from curve 1 yielded the weight loss vs. time curve (curve 3) for the initial rapid reaction. The data for this reaction were replotted as a first-order reaction (Fig. 13). Linear plots of ln C vs. time were obtained for all the samples listed in Table 7, confirming the first-order relationship and enabling calculation of the rates of the reaction.

The data in Table 7 show that the extent of the initial rapid reaction

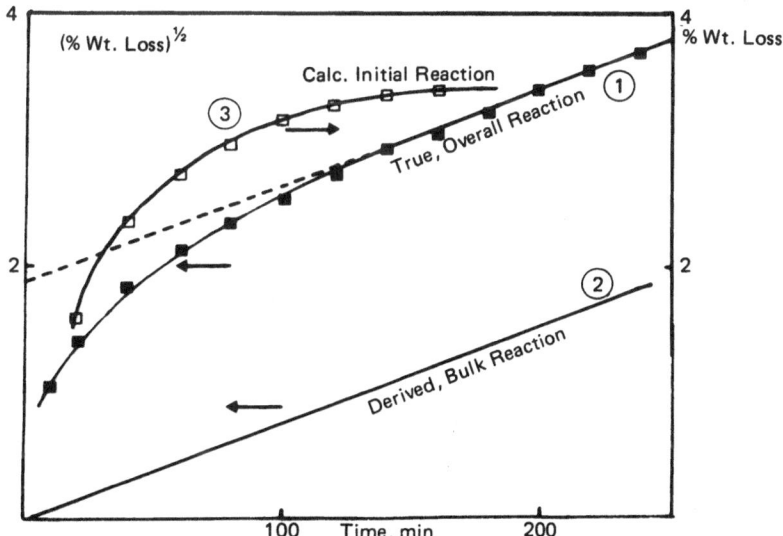

FIGURE 12. Vacuum pyrolysis of rayon: (1) overall reaction, (wt. loss)$^{1/2}$ vs. time; (2) bulk decomposition reaction; (3) initial rapid reaction, wt. loss vs. time.[37]

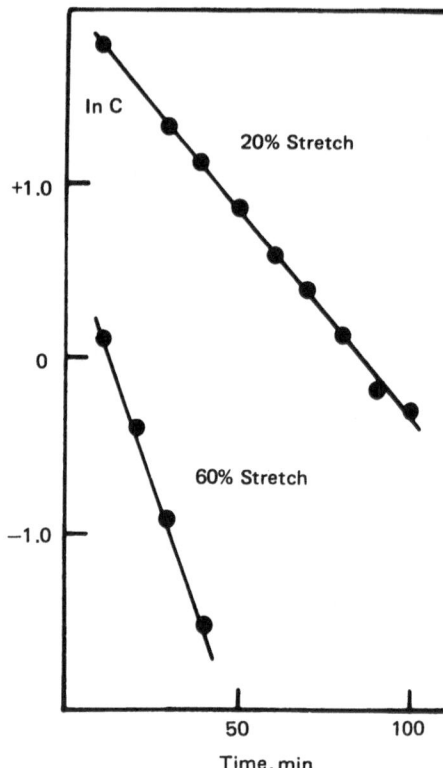

FIGURE 13. Initial reaction, first-order kinetic plot.[37]

TABLE 7

Influence of Orientation on Kinetics of Pyrolysis of Rayon (crystallinity ratio of all samples: 0.422 ± 0.026; D.P. of all samples: 317 ± 12; change in orientation obtained by stretching)[a]

			Initial rapid reaction		Bulk reaction rate	Char (%)	
% stretch	f_o	f_{LOR}	% decomposed	$k \times 10^2$ min^{-1}	$(Wt. loss)^{1/2}$ $\times 10^3$ min^{-1}	By TGA	Calculated
20	0.41	0.32	7.7	2.36	6.33	23.5	23.9
30	0.48	0.36	5.4	2.34	6.50	23.2	20.2
40	0.54	0.39	3.4	3.23	7.67	15.3	17.0
50	0.61	0.52	2.9	3.18	9.00	16.2	16.2
60	0.69	0.65	2.0	5.56	12.20	13.2	14.8

[a] Taken from Ref. 37.

decreases while its rate increases with increasing orientation. In a poorly oriented cellulose, the polymer chain segments in the LOR's have a relatively high freedom of motion at temperatures above the LOR glass-transition range of 220–230° C. They can thus occupy positions favorable to cross-linking. In the low orientation range, therefore, the extent of cross-linking, and consequently the initial weight loss, will increase with decreasing orientation. Since at low orientations the distances between chains are comparatively large, the rates will be relatively slow. Upon increasing orientation, fewer opportunities for cross-linking will occur due to steric hindrances. However, since the chains are closer to each other, the rate of cross-linking will tend to be higher. These relationships are illustrated in Figs. 14 and 15.

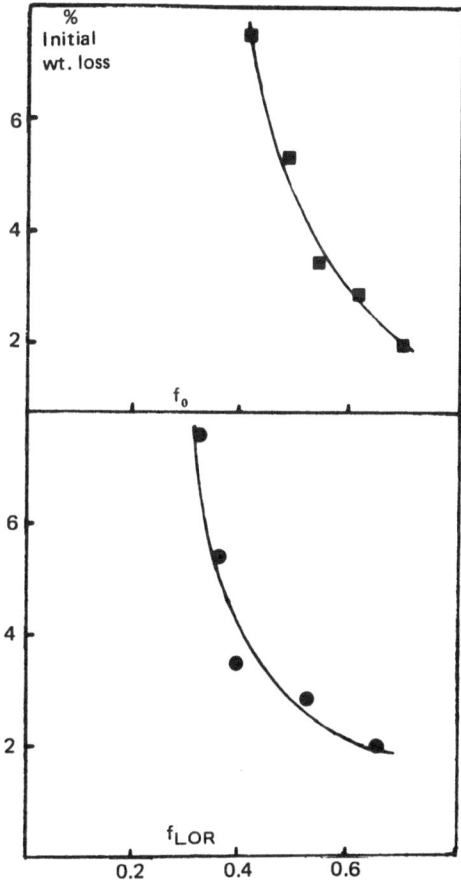

FIGURE 14. Percentage initial weight loss vs. overall orientation (f_o) and LOR orientation (f_{LOR}).[37]

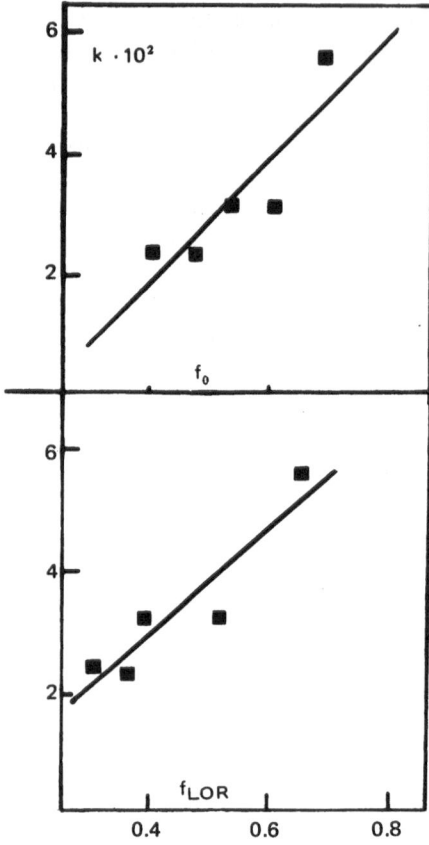

FIGURE 15. First-order rate constant for initial reaction vs. overall orientation (f_o) ($r = 0.883$) and LOR orientation (f_{LOR}) ($r = 0.922$).[37]

The assumption that the initial rapid reaction is due to dehydrative cross-linking is reinforced by the relationship found between the initial weight loss and char formation (Table 7 and Fig. 16). Since it is accepted that cellulose dehydration leads to char formation, a correlation is expected between the initial weight loss and the amount of char produced by total pyrolysis of the same cellulose. The authors[37] also showed that the amounts of char obtained experimentally were consistent with char yields calculated from a consideration of the char-forming processes occurring in the LOR's during pyrolysis (Table 7).

A further corroboration of the assumption that the char is formed from the LOR's is obtained from a series of experiments on the pyrolysis of rayons cross-linked with various amounts of formaldehyde (see Fig. 1). The % char is seen to decrease linearly with the increase in reacted formaldehyde. At the same time, the initial weight loss also decreases. The interchain, inter-AGU, and intra-AGU cross-links with formaldehyde, formed in the cross-linking

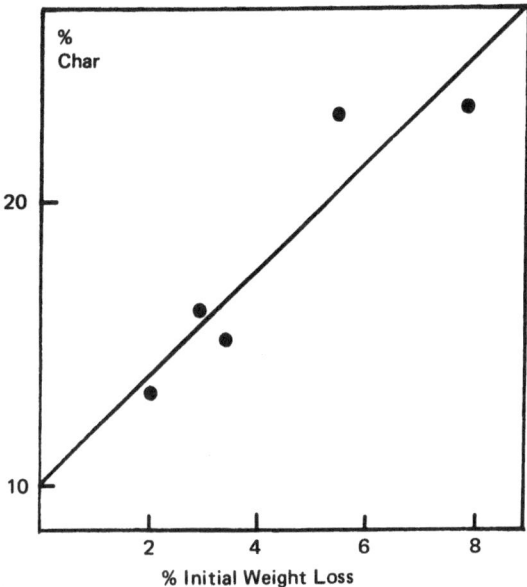

FIGURE 16. Percent char vs. initial weight loss in vacuum pyrolysis. $r = 0.932$.[37]

reaction, prevent dehydration and also immobilize the chains in the LOR's and thus the thermal autocross-linking is also hindered.

4.4. Thermal Analysis

4.4.1. Differential Thermal Analysis (DTA) and Differential Scanning Calorimetry (DSC)

DTA and DSC, which record heat absorption or emission as a function of temperature, have been widely used as a means of studying the thermal decomposition of cellulose and cellulose derivatives. In view of the complexity of the reactions the simplicity of the characteristic thermogram of pure cellulose is surprising. Typically, a strong endotherm is seen near 100°C due to loss of adsorbed water, followed by a broad endotherm between approximately 300 and 350°C. This second endotherm, which accompanies decomposition, has been ascribed to depolymerization and levoglucosan formation.[43] Two other features of the thermogram may be observed. A shallow endotherm has been reported[42] near 200°C and ascribed to the formation of "dehydrocellulose," but this endotherm is slight and difficult to reproduce. A more general characteristic is an exotherm immediately following the major endotherm. Arseneau[44] showed that this exotherm was dependent on the thickness of the sample and suggested that it represented the decomposition of the evolving levoglucosan. DTA and DSC have been compared as to their relative merits in cellulose research[83,84] and have been used to study papers,[85] cross-linking,[86,41] and degrees of substitution of cellulose ethers.[87]

Arseneau[44] also showed that the major endotherm appeared to be comprised of two overlapping curves, the first of which he attributed to the depolymerization and simultaneous levoglucosan formation and the second, larger, one to its volatilization. Basch and Lewin[65] succeeded in resolving the endotherm into two distinct regions for ramie only, thus proving its complexity (Fig. 17).

A detailed study[65] was made of the DSC characteristics of 12 different purified celluloses (see Table 8) in expectation that these could be correlated with the thermal stabilities of these materials, as has been reported with cellulose treated with ammonium phosphate.[57]

The expected correlation was verified for the initial and peak endotherm temperatures; the series of increasing thermal stabilities of rayon, cotton, and ramie show correspondingly increasing endotherm temperatures. The very high endotherm temperatures found for ramie imply high crystallite stability and possibly explain the endotherm peak separation; the two processes occur far enough apart to be seen separately by DSC.

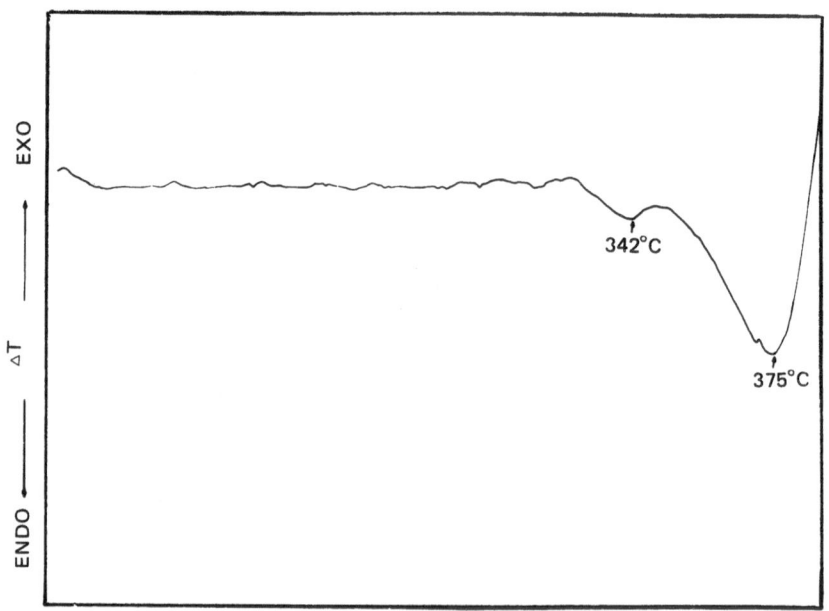

FIGURE 17. DSC thermogram of ramie heated at 10°C/min under vacuum; scale 200–400°C, 20°C/in.[65]

TABLE 8
Properties of the DSC Endotherm[a]

| Sample | Temperature | | Relative area |
	Initial°C	Peak °C	
Cottons[b]	322 ± 3	359 ± 4	74 ± 8
Rayons	310 ± 7	353 ± 2	41 ± 21
Ramie	335	370	32

[a]Taken from Ref. 65.
[b]Including hydrolyzed and mercerized cottons.

The unusually high thermal stability of ramie must be attributed to its very high orientation, as the crystallinities and degrees of polymerization of ramie and native cotton are similar[65] and is related to the mechanism of levoglucosan formation. The formation of levoglucosan from the crystalline regions proceeds in a manner similar to the decomposition of the LOR's at the lower temperatures, by an unzipping of the polymer chains.[54] This unzipping involved a change in conformation of the individual AGU's and is likely to occur on the exposed faces of the crystalline regions in which the stabilizing crystal bonding is partly lacking. If the crystalline regions are highly oriented, as is the case with ramie, there may be increased hydrogen bonding between adjacent regions, sufficient to further stabilize a crystal face and thus retard its pyrolysis.

It has been suggested[88] that high orientations may bring about an agglomeration of crystalline regions from adjacent elementary fibrils leading to thicker regions of higher order. Deuteration studies[89] have recently shown higher values for crystallite thickness and lower values for accessibility of ramie as compared to cotton. Such agglomerations do not possess crystalline perfections, and, therefore, the higher values for crystallite thickness were not found by X-ray diffraction.

The relative endotherm areas (Table 8) are seen to range widely in value and correlate poorly with thermal stability. The explanation probably lies in the exotherm immediately following the endotherm, which masks a variable part of the endotherm area. The high stability of ramie moves the endotherm even farther into the exotherm area and accounts for the anomalously low endotherm area.

4.4.2. *Thermal Gravimetric Analysis (TGA) and Energy of Activation of Pyrolysis*

Many studies of the activation energy of pyrolysis have been reported, despite its limited theoretical significance in a complex series of reactions, with values ranging from 22 kcal/mol[58] to 150 kcal/mol.[60]

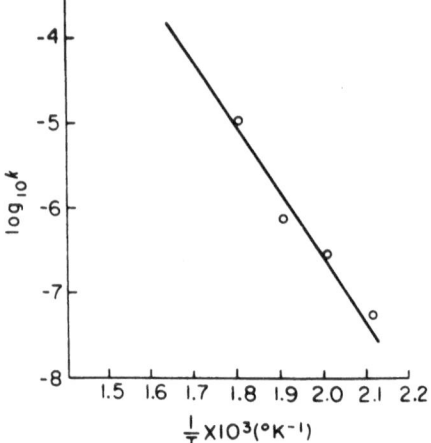

FIGURE 18. Effect of temperature on rate constant in cellulose pyrolysis (*in vacuo*).[64]

It is of interest to note that most of the values for the activation energy which were obtained at temperatures lower than 260° C are close to 30 kcal/mol.[59,33] Fung[64] (see Fig. 18) obtained a value of 35.5 kcal/mol for the activation energy of the pyrolysis by following the decrease in D.P. in the temperature range 200–280° C. Higher values in the vicinity of 50 kcal/mol reported in the literature[59] were obtained in high-temperature pyrolysis,

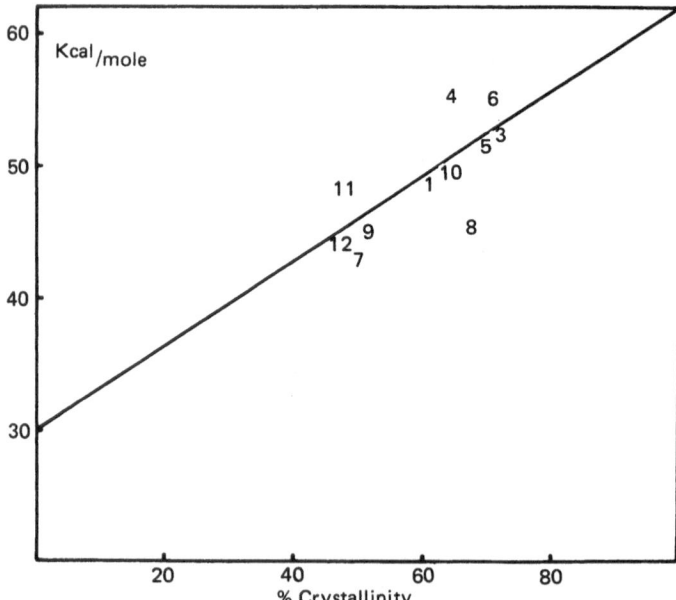

FIGURE 19. Apparent energy of activation vs. % crystallinity. $r = 0.726$.[65]

such as dynamic TGA, in which a significant loss in weight appears only above 300°C. Roberts[90] believes that the lower activation energies are due to catalytic decomposition, while the higher values of 50–60 are the true values pertaining to the complete decomposition of pure cellulose.

A study of E_{act} of different celluloses[65] revealed an approximately linear correlation with the crystallinity of the cellulose. By regression analysis, a coefficient of correlation of 0.726 was obtained. Extrapolating the line yielded a value of 29.6 kcal/mol for the LOR's, while for a totally crystalline cellulose, an E_{act} of over 60 kcal/mol was obtained (see Fig. 19).

These results can be interpreted in terms of two separate stages, LOR pyrolysis and crystallite pyrolysis. For the first stage, LOR pyrolysis, random

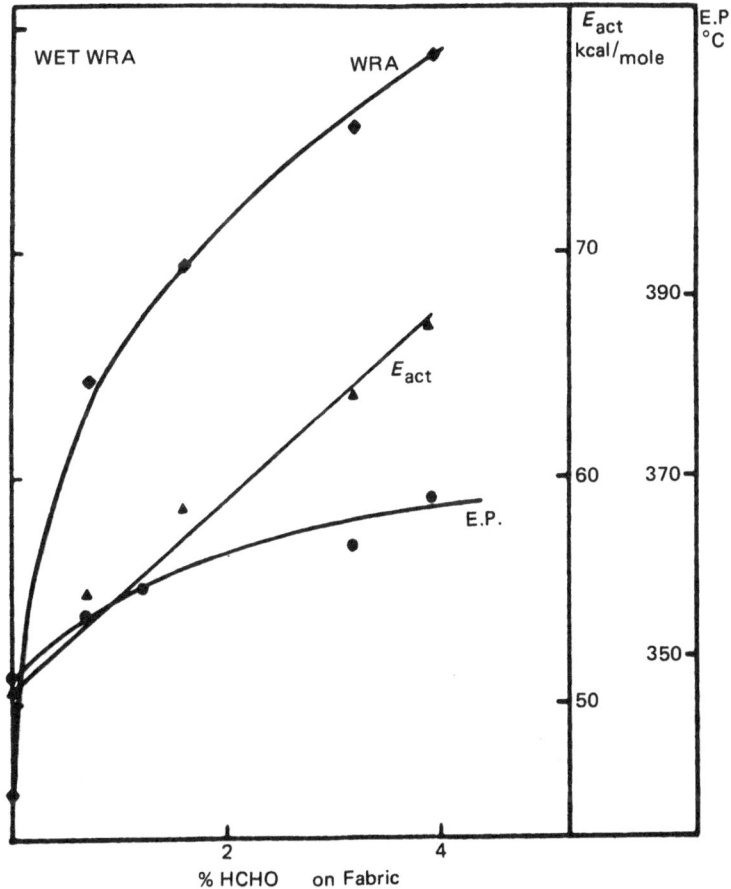

FIGURE 20. E_{act}, DSC endotherm peak, and wet-wrinkle recovery angle vs. % formaldehyde reacted on cotton.[95]

chain cleavage appears to be the rate-determining step,[64] and its activation energy will resemble the values found for hydrolytic chain scission of methyl cellulose (30.15 kcal/mol)[91] and for hydrolysis of cellulose solutions in concentrated acids (29.8, 27.6, and 27.2 kcal/mol).[92–94] The activation energy of 28 kcal/mol found[81] for the thermal autocross-linking falls in the same range.

The second mechanism is due to the pyrolysis of the crystalline regions, and the energy of activation of above 60 kcal/mol is connected with higher thermal stability of the crystallites. The rate-determining step of this reaction

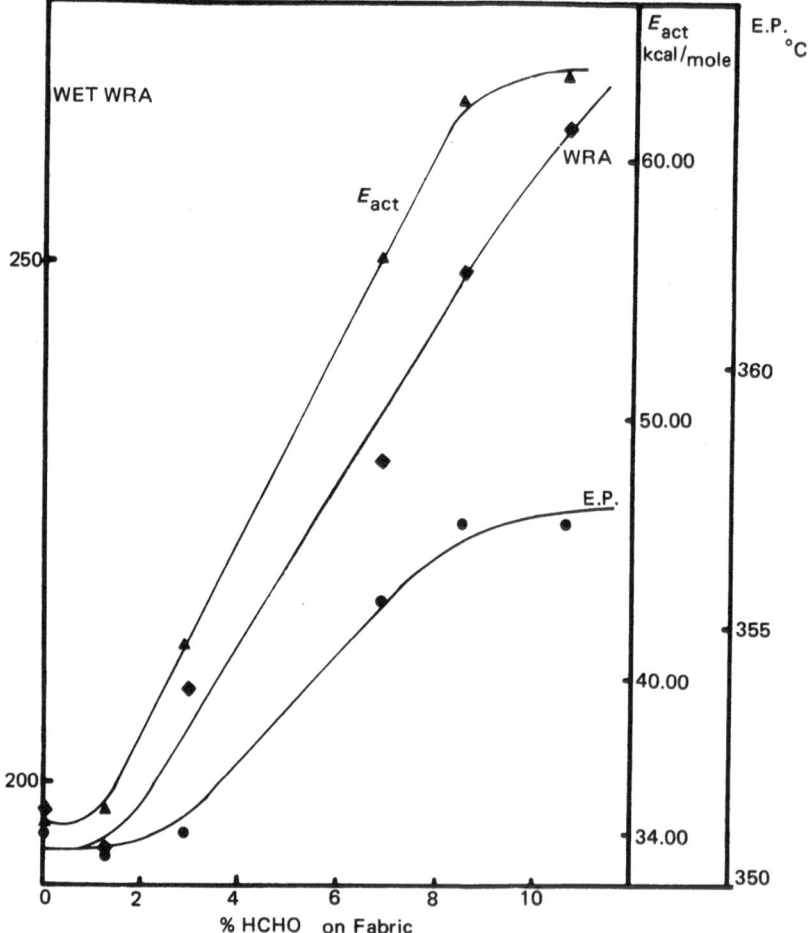

FIGURE 21. E_{act}, DSC endotherm peak, and wet-wrinkle recovery angle vs. % formaldehyde reacted on rayon.[95]

may be the formation of levoglucosan on the exposed faces of the crystalline regions. If this is the case, then a substitution on C_2 or C_6 may bring about an increase in activation energy. This was indeed observed[95] both in formalde-hyde cross-linked cotton and rayon (see Figs. 20 and 21), in which it was found that the activation energy increases in parallel with the wet-wrinkle recovery angle and with the amount of reacted formaldehyde. The cross-linking brings about a stabilization of the structure which expresses itself also in the increase in the temperature of the endotherm peak and in the decrease in the rate of pyrolysis.

5. Air Pyrolysis

5.1. General

The presence of as little as 0.5% oxygen significantly affects the pyrolysis of cellulose.[63] Higgins[96] followed the pyrolysis at 250°C by infrared spectroscopy (see Fig. 22) and observed oxidative changes which also involved the crystalline regions. Major[48] found that at 170°C oxidative degradation occurred in the LOR's. Schwenker and Beck,[34] however, stated that gas chromatographic analysis did not show significant differences between pyrolyses conducted under nitrogen or oxygen. Levoglucosan remains a major pyrolysis product in air pyrolysis.[23]

5.2. Kinetics and Fine Structure

Relatively few kinetic studies of air pyrolysis have been reported. Millet and Goedkin[97] examined the weight loss kinetics of cotton linters in air from 180 to 260°C and found first-order kinetics. This was verified by Basch and Lewin[98] in a study of several different celluloses (see Fig. 23). They found the accessibility to oxygen to be the dominant factor controlling the rate of pyrolysis. Reproducible results were obtained only after samples had been cut very finely to a powder. The plot of the first-order rate constants for the air pyrolysis of the samples is seen in Fig. 24 to be linearly related to the overall accessibility, i.e., the LOR's plus the accessible portion of the crystalline regions, divided by the orientation.

The role of the oxygen appears to be that of a catalyst for the decom-position similar to impurities or to acids in the pyrolysis of cellulose.

The major effect of the orientation in the case of air pyrolysis is in the control of the amounts of oxygen penetrating into the polymer. Since the rate of pyrolysis in air is much higher than in vacuum, it is not surprising that this retarding influence of the orientation is much more pronounced than the accelerating effect noticed in vacuum pyrolysis.

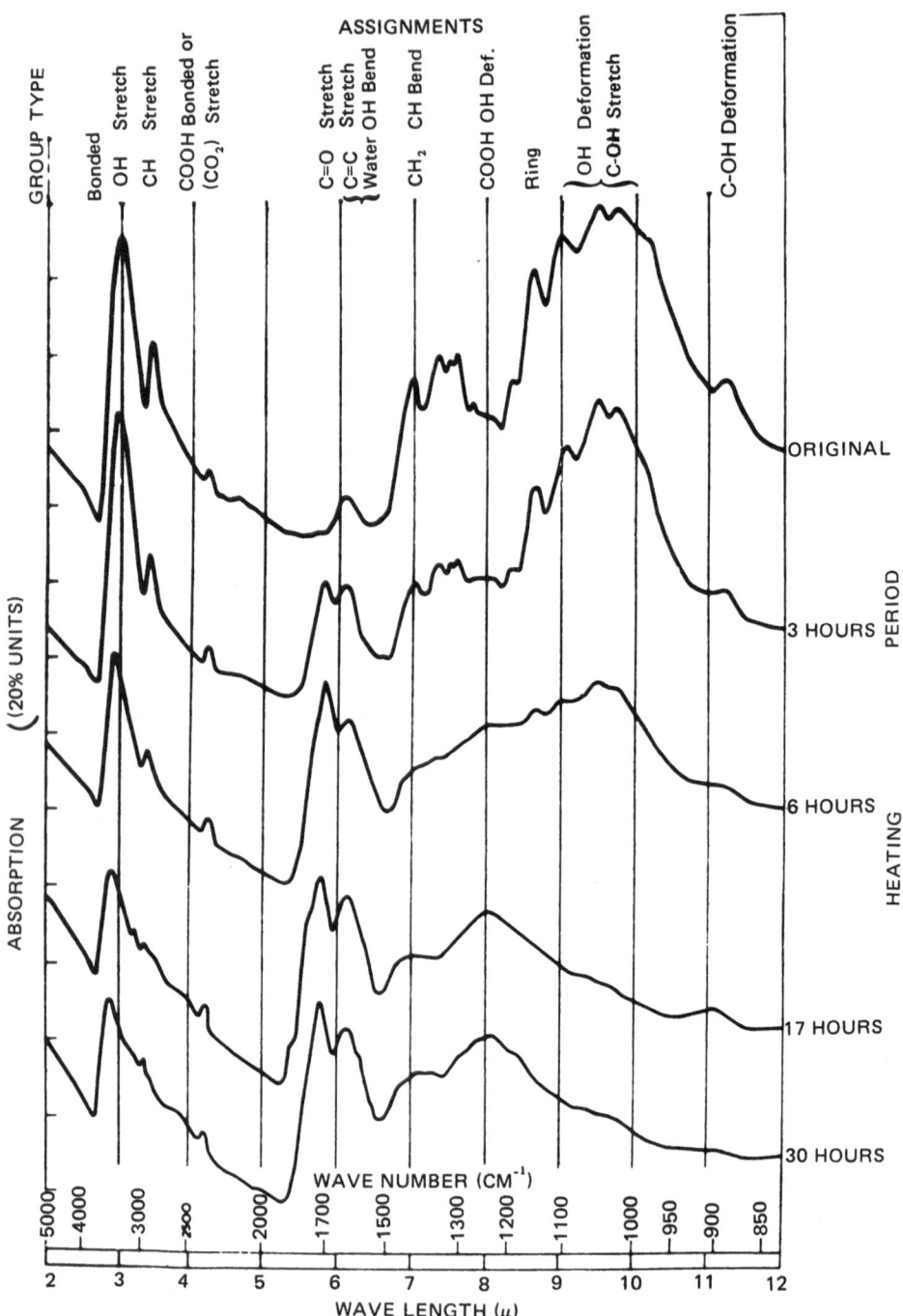

FIGURE 22. Infrared spectra typical of various stages of degradation of cellulose in air at 250°C.[96]

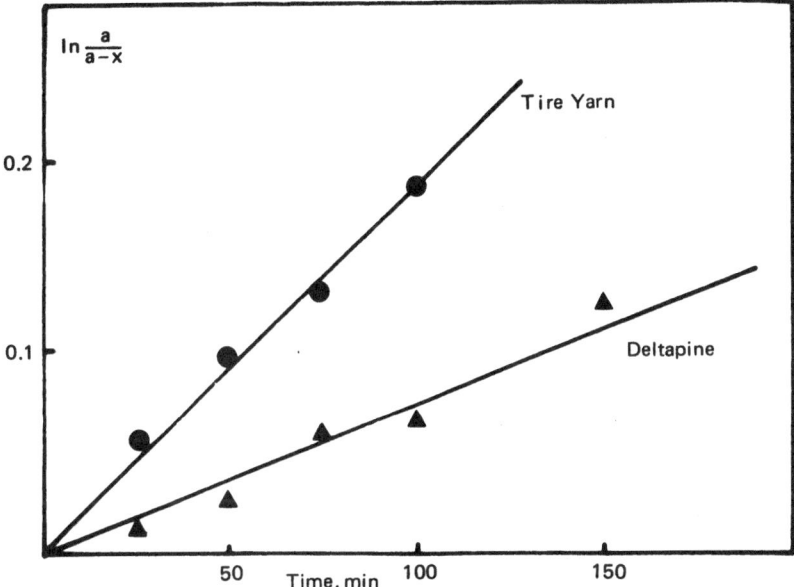

FIGURE 23. Kinetics of pyrolysis in air at 251°C: $\ln[a/(a-x)]$ vs. time.[98]

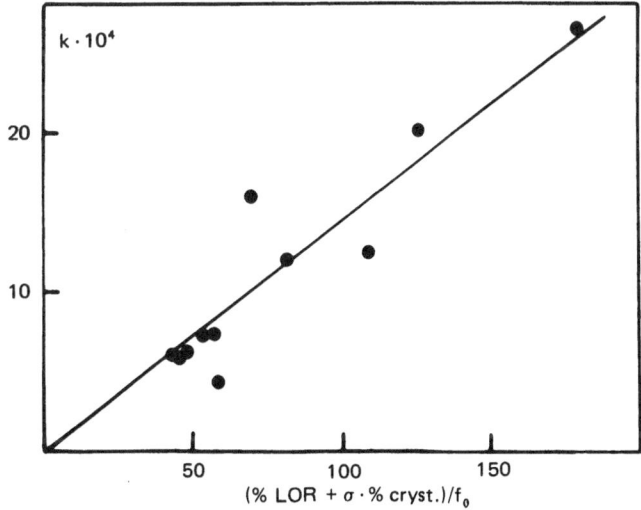

FIGURE 24. First-order rate constant vs. $[\%\ \text{LOR's} + \sigma(\%\ \text{crystallinity})]/f_o$ for air pyrolysis at 251°C; coefficient of correlation $r = 0.923$.[98]

The basic reaction occurring in the system is thus the direct attack of the oxygen molecules upon the cellulose. It would be expected that, similarly to other reactions with cellulose, the air pyrolysis should depend on the rate of diffusion of the oxygen into the polymer, which in turn is known to decrease with the increase of both crystallinity and orientation.

5.3. Levoglucosan Formation, Fine Structure, and Combustibility

The concentration of levoglucosan in the tar formed during high-temperature air pyrolysis has been shown[98] by a novel infrared technique[99] to be directly related to the product of crystallinity and orientation (see Fig. 25).

This is in line with the conclusion of Golova *et al.*[27] that the packing density of the macromolecules is mainly responsible for the yield of levoglucosan. Increasing orientation will tend to favor the greater levoglucosan formation due to decreased accessibility of the crystallite faces to oxygen.

Levoglucosan yields are strongly influenced by even traces of impurities present in or added to the cellulose.[54] It has been shown[100] that acidic impurities can either increase or decrease the percentage of levoglucosan in the tar, depending on their concentration, while basic impurities depress levoglucosan formation at all levels (see Table 9). Flammability of the cellulose is seen to vary, within narrow limits, with levoglucosan formation. These factors explain the observed increase in flammability of cellulosic textiles when treated with low concentrations of flame retardants.[101,102]

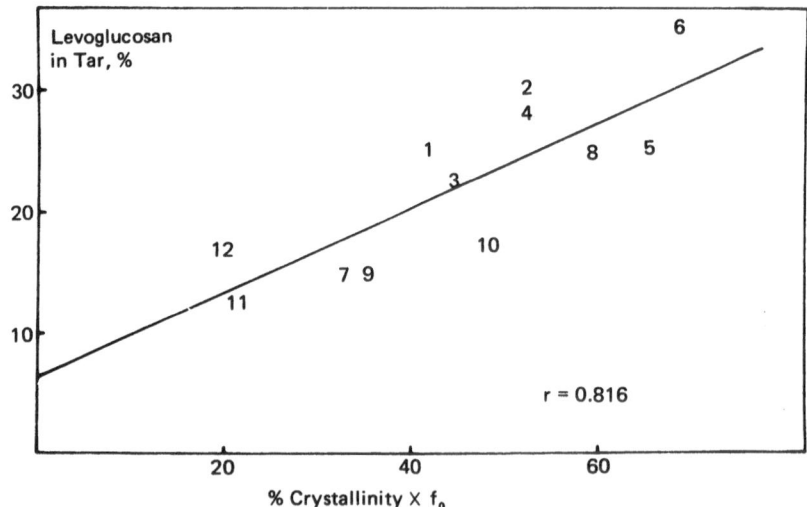

FIGURE 25. Percent levoglucosan in tar formed during high-temperature air pyrolysis vs. % crystallinity $\times$ orientation (f_o). $r = 0.816$.[99]

TABLE 9
Effects of Additives upon Levoglucosan Formation and
Flammability of Cotton[a]

Additive	Concentration (%)	Levoglucosan in tar (%)	LOI
Blank	—	22	18.10
$(NH_4)_2SO_4$	0.1	41	17.40
$(NH_4)_2SO_4$	0.2	28	18.10
$(NH_4)_2SO_4$	1.0	14	19.76
$NaHCO_3$	0.01	16	18.73
$NaHCO_3$	0.1	13	19.15
$NaHCO_3$	1.0	0	20.50

[a]Taken from Ref. 100.

The continued flammability of cellulose in the presence of sufficient flame retardant to depress or eliminate levoglucosan is attributed to the formation of other volatile organic compounds. Cis-4,5-epoxy-2-pentenal has been shown to be formed in appreciable yield in the pyrolysis of phosphoric-acid-treated cellulose.[103]

Cross-linking with formaldehyde also influences both levoglucosan yields and flammability.[41] As seen in Table 10, the levoglucosan con-

TABLE 10
Thermal Properties of Cross-Linked Cellulose[a]

% HCHO	WRA wet	E_{act} (kcal/mol)	Levoglucosan yield (%)	LOI
		Cotton (bleached Pima)[b]		
0.0	130	50.6	25.5	17.7
0.7	221	54.9	24.9	17.7
1.6	247	58.6		17.8
3.2	278	63.6		17.9
3.9	294	66.6	21.4	18.0
		Rayon[c]		
0.0	195	34.4	15.7	18.2
1.3	193	34.8	16.1	18.1
2.9	209	41.2	17.8	18.0
6.9	231	56.1		17.8
8.5	249	62.2		17.3
10.6	263	63.2	20.8	17.1

[a]Taken from Ref. 41.
[b]Cotton; weight = 3.89 oz/yd; crystallinity ratio = 0.715; f_a = 0.9; D.P. = 3082.
[c]Rayon; weight = 4.44 oz/yd; crystallinity ratio = 0.455; f_a = 0.44; D.P. = 318.

tent decreases with increase in amount of formaldehyde reacted with cotton. This decrease is explained by the lower tendency of the crystallite faces with substituted C_2 and C_6 to undergo the unzipping reaction. It is possible that the cross-links bring about an additional pyrolysis path, different from that leading to levoglucosan and necessitating a higher activation energy. It is of interest to note that in the case of rayon (see Table 10) the levoglucosan yield increases with the extent of cross-linking. It appears that in this case the effect of cross-linking was in the direction of increasing crystallinity and that this effect was more pronounced than the opposite effect obtained in cotton.

An increase in crystallinity due to formaldehyde cross-linking has indeed been found in this study for rayon and Vincel 28: Cross-linking of rayon with 10.6% HCHO increased the X-ray C.R. from 0.455 of the uncross-linked fiber to 0.515; in the case of Vincel, an increase from 0.631 to 0.674 was obtained upon cross-linking with 7.12% HCHO. Increases in crystallinity have been previously observed upon acid hydrolysis of celluloses.[2]

6. Flame Retardancy

6.1. Mechanism and Flame-Retardant Structural Considerations

Cellulosic flame retardants may operate through one of two basic mechanisms, solid phase or vapor phase flame inhibition.

Halogens are known to be efficient flame-retardant agents for cellulose and other polymers[104] and act mainly in the vapor phase via free-radical inhibition. Basch et al.[105,106] examined the pyrolysis of brominated wood in detail by chemical, spectroscopic, and thermal analysis and determined the effects of bromine upon the cellulose and the distribution of the bromine in the products of pyrolysis.

Bromination decreased the thermal stability of the wood and increased the amounts of the aqueous and gaseous phases at the expense of the solid residue; the yield of tar was not significantly affected (see Table 11). Upon pyrolysis, considerable amounts of HBr were generated, which penetrated the crystalline regions, increasing levoglucosan formation, lowering the D.P. of the cellulose, and disrupting the crystallites (see Table 12).

Flame retardants based on phosphorus, sulfur, boron, and other acid-forming materials act via dehydration to produce water and char at the expense of flammable tars.[107] Most commercial treatments for cellulose are based on phosphorus, and consequently the effect of the structure of phosphorus-based materials on the mechanism and efficiency of flame retardancy has been widely studied. Hendrix et al.[108] showed that acidic phosphates act via phosphorylation, presumably at the C-6 hydroxyl of the

TABLE 11

Distribution of Pyrolytic Products of Wood Pyrolysis at 250°C for 3 hr under Vacuum[a]

Sample (% Br$_2$)	% residue	% aqueous	% tar	% CO$_2$ (by wt.)
0.00	82.3	—	10.0	0.9
1.85	73.7	12.8	15.2	1.4
4.93	67.8	20.4	12.1	1.4
7.44	65.7	21.0	11.7	1.7
10.61	62.8	24.3	10.9	2.1
17.25	51.7	33.2	10.9	2.8

[a]Taken from Ref. 105.

TABLE 12

Effects of Bromine upon Cellulose Pyrolysis of Wood at 250°C for 3 hr under Vacuum[a]

Sample (% Br$_2$)	% levoglucosan in tar	D.P.	Relative crystallinity (%)
Blank—before pyrolysis	—	1220	34
Blank—after pyrolysis	0.5	166	31
1.85	2.3	76	24
4.93	8.2	41	19
7.44	8.7	86	14
10.61	8.3	32	10
17.25	4.7	[b]	6

[a]Taken from Ref. 105.
[b]Not measurable.

anhydroglucose unit. Consequently structures which increase or favor esterification are likely to be more efficient in promoting flame retardancy. Thus the temperature of flame-retardant decomposition[109] and the ease of esterification[110] or transesterification[111] have been shown to be significant in flame-retardant efficiency. The degree of phosphorus retention in char after burning has been shown to be closely related to the flame-retardant efficiencies of the phosphorus compound as determined by static oxygen bomb calorimetry.[112] This same study also indicated that the nature of the chars which are formed seems to be independent of the chemical structure of the phosphorus compound.

A theoretical and commercially significant feature of phosphorus-based flame retardants is the synergism exhibited with nitrogen. The degree and extent of the synergism depends on the chemistry of the system.[109] It was shown that amide and amine nitrogen increase flame resistance, while nitrile nitrogen detracts and decreases the percentage of phosphorus left in the char.[113] The mechanism of this synergism has been attributed to formation of phosphoramides during pyrolysis, which lower decomposition temperature, increase ester formation, and cause cross-linking.[110] Barker[114] has indeed shown that P–N compounds are more efficient than P–O compounds and has attributed this greater efficiency to the higher polarity, Lewis acidity, and acid catalysis potential of the P–N linkage. The heat of combustion of cellulose, which decreases with increasing amounts of phosphorus in cellulose,[115] showed an inverse relationship with char yield after pyrolysis[116] and was linearly related to N/P ratios.

6.2. Flame Retardancy and Fine Structure of Cellulose

While the possible effects of fine structure have been considered in the pyrolysis of cellulose, they have been virtually ignored insofar as combustion and flameproofing are concerned. This is no doubt due, in part at least, to the obvious effects that fabric construction, moisture content, density, and air permeability have on these phenomena. These effects have been examined quantitatively by Willard and Wondra[117] and Hendrix et al.[118-120]. Furthermore, the reported LOI (limiting oxygen index) values for cotton and rayon are very close if not identical,[121] and the situation is further complicated by the strong dependence of LOI values, and hence combustibility, on cellulose purity.[119]

Tesoro et al.[122] noted that higher amounts of phosphorus- and nitrogen-containing compounds were needed to flameproof rayon as compared with cotton, and they ascribed the differences to the lighter weight of the rayon. However, they also found differences in the phosphorus/nitrogen synergism ratios for the two fabrics which they could not explain. More boron-containing flame retardant is also reported to be needed for rayon as compared with cotton.[123]

Aenishanslin et al.[124] reported (Table 13) considerable differences in the amounts of phosphorus and nitrogen needed to flameproof ramie, cotton, rayon, and a polynosic. They ascribed these differences to differences in orientations of the various fabrics, even though the polynosic required as much P and N as the textile rayon and polynosics are generally more highly oriented. They did not consider other possible factors, such as crystallinity.

Basch and Lewin[125] determined the amount of a number of flame-retardant chemicals and treatments needed to fireproof cotton and rayon. The results (see Table 14) showed that sulfur-based chemicals are about equally effective on both cotton and rayon. The phosphorus-based treatments were far less effective on rayon than on cotton. These results were verified

TABLE 13

Minimum Amounts of P and N Needed to Flameproof Different Cellulosics
with Monomethyloldimethylphosphonopropionamide[a]

Fibers	Weight of the fabric (g/m²)	Minimum phosphorus (% P)	Minimum nitrogen (%N)
Ramie	215	1.37	0.62
	175	1.37	0.62
Cotton	250	1.65	0.74
	150	1.65	0.74
Polynosic	78	2.45	1.10
Rayon	82	2.54	1.15

[a] Taken from Ref. 124.

TABLE 14

Percentages of S or P Needed to Prevent Flame Propagation below a 90°
Burning Angle[a]

Treatment	Fabric	Number of treatments tested	Average % S or P	Average deviation
Sulfur	Cotton	10	1.66	0.19
	Rayon	4	1.88	0.19
Phosphorus	Cotton	8	1.56	0.14
	Rayon	4	2.37	0.24

[a] Taken from Ref. 125.

by LOI measurements made on sulfated and phosphorylated fabrics (see Table 15).

An insight into the different efficiencies of sulfur and phosphorus on cellulose was provided by DSC thermograms of treated fabrics (Figs. 26–31). The sulfated cotton, rayon, and ramie yielded similar thermograms (see Figs. 26, 27, and 28), whose major characteristics were a deep endotherm at 235° C followed by a strong exotherm. However, the thermograms of the phosphorylated cotton and ramie were distinctly different from those of phosphorylated rayon. While the former (Figs. 29 and 30) yielded a slight endotherm near 240° C and a strong endotherm near 290° C, the latter showed only a single deep endotherm near 245° C (Fig. 31). The DSC results thus parallel the combustion data in that they show that while sulfated cotton, rayon, and ramie decompose similarly, phosphorus-treated cotton and ramie decompose quite differently from phosphorus-treated rayon.

TABLE 15
Percentages of S or P Needed to Reach an LOI of 27[a]

Treatment	Fabric			
	Pima	Deltapine	Rayon	Ramie
Sulfation with ammonium sulfamate	1.38	1.25	1.24	1.24
Phosphorylation with urea–phosphoric acid	1.45	1.30	1.72	1.21

[a] Taken from Ref. 125.

Consideration of the mechanisms by which acid-type flame retardants affect the decomposition of cellulose could explain the differing efficiencies of sulfur and phosphorus on cotton and rayon. It is generally agreed[107] that acid-type flame retardants act via dehydration to produce water and char at

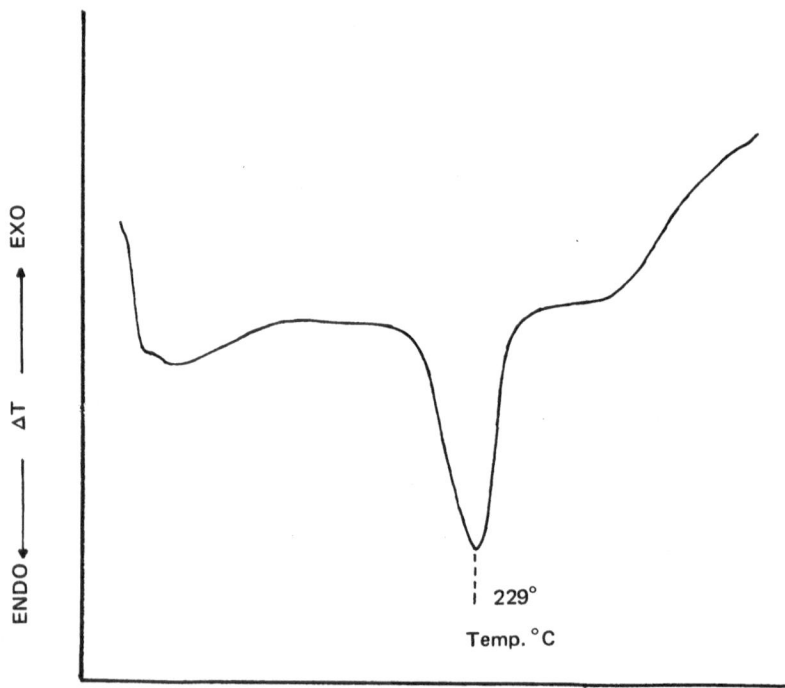

FIGURE 26. DSC curve of cotton sulfated with sulfuric acid.[125]

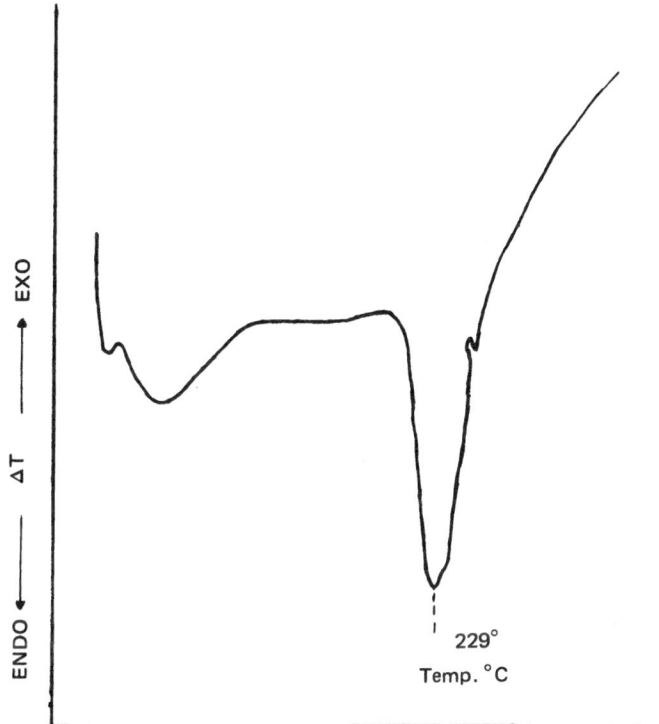

FIGURE 27. DSC curve of rayon sulfated with ammonium sulfamate.[125]

the expense of flammable tars. This dehydration can proceed either by carbonium ion catalysis according to the scheme

$$R_2CH-CH_2OH \xrightarrow{H^+} R_2CH-CH_2OH_2^+ \longrightarrow R_2CH-CH_2^+ + H_2O \\ R_2C{=}CH_2 + H^+ \tag{1}$$

or by esterification and subsequent decomposition of the ester to yield acid, olefin, and water, as follows:

$$R_2CH-CH_2OH + HOR \text{ (acid)} \longrightarrow R_2CH-CH_2OR + H_2O \\ R_2C{=}CH_2 + HOR \text{ (acid)} \tag{2}$$

Reaction sequence (1), once initiated, would be expected to proceed rapidly through all its stages, due to the instabilities of the carbonium ion

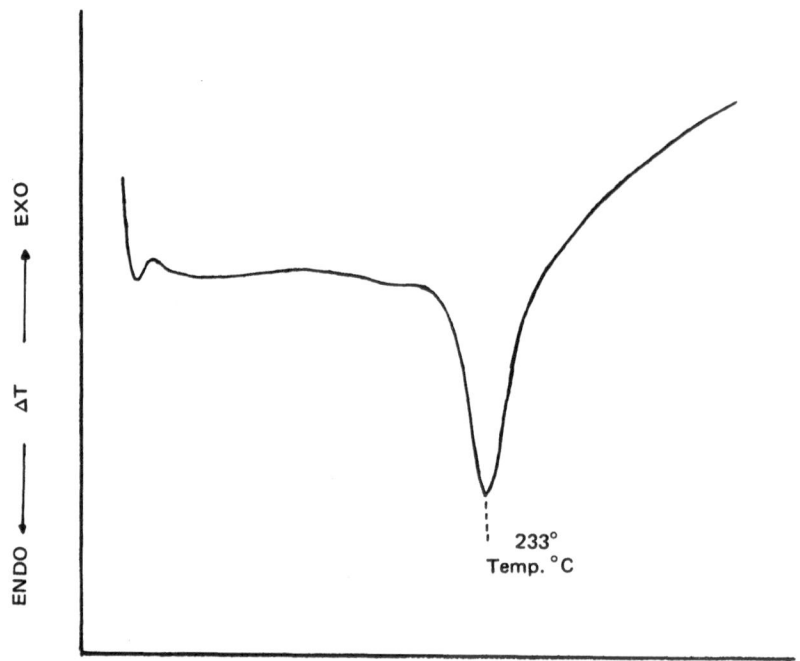

FIGURE 28. DSC curve of ramie impregnated with ammonium bisulfate.[125]

intermediates. It would be favored by the presence of a very strong acid, and the DSC picture of a single, strong endotherm for the decomposition of sulfated cellulose is consistent with this mechanism. Reaction sequence (2) could proceed relatively slowly and even in discreet steps, depending on the stability of the ester. It has been shown that cotton treated with phosphoric acid or ammonium phosphate decomposes primarily by reaction sequence (2).[108]

It is possible that these differing behaviors of sulfur vs. phosphorus and cotton vs. rayon can be explained by the known properties of the reagents and the fabrics. When phosphorylated cotton is heated, the cellulose ester begins to pyrolyze near 240° C, with simultaneous dehydrative decomposition of the accessible regions which had been partly esterfied and formation of free phosphoric or polyphosphoric acid. This free acid will react with the crystalline regions only at higher temperatures to form char and water. In the case of rayon, however, the pyrolytic decomposition of the bulk of the cellulose occurs at lower temperatures where part of the acid is still esterified and unavailable to further direct the pyrolysis to nonflammable materials. Thus, relatively larger amounts of phosphorus are needed to impart a given degree of flame retardancy. The cellulose esters of sulfuric acid are more labile than those of phosphoric acid and will cleave more rapidly and more completely. The released acid, with rising temperature, will attack the remaining cellulose more rapidly than phosphoric acid by catalytic proton dehydration due to greater ionization and mobility and possibly higher acid strength. As a result, all of the acid is available even at the lower decomposition temperature of rayon, and no extra amounts of sulfur are needed for rayon as compared with cotton. This picture would also suggest that the acid catalysis is not greatly hindered by crystallinity; in other words, the acid can swell and penetrate the crystalline regions.

The ability of strong acids to penetrate and disrupt the crystalline lattice has been indicated previously in a study of the pyrolysis of brominated wood.[105] The differences in the actions of sulfur- and phosphorus-based acids would seem to be in the relative rates by which the two mechanisms catalytic dehydration and deesterification, operate in each case.

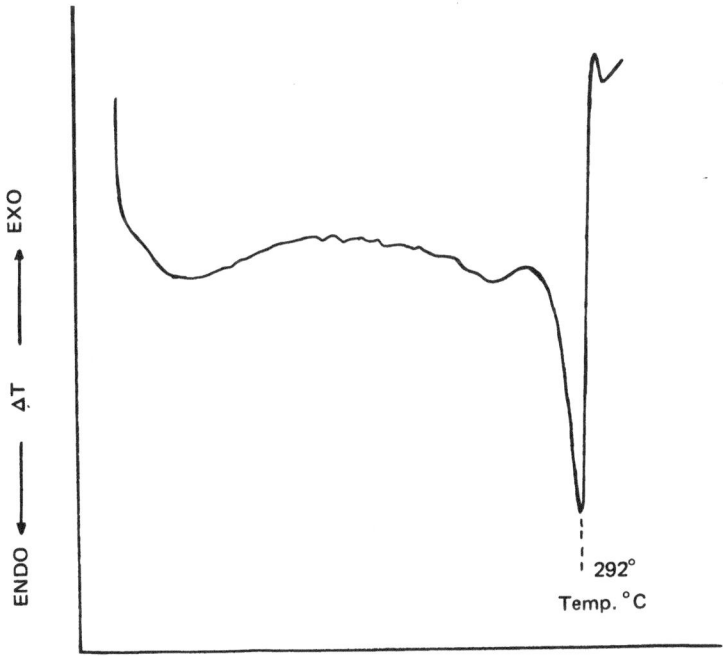

FIGURE 29. DSC curve of cotton phosphorylated with urea–phosphoris acid.[125]

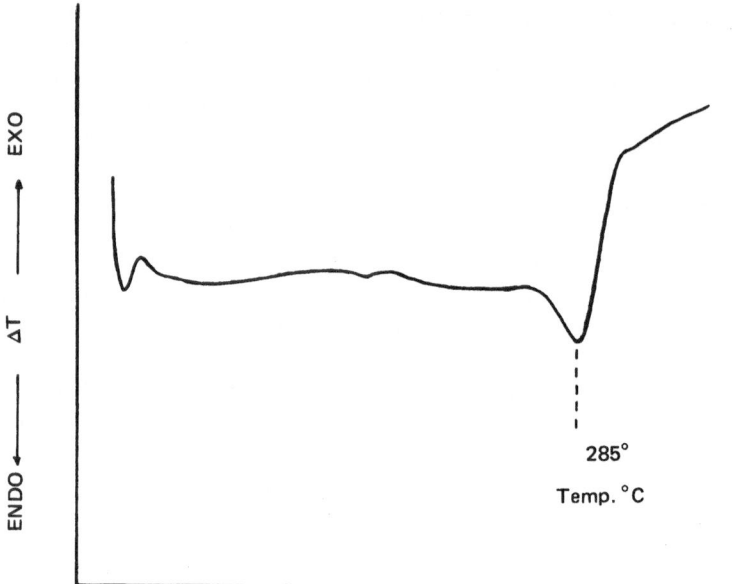

FIGURE 30. DSC curve of ramie impregnated with diammonium phosphate.[125]

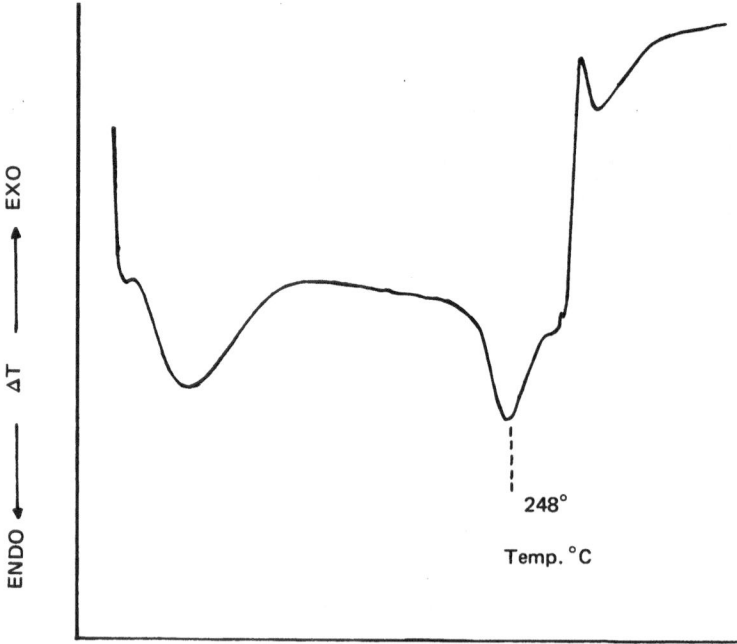

FIGURE 31. DSC curve of rayon phosphorylated with urea–phosphoric acid.[125]

7. References

1. P. H. Hermans, *Physics and Chemistry of Cellulosic Fibres*, American Elsevier, New York (1949).
2. A. Sharples, *J. Polym. Sci.* **13**, 393 (1954).
3. F. H. Forziati and J. W. Rowen, *J. Res. Nat. Bur. Stand.* **46**, 38 (1951).
4. M. Tsuboi, *J. Polym. Sci.* **25**, 159 (1957).
5. A. Basch, T. Wasserman, and M. Lewin, *J. Polym. Sci. Polym. Chem. Ed.* **12**, 1143 (1974).
6. H. J. Marrinan and J. Mann, *J. Appl. Chem.* **4**, *204 (1954)*.
7. R. Jeffries, D. M. Jones, J. G. Roberts, K. Selby, S. C. Simmens, and J. O. Warwicker, *Cellul. Chem. Technol.* **3**, 255 (1969).
8. A. Venkateswaren, *J. Appl. Polym. Sci.* **13**, 2469 (1969).
9. A. N. J. Heyn, *J. Cell Biol.* **29**, 181 (1966).
10. H. Z. Farnberg and N. V. Mikhailon, *Polym. Sci. USSR* **8**, 1805 (1966).
11. R. St. John Manley, *J. Polym. Sci. Part A-2*, **9**, 1025 (1971).
12. M. M. Y. Chang, *J. Polym. Sci. Polym. Chem. Ed.* **12**, 1349 (1974).
13. A. K. Kulshreshtha and N. E. Dweltz, *J. Polym. Sci. Polym. Phys. Ed.* **11**, 487 (1973).
14. S. Y. Lin, *Fibre Sci. Technol.* **5**, 303 (1972).
15. F. Shafizadeh, *Pure Appl. Chem.* **35**, 195 (1973).
16. S. Watanabe, J. Hayashi, and T. Akahori, *J. Polym. Sci. Polym. Chem. Ed.* **12**, 1065 (1974).
17. K. H. Gardner and J. Blackwell, *Biopolymers* **13**, 1975 (1974).
18. G. V. Schulz, *Angew. Chemi.* **13**, 417 (1974).
19. A. Pictet and J. Sarasin, *Helv. Chim. Acta* **1**, 87 (1918).
20. S. L. Madorsky, V. E. Hart, and S. J. Strauss, *J. Res. Nat. Bur. Stand.* **56**, 343 (1956); **60** 393 (1958).
21. F. H. Holmes and C. J. G. Shaw, *J. Appl. Chem.* **11**, 210 (1961).
22. F. A. Byrne, D. Gardiner, and F. H. Holmes, *J. Appl. Chem.* **16**, 81 (1966).
23. R. F. Schwenker, Jr., and E. Pascu, *Ind. Eng. Chem.* **2**, 83 (1957).
24. R. F. Schwenker, Jr., and E. Pacsu, *Ind. Eng. Chem.* **50**, 91 (1958).
25. D. P. Golova, A. M. Pakhomov, and E. A. Andrievskaya, *Proc. Acad. Sci. USSR, Chem. Sci. Div.* **112**, 3 (1957).
26. D. P. Golova, R. G. Krylova, *Dokl. Akad. Nauk SSR* **116**, 419 (1957).
27. D. P. Golova, A. M. Pakhomov, and I. Nikolaeva, *Proc. Acad. Sci. USSR, Chem. Sci. Div.* **112**, 533 (1957).
28. Y. Halpern, R. Ritter, and A. Broido, *J. Org. Chem.* **38**, 204 (1973).
29. A. Ohnishi, E. Takagi, and K. Kato, *Chem. Lett.*, Chem. Soc. Japan **1974**, 1361.
30. A. Broido, M. Evett, and C. C. Hodges, *West. States Sect. Combust. Inst. Pap.*, No. WSS/CI 74-33 (1974); *Chem. Abs.* **82**, 74670h (1975).
31. Y. Halpern and S. Patai, *Isr. J. Chem.* **7**, 673 (1969).
32. F. Shafizadeh and Y. L. Fu, *Carbohydr. Res.* **29**, 113 (1973).
33. E. J. Murphy, *J. Polym. Sci.* **58**, 649 (1962).
34. R. F. Schwenker and L. R. Beck, Jr., *J. Polym. Sci. Part C, Polym. Symp.* **1** (2), 331 (1963).
35. A. E. Lipska and G. E. McCasland, *J. Appl. Polym. Sci.* **15**, 449 (1971).
36. S. Glassner and A. R. Pierce, *Anal. Chem.* **37**, 525 (1965).
37. A. Basch and M. Lewin, *J. Polym. Sci. Polym. Chem. Ed.* **12**, 2053 (1974).
38. A. Broido and M. A. Nelson, *Combust. Flame* **24**, 263 (1975).
39. J. L. Eventova, A. P. Rudenko, I. I. Kulakova, M. M. Kanovich, V. O. Gorbachera, A. A. Konkin, N. S. Volkova, and N. F. Erofeeva, *Khim. Volokna* **4**, 29 (1974); *Chem. Abs.* **82**, 32258Z (1975).
40. B. Miller and T. M. Gorrie, *J. Polym. Sci. Part C*, **36**, 3 (1971).
41. H. Rodrig, A. Basch, and M. Lewin, *J. Polym. Sci. Polym. Chem. Ed.* **13**, 1921 (1975).

42. F. J. Kilzer and A. Broido, Speculations on the Nature of Cellulose Pyrolysis, WSS/CI Paper 64-4, U.S. Department of Agriculture, Washington, D.C. (1964).
43. G. D. M. McKay, Mechanism of Thermal Degradation of Cellulose, Forestry Branch (Canada) Dept. Public No. 1201 (1967).
44. D. F. Arseneau, *Can. J. Chem.* **49**, 632 (1971).
45. J. Derminot and C. Rabourdin-Belin, *Bull. Inst. Text. Fr.* **156**, 721 (1971).
46. R. J. McCarter, *Text. Res. J.* **42**, 709 (1972).
47. M. S. Barns, *Carbohydr. Res.* **34**, 169 (1974).
48. W. D. Major, *Tappi* **41**, 530 (1958).
49. F. Shimazu and C. Sterling, *J. Food Sci.* **31**, 548 (1966).
50. K. Kato and H. Komorita, *Agric. Biol. Chem.* **32**, 21 (1968).
51. B. Philipp, J. Baudisch, and C. H. Ruscher, *Tappi* **52**, 693 (1969).
52. N. Hurdoc and J. A. Schneider, *Bul. Inst. Politech. Iasi* **16**, 357 (1970).
53. P. K. Chatterjee and C. M. Conrad, *Text. Res. J.* **36**, 487 (1966).
54. S. Patai and Y. Halpern, *Isr. J. Chem.* **8**, 655 (1970).
55. M. Weinstein and A. Broido, *Combust. Sci. Technol.* **1**, 287 (1970).
56. Y. Halpern and S. Patai, *Isr. J. Chem.* **7**, 673 (1969).
57. K. Akita and M. Kase, *J. Polym. Sci. Part A* **5**, 833 (1967).
58. K. Kato and N. Takahashi, *Agric. Biol. Chem.* **31**, 519 (1967).
59. W. K. Tang, U.S. Forestry Service Research Paper FPL 71, Forestry Production Laboratory, Madison, Wis. (1967).
60. M. V. Ramiah and D. A. I. Goring, *Cellul. Chem. Technol.* **1**, 277 (1967).
61. M. V. Ramiah, *J. Appl. Polym. Sci.* **14**, 1323 (1970).
62. W. K. Tang and W. K. Neill, *J. Polym. Sci. Part C*, **6**, 65 (1964).
63. A. E. Lipska and W. J. Parker, *J. Appl. Polym. Sci.* **10**, 1439 (1966).
64. D. C. P. Fung, *Tappi* **52**, 319 (1969).
65. A. Basch and M. Lewin, *J. Polym. Sci. Polym. Chem. Ed.* **11**, 3071 (1973).
66. I. N. Andreeva, K. B. Ryzhov, S. P. Pankov, and V. A. Kargin, *Vysokomol. Soedin. Ser. B* **13**, 274 (1971).
67. A. Broido, A. J. Son, A. C. Ouano, and E. M. Barrall, *West States Sect. Combust. Inst.* (Pap.), No. 71-4 (1971).
68. M. V. Shablygin, P. M. Pachomov, K. H. Temnikov, and R. M. Levit, *Khim. Volokna* **16**, 72 (1974).
69. F. Shimazu and C. Sterling, *J. Food Sci.* **31**, 548 (1966).
70. B. Philipp, J. Baudisch, and A. Gaudy, *Faserforsch. Textiltech.* **18**, 9 (1967).
71. I. Rusznak and I. Tanczos, *J. Polym. Sci. Polym. Symp.* **42**, 1475 (1973).
72. S. Seitsonen and J. Mikkoner, *Polym. J.* **5**, 263 (1973).
73. P. K. Chatterjee, *J. Macromol. Sci. Chem.* **8**, 191 (1974).
74. E. L. Back, *Das Papier* **27**, 475 (1973).
75. M. Kimura, T. Hatakeyama, and J. Nakano, *J. Appl. Polym. Sci.* **18**, 3069 (1974).
76. R. H. Atalla and S. C. Nagel, *Science* **185**, 522 (1974).
77. R. H. Atalla and S. C. Nagel, *J. Polym. Sci. Polym. Lett. Ed.* **12**, 565 (1974).
78. A. Sueoka, J. Rayashi, and S. Watanake, *Nippon Kagaku Kaishi* **7**, 1345 (1973); *Chem. Abs.* **79**, 93629 (1973).
79. A. Basch and M. Lewin, *J. Polym. Sci. Polym. Lett. Ed.* **8**, 493 (1975).
80. E. L. Back and L. O. Klinga, *Svens. Papperstidn.* **66**, 745 (1963).
81. E. L. Back, *Transaction of the Cambridge Symposium*, Technical Section of the British Paper and Board Makers Association, Cambridge, England, (1965).
82. B. Philipp and W. Stohr, *Faserforsch. Textiltech.* **24**, 143 (1973).
83. R. M. Perkins, G. L. Drake, Jr., and W. A. Reeves, *J. Appl. Polym. Sci.* **10**, 1041 (1966).
84. C. H. Mack and D. J. Donaldson, *Text. Res. J.* **37**, 1063 (1967).

85. R. L. Hebert, *Tappi* **52**, 1183 (1969).
86. R. D. Gilbert and J. H. Rhodes, *J. Polym. Sci. Part C*, **30**, 509 (1970).
87. P. K. Chatterjee and R. F. Schwenker, *Tappi* **55**, 111 (1972).
88. A. Krassig, and W. J. Kitchen, *J. Polym. Sci.* **51**, 123 (1961).
89. A. M. Scallen, *Text. Res. J.* **41**, 647 (1971).
90. A. F. Roberts, *Combust. Flame* **14**, 261 (1970).
91. G. C. Gibbons, *J. Text. Inst.* **43**, T25 (1952).
92. K. Freudenberg, W. Kuhn, W. Durr, F. Boxz, and G. Steinbrunn, *Ber.* **63 B**, 1510 (1930).
93. K. Freudenberg and G. Blomquist, *Ber.* **68B**, 2070 (1935).
94. G. V. Schulz and J. J. Lohmann, *J. Prakt. Chem.* **157**, 238 (1941).
95. M. Lewin, A. Basch, and Ch. Rodrig, in *Proceedings of the International Symposium on Macromolecules, Rio de Janeiro, July 26–31, 1974* (E. B. Mano, ed.), pp. 225–250, Elsevier, Amsterdam (1975).
96. H. G. Higgins, *J. Polym. Sci.* **28**, 645 (1958).
97. M. A. Millet and V. L. Goedken, *Tappi* **48**, 367 (1965).
98. A. Basch and M. Lewin, *J. Polym. Sci. Polym. Chem. Ed.* **11**, 3095 (1973).
99. A. Basch and M. Lewin, *J. Fire Flammability* **4**, 92 (1973).
100. A. Basch and M. Lewin, *Text. Res. J.* **43**, 693 (1973).
101. B. Miller and R. Turner, *Text. Res. J.* **42**, 629 (1972).
102. S. Coppick, in *Flameproofing Textile Fabrics* (R. W. Little, ed.), pp. 50–54, A.C.S. Monograph Series, Van Nostrand Reinhold, New York (1947).
103. L. K. M. Lam, D. C. P. Fung, Y. Tsuchiya, and K. Sumi, *J. Appl. Polym. Sci.* **17**, 391 (1973).
104. Z. E. Jolles, *Bromine and its Compounds*, Ernest Benn Ltd., London (1966), p. 661.
105. A. Basch, Y. Halpern, T. Wasserman, and M. Lewin, *Cellul. Chem. Technol.* **5**, 533 (1971).
106. A. Basch, B. Hirschmann, and M. Lewin, *Cellul. Chem. Technol.* **7**, 255 (1973).
107. J. W. Lyons, The Chemistry and Uses of Fire Retardants, Wiley-Interscience, New York (1970), p. 19.
108. J. E. Hendrix, J. E. Bostic, Jr., E. S. Olsen, and R. H. Barker, *J. Appl. Polym. Sci.* **14**, 1701 (1970).
109. W. A. Sanderson, W. A. Mueller, and R. Swidler, *Text. Res. J.* **40**, 217 (1970).
110. J. E. Hendrix, G. L. Drake, Jr., and R. H. Barker, *J. Appl. Polym. Sci.* **16**, 257 (1972).
111. W. C. Arney, Jr., and W. C. Kuryla, *J. Fire Flammability* **3**, 183 (1972).
112. M. J. Drews and R. H. Barker, *J. Fire Flammability* **5**, 116 (1974).
113. W. A. Reeves, R. M. Perkins, B. Piccolo, and G. L. Drake, Jr., *Text. Res. J.* **40**, 223 (1970).
114. R. H. Barker, *Textile veredlung* **8**, 180 (1973).
115. K. Yeh and R. H. Barker, *Text. Res. J.* **41**, 932 (1971).
116. B. F. Gilliland and B. F. Smith, *J. Appl. Polym. Sci.* **16**, 1801 (1972).
117. J. J. Willard and R. E. Wondra, *Text. Res. J.* **40**, 203 (1970).
118. J. H. Hendrix, J. V. Beninate, G. L. Drake, Jr., and W. A. Reeves, *Text. Res. J.* **41**, 854 (1971).
119. J. H. Hendrix, G. L. Drake, Jr., and W. A. Reeves, *J. Fire Flammability* **3**, 38 (1972).
120. J. H. Hendrix, G. L. Drake, Jr., and W. A. Reeves, *J. Am. Assoc. Text. Chemists Colorists* **5**, 144 (1973).
121. J. L. Isaacs, *J. Fire Flammability* **1**, 36 (1970).
122. G. C. Tesoro, H. B. Sello and J. J. Williard, *Text. Res. J.* **39**, 180 (1969).
123. M. Abdul Kasem and H. R. Richards, *Ind. Eng. Chem. Prod. Res. Dev.* **11**, 114 (1972).
124. R. Aenishanslin, M. Bigler, Ch. Guth, P. Hoffman, and H. Nachbur, SIRTEC—1er Symposium International de la Recherche Textile Cotoniere Paris, *Inst. Text. Fr.* p. 703 (1969).
125. A. Basch and M. Lewin, *Text. Res. J.* **45**, 246 (1975).

2

Synergism and
Flame Retardancy

Y. P. Khanna and Eli M. Pearce

1. Introduction

This chapter is a review of the concept of synergism and its applicability to the flame retardation of polymers. Literature available on this subject has been analyzed to arrive at some understanding of the phenomenon and its mechanistic basis.

The most important commercial fire-retardant synergistic combinations (Sb–X, N–P, P–X, etc.) are dealt with in regard to the materials involved, application to specific polymer systems, and the mechanisms accompanying such reactions. The synergistic effect is brought about by gas phase and/or condensed phase reactions—one or the other may dominate depending on the polymer substrate, flame retardant, and the test conditions. This review suggests that condensed phase reactions leading to the formation of char (carbonaceous products) or other surface residues of low combustibility have special promise.

2. Synergism

The effect of a mixture of two or more flame retardants may be additive, synergistic, or antagonistic. When the effect of two flame retardants

---

Y. P. Khanna and Eli M. Pearce • Polytechnic Institute of New York, Brooklyn, New York.

is the sum of the effects of the two components taken independently, it is called an *additive* effect. *Synergism* is the case in which the effect of two or more components taken together is greater than the sum of their individual effects. An *antagonistic* effect is one in which the flame retardance is less than that predicted on an additive basis.

The development of synergistically efficient flame retardants leads to less expensive polymer systems with minimal effects on other desirable properties. Although the concept of synergism is useful, mechanistic interpretation of the phenomenon is still at an early stage.

3. Synergistic Reactions in Fire Retardation

Most of the currently used commercial fire-retardant additives are based on the following elements: Sb, Cl, Br, P, N, B, Al, Mg, Zn, and Sn. Although some of the synergistic interactions may include many combinations of these elements, we shall consider in detail only the three most important combinations, Sb–X (Cl, Br), P–X (Cl, Br), and N–P.

3.1. Antimony–Halogen Synergism

The addition of antimony (III) oxide to halogen-containing polymers is one of the classical illustrations of synergism observed in flame retardation. The role of antimony in enhancing the effectiveness of halogen-based flame retardants was first explained for cellulosic fabrics treated with chlorinated paraffins and Sb_2O_3.[1] The Sb–X synergism has also been indicated for polyester resins,[2–4] polystyrene resins,[5] polyolefins,[6] and polyurethanes, polyacrylonitrile, and polyamides.[7–8]

3.1.1. Antimony Compounds[9]

The most extensively used antimony compound is the trioxide (Sb_2O_3). In addition, several other antimony compounds have been patented and/or mentioned in the literature as flame retardants capable of exhibiting synergism when used in the presence of halogen-containing materials.

The tetra- and pentaoxides of antimony have also been investigated for their synergistic effect in comparison to the trioxide.[10–12] For example, Pitts[9] evaluated the three oxides for synergism with chlorine in a 4-phr H_2O, 1.4 lb/ft^3 density-free rise-flexible urethane formulation; the ASTM D1692-68 flammability tests indicated that Sb_2O_3 is much more effective than either of the other two.

Studies on the synergism exhibited with other elements of Group V of the periodic table, namely As and Bi, show that these elements also exhibit

TABLE 1
Various Metal Oxides and Compounds of Antimony as Partial or Total
Replacements for $Sb_2O_3^a$

Partial or total substitutes for Sb_2O_3	Polymer substrate	Ref.
$Al_2O_3 \cdot 3H_2O$	Rubber products	14, 15
	Polyesters	16
Fe_2O_3(or SnO_2) + Sb_2O_3	Cotton	17, 1
10% Sb_2O_3, SnO, SnO_2, or PbO	Nylon + 6% chlorine	18
ZnO	Coatings containing chloroparaffins	19
TiO_2 + Sb_2O_3	Polyurethanes + halogen	20
Fe_2O_3 + TiO_2 + CuO	Flexible urethane + chlorine	9
$2ZnO \cdot 3B_2O_3 \cdot 3\frac{1}{2} H_2O$	Polyvinyl chloride	21
	Polyesters	
	Chlorinated polyesters	22
$SbCl_3$	Fire-resistant coatings	23
Sb_2S_3	Polyvinyl chloride	24
	Rigid urethane foam	25
SbOCl	Flexible urethane foam	9
$NaSbO_3$	Polyvinyl cloride	
	Polyethylene + chlorinated paraffin	26
$KSbC_4H_2O_6 \cdot H_2O$	Polyesters + chloroparaffins	27

^{a}Compiled from Ref. 9.

synergism with chlorine and bromine similar to that of Sb. A comparison of the synergistic flame-retardant effect of the trioxides of Sb, As, and Bi in chlorinated polyethylene[13] indicate an effective order of $Sb_2O_3 > As_2O_3 > Bi_2O_3$.

There is no question about the practical use of Sb_2O_3; however, its high cost and dependency on import have encouraged the development of some substitutes which could serve at least as a partial replacement in some specific polymer systems. Table 1 summarizes some of the antimony compounds that are in part or totally capable of replacing Sb_2O_3.

3.1.2. Halogen Compounds[9]

Theoretically any halogen-containing compound which can decompose to give a hydrogen halide can be used with an antimony compound to produce a synergistic effect. However, the overall effect depends on the halogen content, its ease of liberation, and the type of halogen.

Lyons[7] concludes that on an equal molar basis the ease of halogen liberation is I > Br > Cl > F. In practice, iodine-based compounds can be eliminated on the basis of high cost and their instability to heat and light. Also, fluorine compounds have little practical importance since the fluorine

TABLE 2

Halogenated Compounds Exhibiting Synergistic Flame Retardancy with Sb_2O_3[a]

Compound		Substrate	Ref.
Name	Structure		
Polyvinylidene fluoride	$(CH_2CF_2)_n$	Polyurethanes	28
Hexachloroethane	C_2Cl_6	Polyesters	29
Polyvinyl chloride	$(CH_2CHCl)_n$	Polystyrene	30
Chlorendic anhydride		Polyesters	31
Tetrachlorophthalic anhydride		Polyesters	32
Chlorendic anhydride		Epoxies Polyesters Polyurethanes	33 34 34
Pentabromotoluene		Polyesters	16
Hexabromobiphenyl		Propylene Styrenics	35 34
Tetrabromophthalic anhydride		Epoxies Polyesters Polyurethanes	36 34 34
Tribromophenol		Rigid polyurethanes	37

TABLE 2 (*Continued*)

Compound		Substrate	Ref.
Name	Structure		
Tetrabromobisphenol-A		Epoxies	16
		Polycarbonates	34
		Polyesters	34
Dechlorane Plus® (Hooker)		Polyolefins	34
Dechlorane 604® (Hooker)		Styrenics	34

[a]Compiled from Refs. 9 and 34.

is so strongly bound to the carbon that apparently the potentially active HF is not released at the appropriate time for synergist formation.[7]

The most important halogens in practical use for flame retardancy are bromine and chlorine. Table 2 summarizes some of the halogenated compounds that have been proposed as synergists when used in combination with Sb_2O_3.

3.1.3. Mechanism of Sb–X Synergism

In polymer systems containing Sb_2O_3 and halogen, it has often been suggested that fire retardance is mainly due to gas phase flame inhibition by the volatile antimony trihalide.[7,8,38–40] Figure 1 shows that the maximum flame retardancy with the Sb_2O_3–halogen combination is observed when the mole ratio Sb/X is 1:3.

In addition to influencing the flame inhibition, it has been shown that the Sb_2O_3–halogen combination can also influence the condensed phase chemistry and enhance char formation to an extensive degree in some polymers.

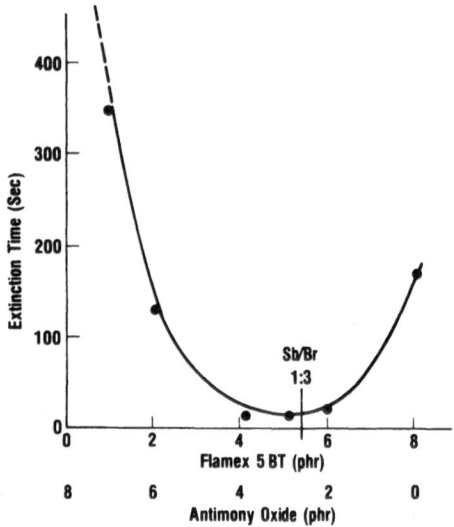

FIGURE 1. Synergistic antimony–halogen flame retardancy in polyester laminates.[9]

Hastie[41] has described the criteria for the solid phase and the gas phase modes of flame inhibition (Table 3). Let us consider these two modes of flame inhibition to explain Sb–X synergism. The HCl formed by the degradation of the chlorinated organic material reacts with Sb_2O_3 to yield SbOCl, and

TABLE 3

Determinants of Solid Phase and Gas Phase Modes of Flame Inhibition[a]

Mechanism of flame inhibition	
Solid phase	Gas phase
1. Increase in char formation	1. Flame retardant is lost from the substrate.
2. Flame retardant is retained in the substrate	2. Flammability is insensitive to the structure of the substrate
3. Flame retardant has no effect in the gas phase	3. Flame retardancy is sensitive to the oxidant phase
4. Flame retardancy is governed by the structure of the substrate	4. The composition of the volatiles doesn't change or changes only slightly
5. Flammability is insensitive to the oxidant phase, e.g., O_2 or N_2O	
6. Composition of the volatiles from the pyrolysis of the substrate changes in the presence of the flame retardant	

[a] Based on Ref. 41.

subsequent reactions explain the increased flame retardancy observed by earlier workers.[1] The formation of an intermediate by the combination of Sb_2O_3 and chlorine is also confirmed by thermoanalytical data[42]; e.g., the individual TGA's of Sb_2O_3 and a chlorinated paraffin showed a 0 and a 67% weight loss, respectively, in the same temperature range. Assuming there was no interaction between the components of this system, a 34% weight loss would be expected from a 50:50 mixture. Instead, a 72% weight loss was observed in the temperature range under consideration.

In a typical substrate of $R \cdot HCl$ (chlorine source) + Sb_2O_3 polymer, the formation of SbOCl is illustrated by the following equations:

$$R \cdot HCl \xrightarrow{\sim 250°C} R + HCl$$

$$2HCl + Sb_2O_3 \xrightarrow{\sim 250°C} 2SbOCl + H_2O$$

Working under the assumption that Sb_2O_3 is the primary product of the interaction between Sb_2O_3 and a chlorine source, Pitts et al.[43] have studied the thermal decomposition of SbOCl (Fig. 2). They proposed the

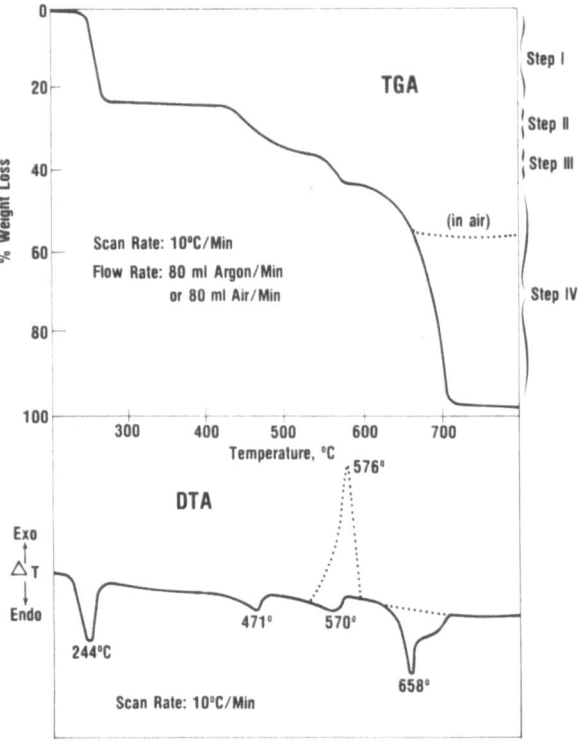

FIGURE 2. Thermal decomposition of antimony oxychloride (SbOCl).[43]

following reaction sequence:

Step I: $5\,SbOCl(s) \xrightarrow{245\text{--}280^\circ C} Sb_4O_5Cl_2(s) + SbCl_3(g)$

Step II: $4\,Sb_4O_5Cl_2(s) \xrightarrow{410\text{--}475^\circ C} 5\,Sb_3O_4Cl(s) + SbCl_3(g)$

Step III: $3\,Sb_3O_4Cl(s) \xrightarrow{475\text{--}565^\circ C} 4\,Sb_2O_3(s) + SbCl_3(g)$

Step IV: $Sb_2O_3(s) \xrightarrow{685^\circ C} Sb_2O_3(l)$

Accordingly, the thermal decomposition of SbOCl results in the evolution of gaseous $SbCl_3$ in three endothermic steps, through the temperature range 245–565°C.

The endothermic release of $SbCl_3$, in combination with the flammable degradation products, is considered to be responsible for the flame-retardance effect observed in flexible urethane foams containing SbOCl.[9] The thermal decomposition of SbOCl occurs within an appropriate temperature range which happens to match the degradation temperature range of the foam itself. Metal oxides such as CaO and ZnO were found to have a negative effect when used as partial replacements for Sb_2O_3. The thermogravimetric analysis (TGA) data reveal that both of these oxides raise the decomposition temperature of SbOCl by 25–50°C. On the other hand, Fe_2O_3 and CuO proved to be quite useful even in catalytic amounts. Interestingly, the TGA study indicates that both Fe_2O_3 and CuO lower the decomposition temperature of SbOCl by 50–100°C (Fig. 3). However, the TGA curves of the foams containing these oxides and Sb_2O_3 showed no significant differences. These observations suggest that both Fe_2O_3 and CuO enhance the flame retardancy by lowering the energy barrier for the transformation of SbOCl to

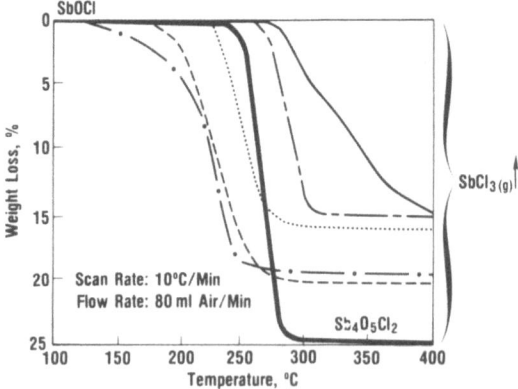

FIGURE 3. Thermogravimetric analysis curves of antimony oxychloride-metal oxide mixture. (———) SbOCl, (— • —) 8SbOCl + 2CuO, (— — —) 8SbOCl + 2Fe$_2$O$_3$, (-------) 10SbOCl + 5TiO, (- —— -) 10SbOCl + 5CaO, (———) 10SbOCl + 5ZnO.[9]

$SbCl_3$. Therefore, there is a strong support that in Sb_2O_3–halogen synergistic reactions, $SbOCl$ is formed as an intermediate which on thermal decomposition evolves gaseous $SbCl_3$, the actual flame retardant. Further evidence for this is as follows:

1. The flame extinction time for polyester laminates shows a minima at an SblCl ratio of 1:3 (Fig. 1).
2. For a flexible foam formulation, the flame retardancy improved significantly at first, but only little further improvement was observed after reaching an SblCl ratio of 1:3 (Fig. 4).
3. In chlorinated polyethylene containing Sb_2O_3, Fenimore and Jones established from oxygen index studies that maximum values are obtained at theoretical Sb/Cl ratios of 1:3 or greater.[44]
4. Our work[45] on the flammability of styrene–vinyl benzyl chloride copolymers also indicates that a maximum on the oxygen index is observed at an Sb/Cl ratio of 1:3.

Assuming that SbX_3 is the actual flame retardant, a number of explanations for this have been put forward.[7,8,38–40]

Antimony trichloride is considered to be a "free-radical trap" in the gas phase of the burning system. Another explanation suggests that trivalent antimony facilitates the generation of halogen radicals which interfere with the concentration of the flame-propagating radicals, such as $H\cdot$, $HO\cdot$, $O\cdot$, and perhaps $CH_3\cdot$ and $HOO\cdot$.[5] Others believe that $SbCl_3$ delays the rate of escape of halogen from the flame zone and thereby increases the chances of interferences with the flame-propagating species.[46] Existence of volatile $SbCl_3$ could also act by blanketing the flame,[38] or, more probably,

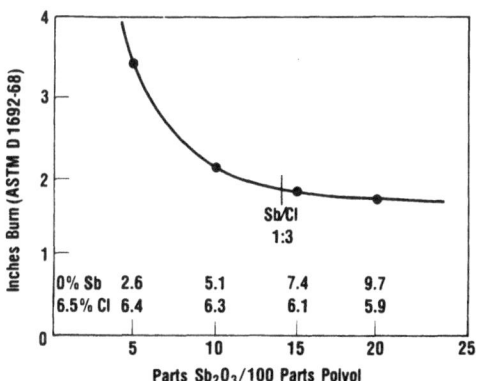

FIGURE 4. Flammability of flexible polyurethane foam containing chlorine and antimony oxide.[9]

condensed liquid or solid SbCl$_3$ particles could reduce the energy of the flame by the wall or surface effect.[39,46,47]

Martin and Price[48] have investigated the flammability of epoxy resins using triphenyl phosphine and triphenyl antimony as flame retardants. Both proved to be effective flame inhibitors without the presence of any halogen. Here the antimony appears to act in retarding the gas phase flame reactions since it behaves differently in the presence of N$_2$O–N$_2$ and O$_2$–N$_2$ atmospheres. Therefore, for this case it is reasonable to assume tnat only antimony may be the real flame quencher.

The bulk of evidence to date leads one to conclude that Sb–X synergism is the result of the formation of an SbOX intermediate which then decomposes to form the volatile trihalide, SbX$_3$, which serves as the means of transporting trivalent antimony into the gas phase where it then functions in its role as a flame inhibitor.

Antimony–halogen flame-retardant combinations also increase the char yield in some polymer systems.[30] For example, when cellulose treated with Sb$_2$O$_3$ and a chlorine compound is heated in air, charring is observed at 280°C—a temperature at which untreated cotton remains unaffected.[49] Thus, it may be postulated that antimony halides can function not only in the gas phase but also in the solid phase by dehydrating, dehydrogenating, and dehydrohalogenating the pyrolyzing polymer phase.

In the flame retardation of polymers containing antimony–halogen fire-retardant additives, either the gas phase or the condensed phase or both of the mechanisms may contribute depending on the flammable substrate. For example, Dechlorane Plus with antimony oxide in polyolefins gives a more substantial char yield than other related polycyclic chlorocarbons,[34]

3.2. Phosphorus–Halogen Synergism

In spite of the practical importance of flame retardants containing both halogen and phosphorus, the explanation for the existence of phosphorus–halogen synergism is still doubtful.

Lyons[7] has summarized the typical levels of phosphorus, halogen, and phosphorus–halogen combinations required to attain self-extinguishing properties in acrylic, cellulose, epoxy, polyester, polyolefin, polystyrene, and polyurethane systems. Lyons figures indicate that for polyolefins, polyacrylates, and epoxy resins, lower amounts of phosphorus and halogen are required in combination than those calculated on a linear additivity model. However, polyesters and polyacrylonitrile follow the linear model predictions, whereas the data on polyurethanes point out a phosphorus–halogen antagonism. Based on a thorough study on flammability of several polymer substrates containing various phosphorus–halogen combinations, Weil[34] has concluded that depending on the substrate and the particular phosphorus–halogen compound combinations used, synergistic, additive, or antagonistic effects may be observed.

3.2.1. Phosphorus–Halogen Compounds

Some of the phosphorus–halogen fire-retardant combinations used commercially are summarized in Table 4.

3.2.2. Mechanism of P–X Synergism

Although the practical importance of flame retardants containing both halogen and phosphorus is well known, there is little evidence in the literature to strongly support proposed phosphorus–halogen synergisms. Moreover, in cases where the existence of the synergism appears to be true, a mechanistic explanation is lacking.

Recently Papa and Proops[59] have studied the flame retardancy of flexible polyurethane foams made from polyols containing bromine and phosphorus to illustrate the concept of synergism. Several complexities were found in terms of phosphorus and bromine content (separately and together) with regard to oxygen index, char yield, and char formation. According to their charring studies, most of the phosphorus originally present in the foam was found in the char. Unexpectedly, elemental analysis of the char revealed the presence of substantial amounts of bromine. Like phosphorus, bromine also enhanced the char formation. Ionic bromine appeared to be the most effective elemental form. Charring studies at 300°C revealed that the chars tended to retain ionic bromine and phosphorus in a 1:1 ratio, whereas chars at 500°C afforded 1:2.5–3 ratio of total bromine to phosphorus. On this basis, Papa and Proops suggest a specific chemical interaction leading to these ratios. More interestingly, foams containing aromatically bonded bromine did not show any constant P/Br ratio in the char, and in some cases there was complete loss of bromine. These studies suggest that the mode of flame retardance and the existence of synergism can be both structure and concentration dependent.

Although there is no strong evidence for a general phosphorus–halogen synergism, hypotheses have been put forward to explain this concept. The formation of flame-retardant phosphorus halides and oxyhalides from phosphorus additives and halogen donors are postulated in the literature.[40,60] The phosphorus–halogen compounds have been considered to be either free-radical quenchers or flame retardants working in some other manner. However, the bond energy considerations appear unfavorable for the formation of P–X compounds from, for example, phosphorus esters and carbon–halogen compounds.[7]

TABLE 4

Some Fire-Retardant Combinations Based on Phosphorus–Halogen Compounds

Phosphorus compounds	Halogen compounds	Phosphorus–halogen compounds	Polymer substrate	Ref.
Phosphoric acid		2,3-Dibromopropyl phosphate	Polypropylene	50
[ClCH₂CH₂O)₃P]			Polystyrene	51
(ClCH₂CH₂PCl₂)		2-Bromoethyl diglycol phosphate	Rigid polyurethane foam	52
	Epichlorohydrin			53
	Tetrachlorophthalic anhydride			54
Phosphorines	Chloroethylene oxide		Unsaturated polyesters	55
Phosphoric acid	Dibromo succinic acid			56
Tricresyl phosphate				57
Cresyl diphenyl phosphate				
Isodecyl diphenyl phosphate				
2-Ethylhexyl diphenylphosphate	Dibromopentyl glycol		Polyvinyl chloride	58
Phosphate containing polyol				
Diethyl-N,N-bis(2-hydroxyethyl aminomethyl phosphonate)				
Tris(dipropylene glycol)phosphite			Flexible polyurethane foam	59

3.3. Nitrogen–Phosphorus Synergism

It is well established that adding nitrogen often reduces the need for phosphorus.[58] Compounds containing both nitrogen and phosphorus have been used for a long time as fire retardants. For example,

1. A combination of phosphoric acid and urea was used in a pad-dry cure phosphorylation process to produce greater flame resistance in cotton. It was realized that the combined effect of phosphoric acid and urea was greater at a lower concentration than either of the two used alone.[61]
2. Some of the durable flame-retardant finishes for cotton employ tetrabis(hydroxyl methyl)phosphonium chloride together with ammonia, urea, or trimethylol melamine.[62]
3. Other examples include ammonium phosphate used for flame-retarding wood, paper, and cotton and N-methylol dialkyl phosphoric propionamide used for children's sleep wear.

In most of the above cases, the role of nitrogen was believed to be influencing the attachment of the phosphorus to the cellulosic. Recently, the interaction of nitrogen and phosphorus has been evaluated on a more quantitative basis.

Willard and Wondra[63] studied the nitrogen and phosphorus interaction in cotton and viscose rayon, treated with various finishes, using the oxygen index (OI) for evaluation. Their results for Pyrovatex CP [i.e., $(CH_3O)_2P(O)CH_2CH_2CONH-CH_2OH$]/trimethylol melamine-treated cotton sheeting, clearly indicate the increasing slopes of OI vs. % P as the % N is increased (Fig. 5) or of OI vs. % N as the % P is increased (Fig. 6). The nearly perfect linear relationships between OI and % P or % N in this study strongly suggests a synergistic interaction between phosphorus and nitrogen.

Willard and Wondra's results also indicate that there is no specific N/P ratio at which the synergistic effect is at a maximum. Instead, depending on the concentration of phosphorus and the structure of the nitrogen compound, synergism, additivity, or antagonism may be observed.

The literature contains only scattered examples of nitrogen–phosphorus synergism in polymers other than cellulose. Moreover, in polymers where the existence of N–P synergism appeared to be true, the mechanism has not been investigated in detail.

Nitrogen compounds such as urea, cynamide, guanidine, and methylol melamines accelerate the phosphorylation of cellulose.[64-67] Based on this, one may expect the N–P synergism to be related to the phosphorylation of cellulose as the first step. Recently, Hendrix et al.[68] have shown that the pyrolysis of cellulose treated with phosphates or phosphoramides proceeds through the formation of cellulose phosphate or phosphoramide esters

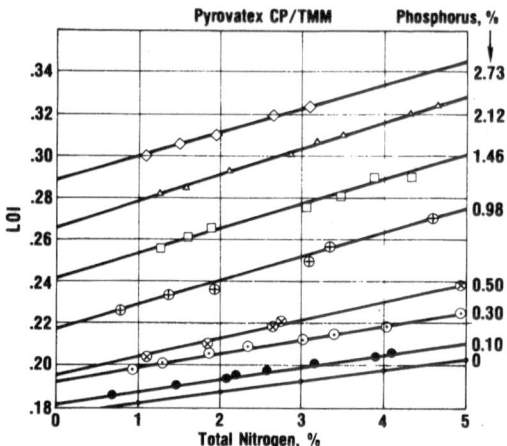

FIGURE 5. Oxygen index as a function of nitrogen content at fixed phosphorus levels for Pyrovatex CP/trimethylol melamine-treated cotton sheeting.[62]

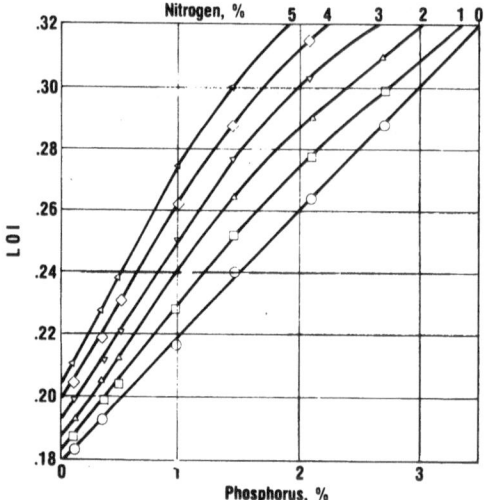

FIGURE 6. Oxygen index as a function of phosphorus content at fixed nitrogen levels for Pyrovatex CP/trimethylol melamine-treated cotton sheeting.[62]

(phosphorylation reaction) followed by subsequent ester pyrolysis to yield dehydrated cellulose char. Formation of phosphoramides during pyrolysis of flame retardants containing P and N constitutes a possible basis for the reported N–P synergism observed in commercial flame retardants. Flame retardancy of the mixed N and P systems is related to the extent to which this intermediate is formed. A thermally stable cross-linked residue is obtained on pyrolysis of the phosphorylated cellulose. Such cross-linking seems to be quite effective in enhancing flame retardation.

3.4. Synergism with Free-Radical Initiators

The free-radical synergist is a more recent development in the flame retardation of polystyrene. By this method, additive requirements have been reduced so that the effects on physical properties are relatively small. Polypropylene and cellulose acetate have also been compounded with free-radical initiators to meet "self-extinguishing" requirements.[62]

Eichhorn[69] has reported the synergism in flame-retarding polystyrene with the following initiators: dicumyl peroxide, azobisisobutyronitrile, *p*-quinone dioxime, quinone dioxime dibenzoate, 2,6-dichloroquinone chlorimide, benzothiazole sulfenamides, and some disulfides. Various hypotheses have been offered to explain the mechanism of such synergistic behavior. The observations that (1) the synergistic effect is decreased by free-radical inhibitors,[69] (2) the synergist reduces the melt viscosity of the burning polymer[70,71] and the flaming material drips readily, and (3) the prevention of dripping either by including glass fibers or testing the samples in a shallow platinum boat suppresses the synergistic action may lead one to conclude that the synergism proposed with free-radical initiators is in fact an artifact arising from a decrease in melt viscosity of the polymer, subsequent dripping, and/or melt flow which translates to an improved flammability value with a specific test method.

3.5. Synergism and Condensed Phase Reactions

As discussed in the foregoing sections, there are two primary mechanisms which give rise to synergism; one plays a role in the gas phase, and the other influences the solid phase. Synergism arising from the solid or condensed phase reactions leads to the formation of char or other surface residues of low combustibility and offers the most promising route to fire resistance. Char formation results in reduced total fuel production since the polymer on degradation leaves much of its carbon as a relatively noncombustible residue rather than yielding volatile combustible fragments.

Our preliminary work on fire retardance through char formation also indicates that solid phase reactions play a significant role in flame retardation[45] In an attempt to flame-retard polystyrene, we copolymerized styrene with vinyl benzyl chloride and studied the copolymers for flammability, charring, and synergism alone and with Sb_2O_3 or ZnO.

Charring studies of the copolymers at 500 and 1100°C reveal that the experimental char yield is always higher than the theoretical char—the latter

TABLE 5

Char Yield and Oxygen Index of Styrene and Vinyl Benzyl Chloride (VBC)
Homo- and Copolymers

Polymer[a]	Composition		% Char at[b]		Oxygen index[c] (%)
	% Cl	Mole % VBC	500°C	1100°C	
Polystyrene	0.0	0	0	0	19
Styrene–VBC copolymer	6.7	22	20	7	43
	13.5	49	47	28	49
	16.6	64	47	28	36
	18.1	72	43	23	36
Polyvinyl benzyl chloride	23.0	100	55	32	50

[a] Prepared by free-radical polymerization using benzoyl peroxide (1 mole %); reaction temperature and time were 80°C and about 1 hr, respectively.
[b] From TGA studies; DuPont 951 TGA, 10-mg sample in Pt pan, 10°C/min heating, argon atmosphere.
[c] Measured using General Electric Oxygen Index Apparatus.

calculated on the basis that polystyrene leaves no residue at these temperatures. These preliminary results are summarized in Table 5.

Both Sb_2O_3 and ZnO showed a synergistic effect when combined with the copolymers. Possibly solid phase as well as gas phase mechanisms are operable. Enhanced cross-linking via ether link formation, alkylation reactions, and others may occur. Some reactions could be accompanied by the elimination of H_2O which may act endothermically or act as a heat transfer agent, thus acting as a flame inhibitor. The exact mechanism for flame retardation in these cases have not been elucidated.

4. Synergism and Future Studies on Fire Retardation

In developing fire-resistant compositions for polymers, synergism allows the utilization of flame-retarding additives at lower concentrations, avoiding the undesirable effects on product performance properties coupled with minimum cost. Thus the concept of synergism appears to be quite important for the future development of flame-retardant polymer compositions. Research efforts which lead to an understanding of the principles of charring should proceed along the following lines:

1. The determination of modifications in polymer structure that assist in char formation.
2. The development of blends or additives that cross-link with the polymer substrate and thus lead to char formation.
3. The study of reagents that can induce or accelerate the charring process.

Probably the most effective way to achieve fire resistance is an approach which combines the benefits of flame inhibition with those of char formation. It appears at this point that a great deal of research still needs to be done to accomplish this goal and that an understanding of the mechanisms involved should go a long way toward accomplishing it.

5. References

1. R. W. Little, *Flameproofing Textile Fabrics*, ACS Monograph Series, No. 104, Van Nostrand Reinhold, New York (1947), p. 240.
2. R. C. Nametz, *Ind. Eng. Chem.* **59**, 99 (1967).
3. G. S. Learmonth, A. Nesbitt, and D. C. Thwaite, *Br. Polym. J.* **1**, 149 (1969).
4. G. S. Learmonth and D. C. Thwaite, *Br. Polym. J.* **1**, 154 (1969).
5. R. F. Lindeman, *Ind. Eng. Chem.* **61**, 70 (1969).
6. C. P. Fenimore and F. J. Martin, *Combust. Flame* **10**, 135 (1966).
7. J. W. Lyons, *The Chemistry and Uses of Fire Retardants*, Wiley, New York (1970).
8. G. L. Tesoro, Synergism During Fire Retardation, presented at Polymer Conference Series on Flammability Characteristics of Polymeric Materials, University of Utah, Salt Lake City (June 1971).
9. J. J. Pitts, *J. Fire Flammability* **3**, 51 (1972).
10. W. M. Layham, U.S. Pat. 3,159,591, (Dec. 1,1964) (to Union Carbide Corp.); **62**, 7639a (1965).
11. P. J. Catenrino and R. M. McGlamery, U.S. Pat. 3,113,118 (Dec. 3, 1963) (to Phillips Petroleum Co.); **60**, 5751 (1964).
12. M. Peters, D. Schleede, H. *Jochinke, and H. Klug*, U.S. Pat. 3,025,262 (March 13, 1962) (to Farbwerke Hoescht A. G.).
13. C. P. Fenimore and F. J. Martin, *Combust. Flame* **10**, 135 (1966).
14. H. Frood and H. P. Alger, Br. Pat. 183,922 (April 30, 1921); *Chem. Abs.* **17**, 228 (1923).
15. K. C. Hecker, *Rubber World* **159** (3), 59 (Dec. 1968).
16. A. G. Walter, *Br. Plast.* **42** (7), 128 (1969).
17. L. A. Jordan and L. A. O'Neill, Br. Pat. 573,472 (Nov. 22, 1945).
18. R. R. Lunt, Jr., U.S. Pat. 3,440,211 (Jan. 23, 1968) (to E. I. DuPont de Nemours and Co.).
19. M. Leatherman, U.S. Pat. 2,472,112 (June 7, 1949); *Chem. Abs.* **43**, 6763q (1949).
20. Raychem. Corp., Neth. Pat. Appl. 6,603,028 (Sept. 1966); *Chem. Abs.* **66**, 56209e (1967).
21. U.S. Borax and Chemical Corp., Belg. Pat. 720,520 (Sept. 9, 1968).
22. J. G. Bower, in *Proceedings of the Flame Retardant Cotton Batting Workshop*, New Orleans, La. (Nov. 15, 1968), p. 60.
23. W. K. Hearn, U.S. Pat. 2,247,633 (July 1, 1939); *Chem. Abs.* **35**, 64737 (1941).
24. A. G. Walker, *Br. Plast.* **42** (7), 128 (1969).
25. C. J. Hilado, W. C. Kuryla, R. W. McLaughlin, and W. R. Proops, *J. Cell. Plast.* **6** (5), 215 (1970).
26. M&T Chemical's Technical Data Sheet No. 128, Thermogard®FR, revised (1969).
27. J. E. Deneich, U.S. Pat. 2,924,532 (Feb. 9, 1960) (to Diamond Alkali Co.) *Chem. Abs.* **54**, 11584h (1960).
28. Raychem. Corp., Neth. Pat. Appl. 6,603,028 (Sept. 1966); *Chem. Abs.* **66**, 56209e (1967).
29. A. A. Samuel, Ger. Pat. 1,155,602 (Oct. 10, 1963); *Chem. Abs.* **60**, 727g (1964).
30. Dynamit-Nobel A. G., Neth. Pat. Appl. 6,411,769 (April 13, 1965).
31. B. T. Hayes, W. J. Read, and L. H. Vaughan, *Chem. Ind. London* **35**, 1162 (Aug. 31, 1957).

32. Y. C. Chae, The Role of Halogenated Fire Retardants in Inhibiting Polymer Flammability, presented at Polymer Conferences Series, University of Utah, Salt Lake City (June 18, 1970).
33. Weyenhaeusen Co. (by J. D. Frichette), Belg. Pat. 629,381 (Oct. 21, 1963); *Chem. Abs.* **60**, 13404h (1964).
34. E. D. Weil, Additivity, Synergism and Antagonism in Flame Retardance, Plastics Institute of America, Stevens Institute of Technology, Hoboken, New Jersey (June 1971).
35. J. Kestler, *Mod. Plast.* **47**, 96 (Sept. 1970); *ibid.*, **44** (1), 102 (1966).
36. Peter Spence & Sons, Ltd. (by L. Williams), Br. Pat. 1,010,204 (Nov. 17, 1965); *Chem. Abs.* **64**, 38023 (1966).
37. Z. E. Jolles, *Plast. Inst. Trans. J. Conf. Suppl.* No. 2, 3–8 (1967); *Chem. Abs.* **61**, 22374k (1967).
38. J. A. Rhys. *Chem. Ind. London* **187** (Feb. 1969).
39. W. G. Schmidt, *Trans. J. Plast. Inst.* 247 (Dec. 1965).
40. I. N. Einhorn, Fire Retardance of Polymeric Materials, presented at Polymer Conference Series on Flammability Characteristics of Polymeric Materials, University of Utah, Salt Lake City (June 15, 1970).
41. J. W. Hastie, *J. Res. Nat. Bur. Stand. Sect. A* **77** (6) (Nov.–Dec. 1973).
42. I. Toubal, Halogen Synergists for Flame Retardation, at Stevens Institute of Technology, Hoboken, New Jersey (June 1971).
43. J. J. Pitts, P. H. Scott, and D. G. Powell, *J. Cell. Plast.* **6**, 35 (1970).
44. C. P. Fenimore and G. W. Jones, *Combust. Flame* **10**, 295 (1966).
45. Y. P. Khanna and E. M. Pearce, unpublished results (1975).
46. J. A. Rhys and R. F. Cleaver, *Plastics Rubber Weekly* 20 (Nov. 13, 1970).
47. M. G. Langer, Evaluation of Fire Retardants by Mass Spectrometric Thermal Analysis, presented at Polymer Conference Series on Flammability Characteristics of Polymeric Materials, University of Utah, Salt Lake City (June 15, 1970).
48. F. J. Martin and H. R. Price, *J. Appl. Polym. Sci.* **12**, 143 (1968).
49. N. J. Reed and E. G. Heighway-Borg, *J. Soc. Dyers Colour.* **74**, 823 (1958).
50. G. J. Listner, U.S. Pat. 3,403,118 (Sept. 24, 1968) (to Johnson & Johnson).
51. K. H. Krause, *Konstst. Rundsch.* 337 (Aug.) 1959; Ger. Pat. 1,046,313 (to Badische Anilin and Soda Fabrik A. G.) (Dec. 1958).
52. A. Heslinge and P. J. Napjus Ger. Pat. 1,166,465 (March 26, 1964) (to Chemische Fabrik Kalk G.m.b.H.) **60**, 16076 (1964).
53. G. W. McCary, Jr., and P. L. Smith, U.S. Pat. 3,419,642 (Dec. 31, 1968) (to Union Carbide Corp.).
54. U. Bahr, K. H. Andres, and G. Braun, Ger. Pat. 1,098,707 (Feb. 2, 1961) (to Barb. Bayer A. G.).
55. F. Rochlitz and M. Vilesek, Ger. Pat. 1,165,262 (March 12, 1964) (to Farb. Hoeschst A. G.).
56. Dynamit Nobel A.G., Belg. Pat. 671,780 (Nov. 3, 1965).
57. Chemische Werke Albent, Neth. Pat. Appl. 6,412,649 (May 3, 1965).
58. V. R. Sawhney, Synergism in Flame Retardance, at Stevens Institute of Technology, Hoboken, New Jersey (June 1971).
59. A. J. Papa and W. R. Proops, *J. Appl. Polym. Sci.* **16**, 2361 (1972).
60. C. J. Milado, *Flammability Handbook for Plastics*, Technomic Publishers, Westport, Conn. (1969).
61. R. W. Little, *Combust. Flame* **12**, 191 (1968).
62. W. C. Kuryla and A. J. Papa, *Flame Retardancy of Polymeric Materials*, Vol. 3, Marcel Dekker, New York (1975).
63. J. J. Willard and R. E. Wondra, *Text. Res. J.* **40**, 203 (1970).
64. J. Bancroft and Sons Co., Br. Pat. 604,197 (1948).
65. F. V. Davis, J. Findley, and E. Rogers, *J. Text. Inst.* **40**, T839 (1949).

66. T. H. Marton, Br. Pat. 634,690 (1950) (to Courtaulds Ltd.).
67. S. J. O'Brien, *Text. Res. J.* **38,** 256 (1968).
68. J. E. Hendrix, G. L. Drake, Jr., and R. H. Barber, *J. Appl. Polym. Sci.* **16,** 257 (1972).
69. J. Eichhorn, *J. Appl. Polym. Sci.* **8,** 2497 (1964).
70. E. V. Govinlock, J. F. Porter, and R. R. Hindersinn, *J. Fire Flammability* **2,** 207 (1971).
71. C. P. Fenimore, *Combust. Flame* **12,** 155 (1968).

3

Ignition of Polymers

Bernard Miller and J. Ronald Martin

1. Introduction

For a polymer to ignite, it must first decompose. While low-molecular-weight condensed materials may produce volatile combustible species through sublimation or evaporation processes, one of the fundamental characteristics of the genus polymer is the lack of a boiling point. Thus, the only mechanism by which volatiles can be produced from a pure polymer must be some form of partial or complete thermal decomposition.

All the available experimental evidence suggests that what we define as ignition (that is, initiation of self-sustaining flaming combustion for an observable time) occurs in the gas phase. The most vivid demonstration of this was reported by Summerfield et al.,[1] who, with the aid of high-speed photography, showed that when a polymer was placed in a stream of high-temperature air ignition occurred only downstream of the solid. This experiment was performed at very high flow rates, and, in consequence, Brownian motion and other diffusional effects were not noticeable factors. It is not difficult to accept the idea that decomposition products of a polymer serve as fuel and that to bring about ignition they must come in contact with oxygen in the air (unless there is some other oxidizing agent present). However, we do not as yet have a clear enough understanding of

Bernard Miller and J. Ronald Martin • Textile Research Institute, Princeton, New Jersey.

the controlling mechanism involved, which may include both chemical and physical processes.

Since thermally induced polymer decomposition is the first step in the ignition process, most experimental and theoretical studies can be classified according to how heat is supplied to the polymer. The three most important modes of supply are by radiation, conduction, and conduction–convection. Radiant heat sources have been preferred by most investigators, since there is a substantial body of established theory and practice regarding how to carry out and analyze radiant heating experiments. Much of the earlier work done on cellulosic materials was reviewed by Welker[2] in 1970; some later writers on this subject have been Alvares and Martin,[3] Shivadev and Emmons,[4] and Welker *et al.*[5,6] All of these investigations dealt with the same basic experiment: a polymer exposed to a radiant heat flux in an environment of unheated air or other oxygen-containing gas. The only other source of heat was, in some cases, a small heated wire or flame not in contact with the polymer (e.g., piloted ignition). A number of independent variables were reported to have an influence on the ignition process, such as applied heat flux, oxygen partial pressure, total gas pressure, etc. Usually, the dependent variables studied were ignition time and/or the apparent rise in surface temperature of the polymer. Accurate measurements of the latter are difficult to obtain, which may account for the contradictions concerning so-called ignition temperatures which appear in the literature. Ohlemiller and Summerfield[7] reported on the use of a laser beam to ignite polymers and pointed out that ignition by radiant heating in unheated air appeared to be much slower than when a hot gas was used as a heat source.

Radiant heating ignition experiments suffer from certain inherent limitations which make them poor sources of useful information. Color changes during pyrolysis will produce changes in a polymer's absorbance, and thus the effective rate of energy absorbed will not be constant. To get around this, some investigators blacken their materials beforehand, which should make their data more reliable but may not tell us how the polymer itself behaves. Even a small amount of additive (e.g., a dye) can affect the thermal decomposition behavior of a polymer. Another major limitation of radiation-induced ignition is that it is very difficult to carry out controlled experiments on thermoplastic polymers. Obviously, maintaining a constant heat flux on a melting or moving target is a difficult problem, and investigators have opted for more intractable targets such as cotton, paper, wood, and thermoset resins. A third objection is a practical one; in evaluating ignition behavior as a hazard factor, the case where all the thermal energy comes from a high-intensity radiant source represents a rather special circumstance. The more likely situation would include the presence of hot gases or solids in contact with the polymer.

Ignition brought about by conductive heat transfer from a hot surface is certainly a realistic circumstance; however, it is extremely difficult to perform systematic experiments of this type. A characteristic shared by many polymers, particularly when in film or fabric form, is a tendency to move (i.e., shrink, curl, twist, etc.) when exposed to heat. Therefore it is difficult to arrange for constant contact between (say) a heated wire and a polymer surface. Data reported for such experiments would be best ignored unless the investigator has taken considerable pains to deal with this problem. Conductive heat transfer from a rapidly moving hot fluid (studied by Summerfield's group[1] can be reproducible but represents a very extreme condition not likely to occur in any actual situation.

Heat transfer through a combination of conduction and convection, such as when a polymer is exposed to heated air, is the most common mode whereby polymers ignite. This includes ignition by exposure to a flame where the main function of the flame is to supply hot gas to heat the polymer.

Experimental studies on ignition of polymers by contact with heated air (usually called autoignition) have been reported by various workers. In the majority of cases, the primary objective was the establishment of the minimum air temperature that would produce ignition. Setchkin[8] appears to have been the first to propose this as a means for characterizing polymers, and his method is still extant at this time.[9] If one reads the original paper describing this method, it is obvious that the author offered it only as a means for determining a procedural property and in no way suggested that the "ignition temperature" obtained in this manner was the temperature of the polymer surface. Unfortunately, misuse of the Setchkin method appears to have fostered the idea that a polymer must be heated to a specific temperature in order for it to ignite. This criterion appears as an accepted fact in many of the published theoretical analyses of polymer ignition.[3,5,10] Ignition temperature data for many polymers have been reported by investigators using a variety of heat transfer modes,[11,12,13] but rarely is there agreement between different authors. In some cases the reported values are ridiculously low or high.

In our opinion, the idea that a polymer must reach a certain temperature in order to ignite has no tenable foundation in theory or experimental observation. A perusal of the available publications on the subject reveals that there is little justification, *a priori* or otherwise, for this assumption. On the other hand, one can demonstrate quite easily that a polymer can be heated well above its "ignition temperature" without undergoing ignition. Such experiments are carried out routinely in any laboratory where thermal analysis studies (e.g., DTA, TG, etc.) are performed. In such instances, ignition does not occur because of the low rates of heating used in conventional thermal analysis. As will be shown, the critical factor for polymer ignition is the rate of energy input (which in turn determines the rate of thermal decomposition).

Ignition brought about by contact with hot air can be a highly reproducible phenomenon if the experiment is carried out with proper appreciation and control of independent variables and if provision is made to determine as the primary response the ignition time (or ignition delay, as it is sometimes called). The most critical variables are sample mass and surface area, sample mounting, air temperature, and the initial contact between sample and air. Once these factors are controlled and accounted for, autoignition time data can be used to reveal a wealth of useful information about the phenomenon of polymer ignition. In the next section we shall describe in detail one way this can be done.

2. Autoignition Studies

2.1. Experimental Technique

To obtain useful autoignition data it is necessary to arrange for the determination of ignition delay time under carefully controlled conditions. A schematic diagram of the apparatus used for this purpose at Textile Research Institute is shown in Fig. 1. The air is heated by means of electrical resistances within the cylindrical oven with provisions made for some preheating when a flowing atmosphere is used. The oven temperature is regulated by a proportional temperature controller and is monitored at a location right next to where the sample will be. The sample is impaled onto the prongs of a sample holder which is attached to a 0.15-m-long ceramic rod. The sample is automatically injected into the oven by means of a pneumatic system; the injection time is less than 0.5 sec and highly reproducible. The moment of injection and moment of ignition are detected from the time-based readout of a thermocouple attached to the sample holder between the sample prongs, near the sample but not touching it. From this trace of thermocouple output versus time (see Fig. 1) two inflection points can be clearly identified: one corresponding to the beginning of heating at the time the sample is injected into the oven and the other to the moment of ignition. The time lapse between these two points represents the ignition delay time, θ. Experimental values of θ reported herewith represent a mean average of five to ten independent determinations of θ with a standard deviation generally of 0.5 sec or less.

The samples are usually single or multiple layers of fabric or film (25 × 13 mm) stapled to a stainless steel screen of the same size. (Samples may also consist of yarns evenly wrapped around the metal screen; this arrangement is especially convenient for studying multicomponent systems, i.e., combina-

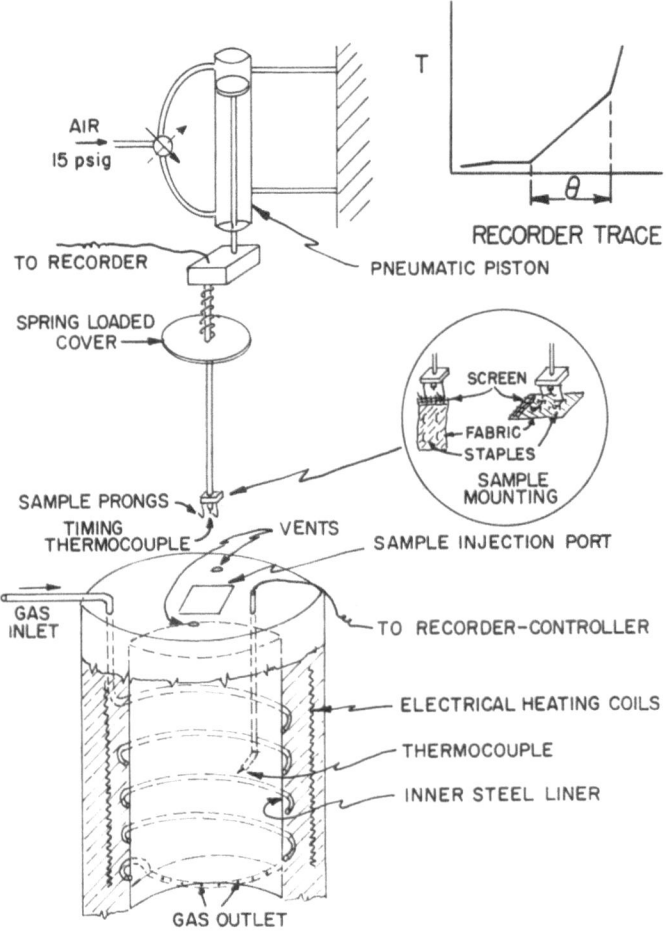

FIGURE 1. Schematic diagram of autoignition apparatus.

tions of two or more polymers.) The screen allows one to work with thermoplastic samples, preserving their general shape and preventing dripping long enough to allow ignition to occur at the desired location. One can impale nonthermoplastic materials by themselves directly onto the prongs of the sample holder; however, nearly all experiments have been carried out with samples in contact with the metal screen so that both melting and nonmelting materials can be compared on the same basis. The screen serves as a minor heat sink which competes with the sample for heat and lengthens the time to ignition. This time increment is usually small and depends on the air temperature.

As shown in Fig. 1, samples can be injected either vertically or horizontally. However, all the data reported herein refer to the vertical arrangement. Experiments can also be performed using air flows from 0 to 10 cm^3/sec without affecting the results; the data are usually obtained at flows of approximately 2 cm^3/sec.

2.2. General Kinetic Model of the Ignition Process

Initially, studies using this technique showed that ignition time depends on a number of experimental parameters, among them the mass of the polymer sample.[14] Therefore, ignition times are usually determined as a function of sample mass with sample area constant (i.e., as a function of thickness); this is accomplished by using multilayered samples. For the areal density range encountered in fabric, film, and yarn samples, the relationship between sample mass and ignition time at a fixed air temperature has been found to be linear for all systems investigated. A typical result is shown in Fig. 2 where ignition times are plotted as a function of sample mass for a cotton fabric at four air temperatures. For each temperature the linear relationship is extrapolated to zero mass to obtain what is termed the intrinsic ignition time, θ_0. This extrapolation can be considered as a means

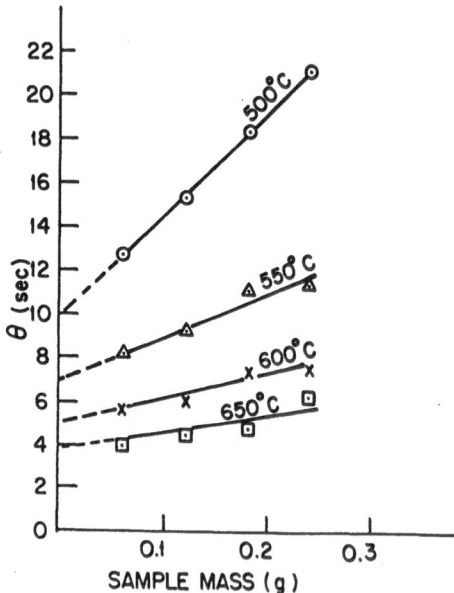

FIGURE 2. Autoignition behavior of cotton fabric stapled to wire screen; ignition time vs. sample mass.

of eliminating the bulk effects of mass, so that the intrinsic ignition time represents the ignition time for an infinitely thin sample, one, in effect, which has been heated instantaneously to its decomposition temperature. This will be elaborated on in the following discussion in which the auto-ignition process is considered to be resolvable into a series of consecutive events.

As a first approach the dependence of ignition time on sample mass was interpreted in terms of two sequential controlling mechanisms in the auto-ignition process: (1) heating of the material up to its initial decomposition temperature and (2) thermal decomposition leading +o the formation of combustible gases which eventually ignite. The contributions of these two processes to the ignition time are shown in Fig. 3 as hypothetical plots of sample temperature and extent of decomposition versus time with sample mass as a parameter. Below the decomposition temperature, sample temperature increases with time in a decaying exponential manner until decomposition is initiated. Above the decomposition temperature, sample temperature still increases in a decaying exponential manner but at a slightly slower rate because of the absorption of heat by endothermic degradation

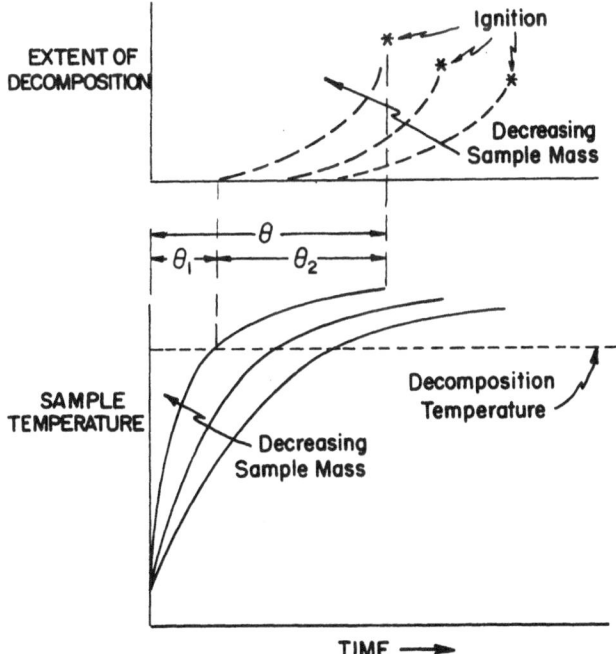

FIGURE 3. Hypothetical plots of the variation of sample temperature and extent of decomposition with time at several values of sample mass.

processes (except for materials such as polyacrylonitrile which have exothermic decompositions[15]). Once the initial decomposition temperature has been exceeded, the degree of degradation increases with time in an accelerated exponential manner (as a result of increasing sample temperature) up to the point of ignition.

As illustrated in Fig. 3, the total time to ignition, θ, should consist of two parts: the time required to heat the sample to its decomposition temperature, θ_1, and the remaining time required for the concentration of combustible gases diffusing from the sample to reach an explosive level and ignite, θ_2. From heat transfer considerations, θ_1 would be expected to increase linearly with sample mass if there were no significant influence of bulk thermal conductivity (that is, the sample was thermally thin). Heat transfer studies of similar systems have shown this to be a valid assumption.[16] On the other hand, θ_2 should be practically independent of sample mass. Therefore, the linear extrapolation of the ignition time–sample mass relationship to zero mass eliminates θ_1 and gives us an ignition time for the material if it were instantaneously heated to its decomposition temperature. From this it follows that at a given temperature θ_0 may be assumed equal to θ_2. The general argument is illustrated in Fig. 4, which shows the relationship between ignition time and sample mass and the contributions of θ_1 and θ_2 at finite sample masses.

This hypothetical resolution of the total ignition time into two sequential steps has been tested experimentally by determining the extent of polymer decomposition (i.e., weight loss) as a function of oven residence time. This enables one to directly determine the time required for decomposition to be initiated; this can then be compared with the value of θ_1 as inferred from a plot of ignition time versus sample mass as shown in Fig. 4. Figure 5 shows the weight retentions for one, two, four, and six layers of cotton fabric sewn

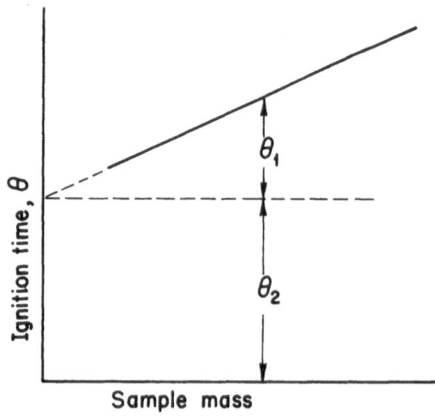

FIGURE 4. General form of the relationship between ignition time and sample mass, showing the mass-dependent and -independent components of the observed time for ignition.

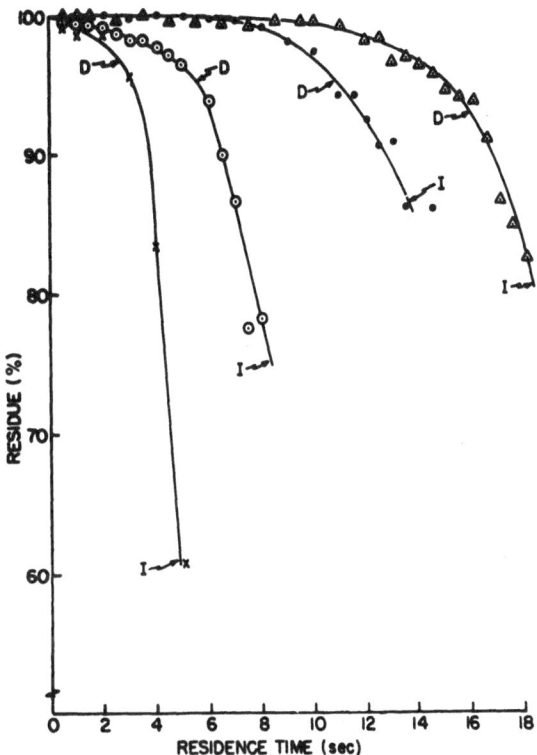

FIGURE 5. Percent weight retention as a function of oven residence time for cotton fabric prior to ignition; D = start of decomposition as determined from ignition time–sample mass relationship; I = ignition. Air temperature = 550°C; $\times$ 0.046 g; $\odot$ 0.100 g; ● 0.194 g; $\triangle$ 0.280 g.

together with cotton thread (no metal screen backing) as a function of residence time in the oven with the temperature set at 550°C. (The weights of the samples are 0.046, 0.100, 0.194, and 0.280 g, respectively.) These data were obtained by interrupted heating experiments in which the samples were withdrawn quickly from the oven after predetermined times, quenched in a stream of cold nitrogen, and then weighed. Ignition time data (Fig. 6) were used to determine values of θ_1, the times after which significant decomposition should begin. These times are indicated by D on the curves in Fig. 5. Although some weight loss does occur before the theoretically predicted decomposition points, these weight losses are no greater than 7%, so that the hypothesis of ignition time resolution into two steps, which is fundamental to subsequent interpretations of autoignition data, can be considered valid.

In Fig. 5, the actual time of ignition for each of the samples is also indicated. From this it is apparent that ignition does not depend on the

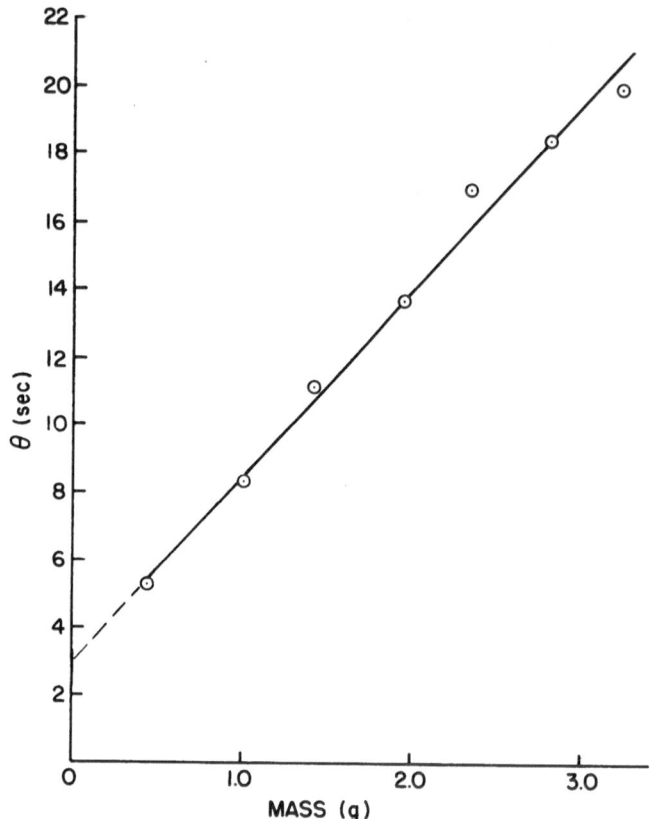

FIGURE 6. Autoignition behavior of cotton fabric used for Fig. 5. Air temperature = 550°C.

decomposition of a specific fraction of the material. In fact, the smaller samples decomposed more before igniting. Considering the extent of weight loss in absolute terms, as shown in Fig. 7, also reveals that ignition did not occur after a common weight loss. More likely, the critical factor is the attainment of a sufficiently high rate of weight loss (i.e., rate of combustible fuel generation). This idea is consistent with the very high rates of decomposition at the instant of ignition for each of the samples indicated in Figs. 5 and 7.

2.3. Autoignition Data for Single-Component Systems

Autoignition data have been obtained for multiple-layered samples of a wide assortment of textile materials. Figures 8 through 20 show relationships between sample mass and ignition time at a number of air temperatures for

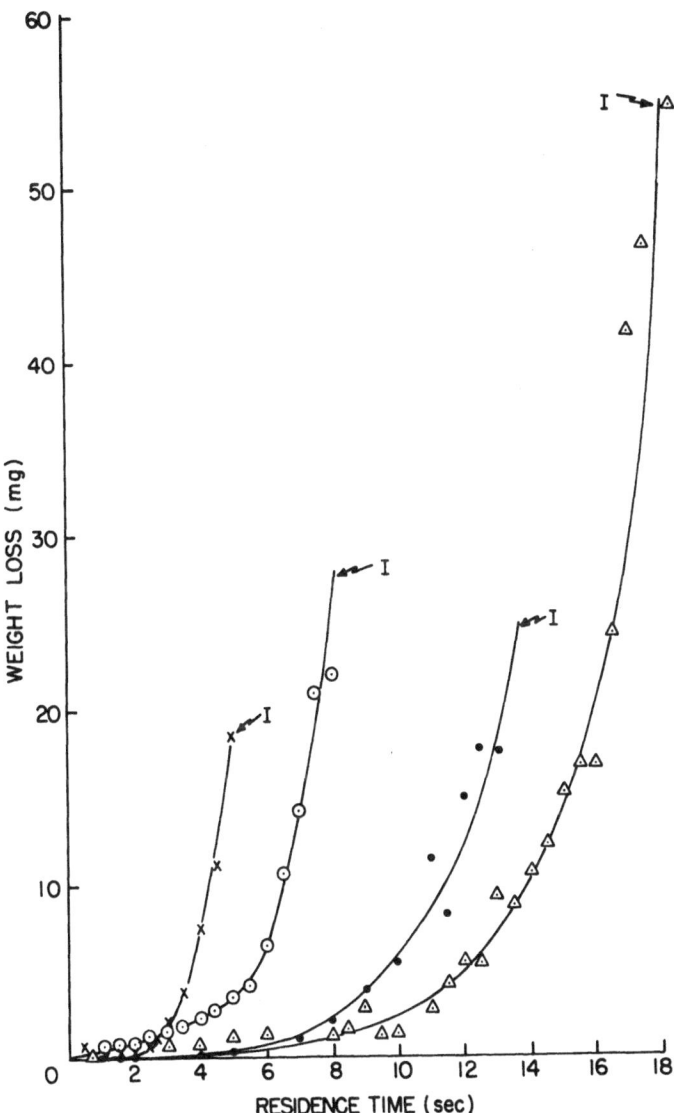

FIGURE 7. Weight loss as a function of oven residence time for cotton fabric used in Figs. 5 and 6 prior to ignition; I = ignition. Air temperature = 550°C; × 0.046 g; ⊙ 0.100 g; ● 0.194 g; △ 0.280 g.

Bernard Miller and J. Ronald Martin

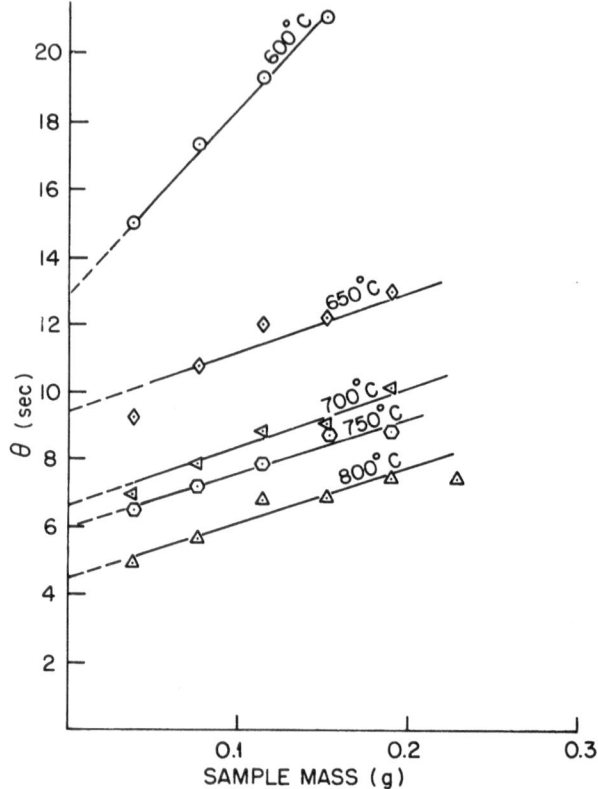

FIGURE 8. Autoignition of polyethylene terephthalate fabric (Dacron®).

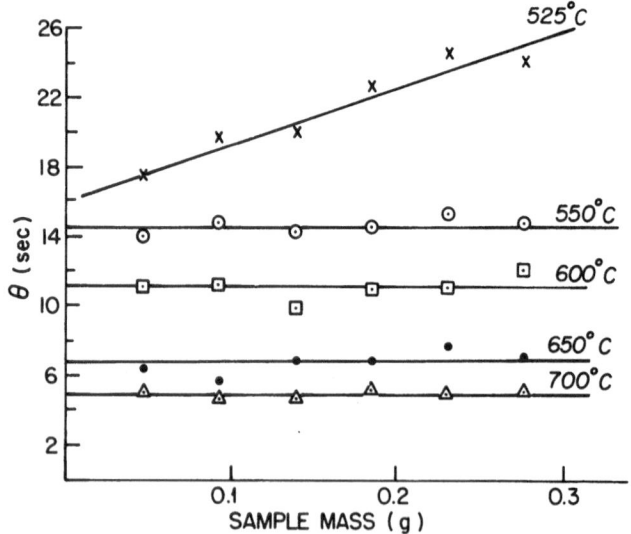

FIGURE 9. Autoignition of polyacrylonitrile fabric (Orlon®).

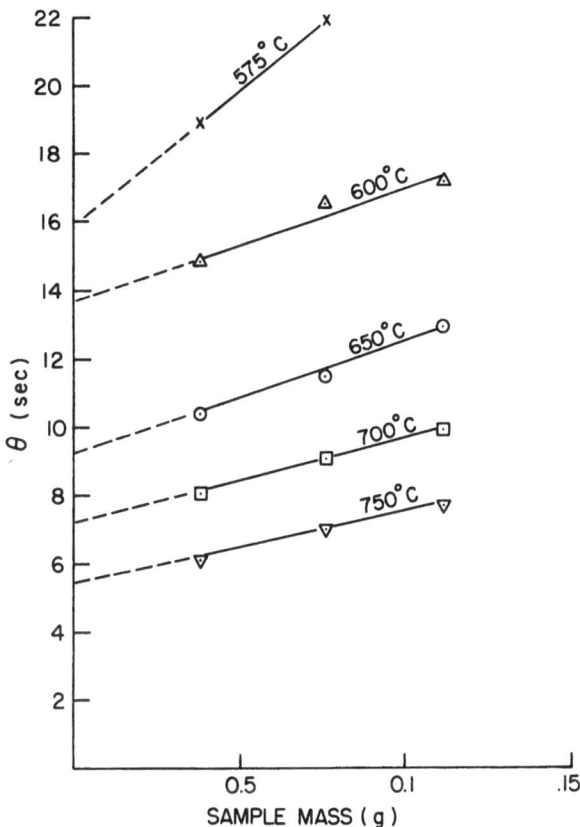

FIGURE 10. Autoignition of polypropylene fabric.

fabrics of polyethylene terephthalate (Dacron®), polyacrylonitrile (Orlon®), polypropylene, cellulose acetate, polyvinyl chloride (Rhovyl®), rayon, wool, Nomex®, nylon 66, novaloid (Kynol®), polybenzimidazole, modacrylic (Dynel®), and Kermel®. Over the mass range investigated, all these materials display a linear relationship between sample mass and ignition time. The data for each air temperature have been extrapolated to zero mass to give an intrinsic ignition time, θ_0.

It is of interest to note that the data for Orlon® (Fig. 9) at all but the lowest temperature investigated (525° C) show an ignition time independent of sample mass over the mass range investigated. All other materials studied show a decreasing dependence of ignition time on sample mass with increasing temperature.

The dependence of intrinsic ignition time on air temperature can be described in the form of Arrhenius plots (log $1/\theta_0$ vs. $1/T$), as shown in

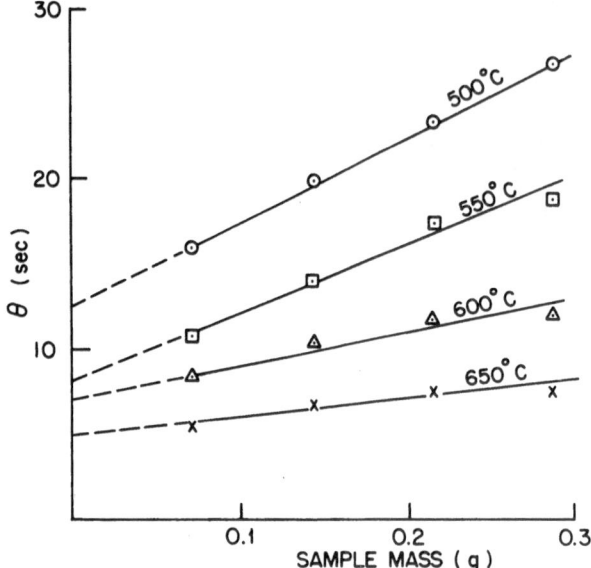

FIGURE 11. Autoignition of cellulose acetate fabric.

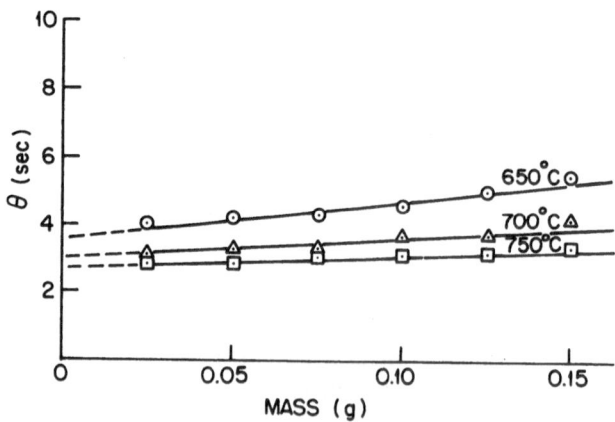

FIGURE 12. Autoignition of polyvinyl chloride (Rhovyl®) nonwoven fabric.

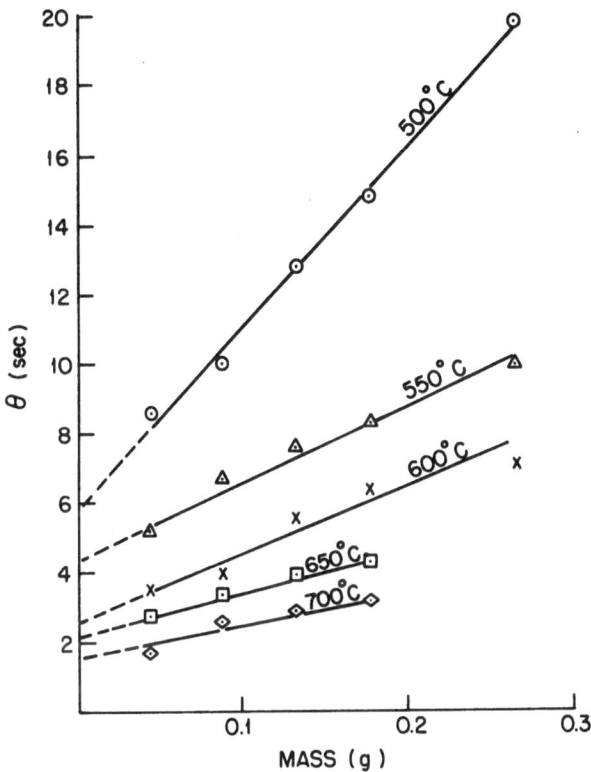

FIGURE 13. Autoignition of rayon challis fabric.

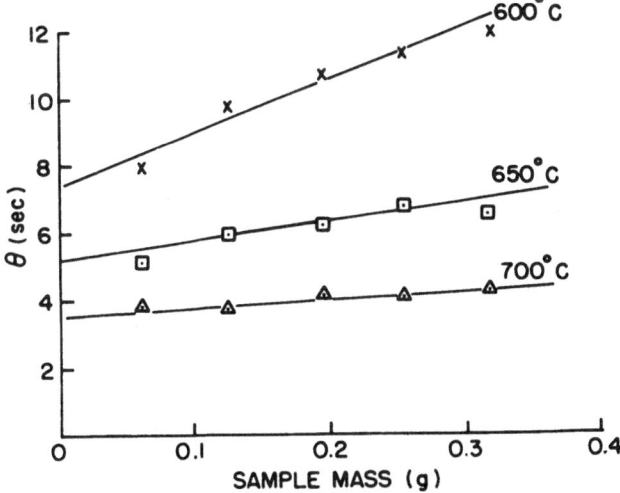

FIGURE 14. Autoignition of wool fabric.

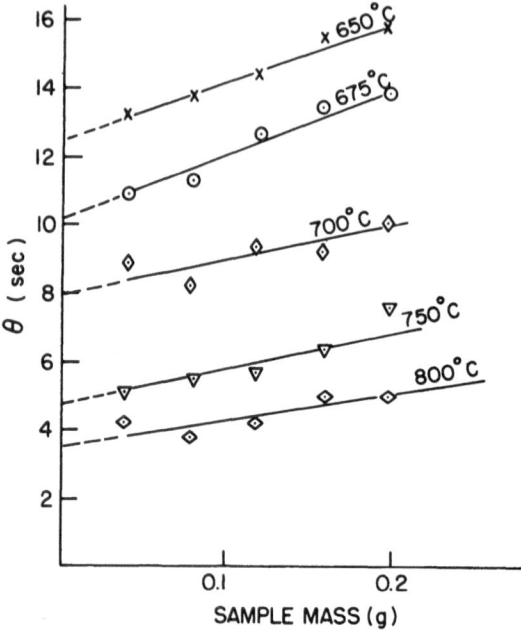

FIGURE 15. Autoignition of Nomex® fabric.

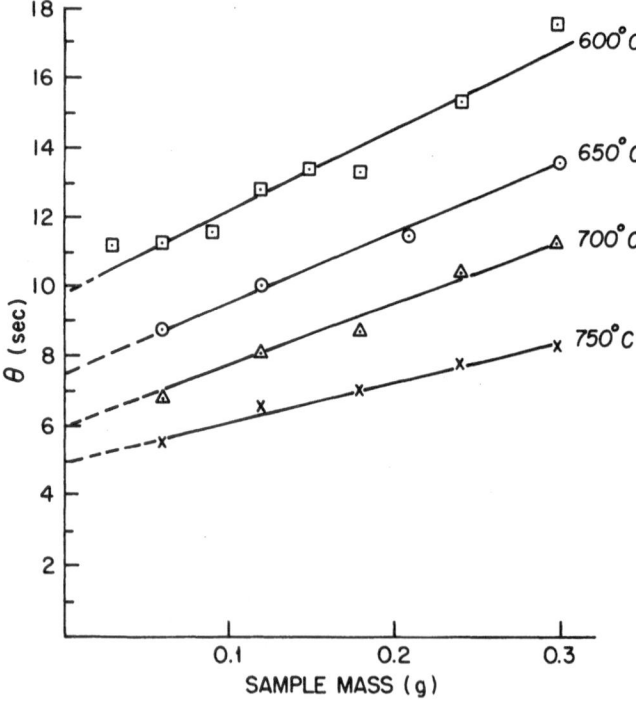

FIGURE 16. Autoignition of nylon 66 fabric.

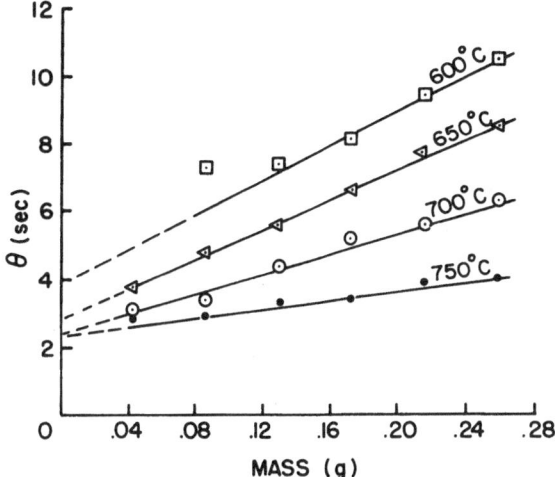

FIGURE 17. Autoignition of novaloid fabric (Kynol®).

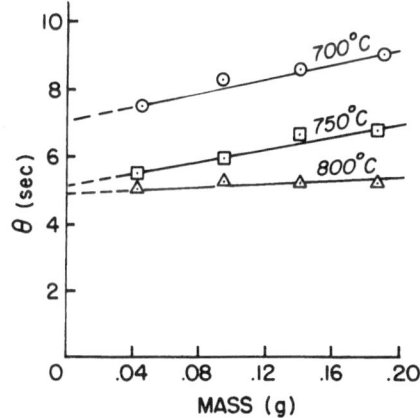

FIGURE 18. Autoignition of polybenzimidazole (PBI) fabric.

Fig. 21 for some of the materials investigated. The relationship is consistently linear with an apparent activation energy (obtained from the slopes) corresponding to 8–10 kcal/mol, except for Nomex® and Kermel®, which display activation energies of 13 and 17 kcal/mol, respectively. These low activation energies coupled with the fact that the same activation energy is obtained for nearly all species investigated suggest that the rate-controlling step is some physical process (heat or mass transfer) which would be expected to have such a low activation energy. It should be kept in mind that these activation energies were obtained using

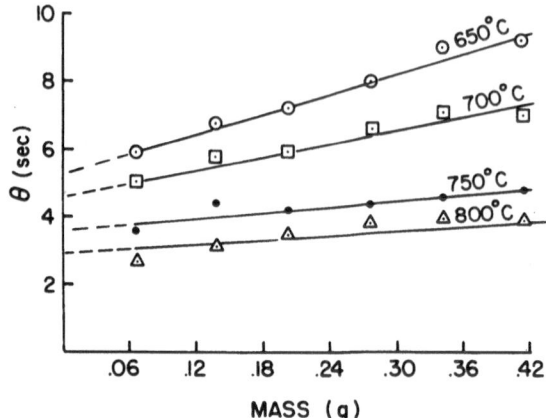

FIGURE 19. Autoignition of modacrylic fabric (Dynel®).

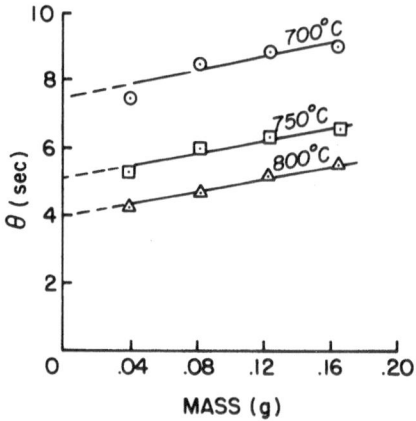

FIGURE 20. Autoignition of Kermel® fabric.

the temperature of the air in the oven and cannot be considered to be related to any solid phase restrictions. Most likely, they reflect gas diffusion processes.

2.4. Effects of Selected Experimental Variables

The previous results have been obtained with 25 × 13 mm (1 × ½ in.) fabric samples stapled to a metal screen and placed vertically in a heated oven. Since these experimental conditions have been arrived at somewhat

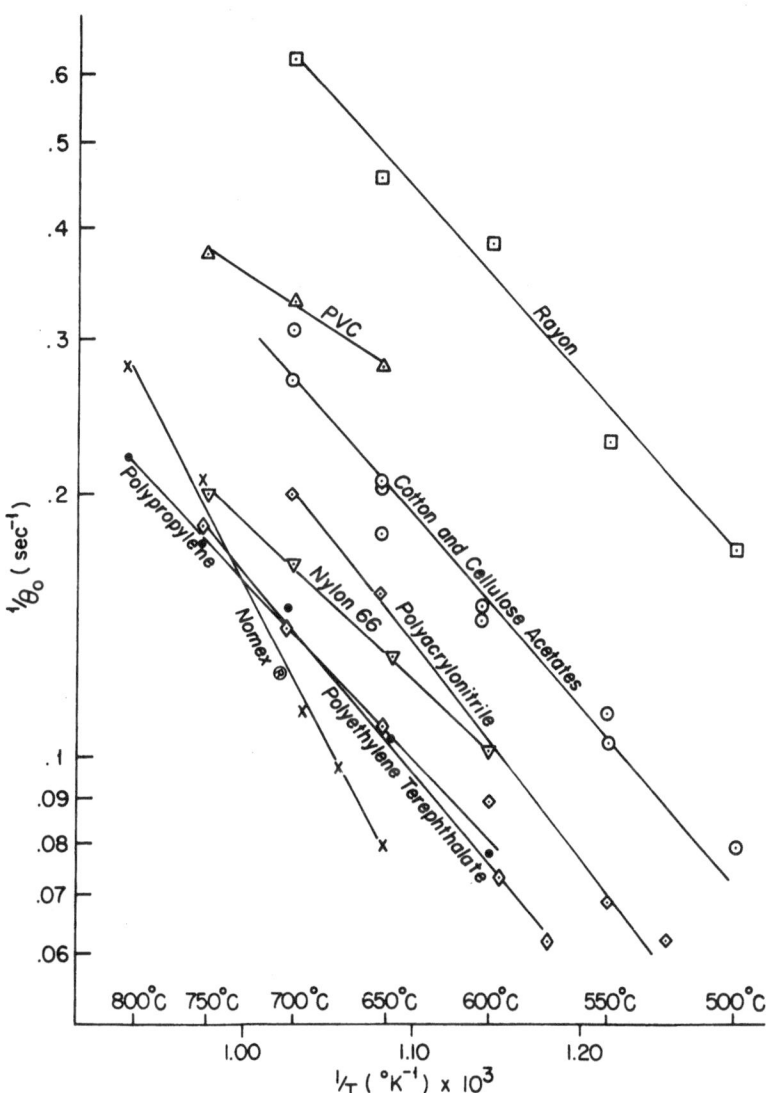

FIGURE 21. Arrhenius plots of intrinsic ignition time (θ_0) as a function of furnace air temperature.

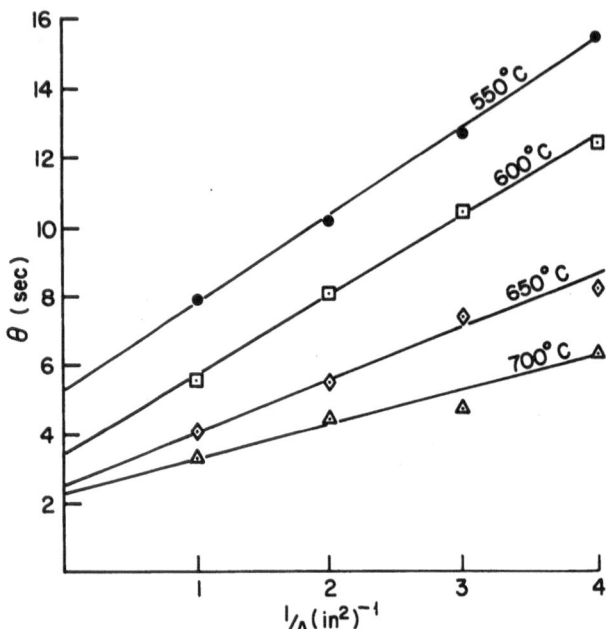

FIGURE 22. Autoignition time of cotton as a function of bulk sample area; sample mass = 0.07 g.

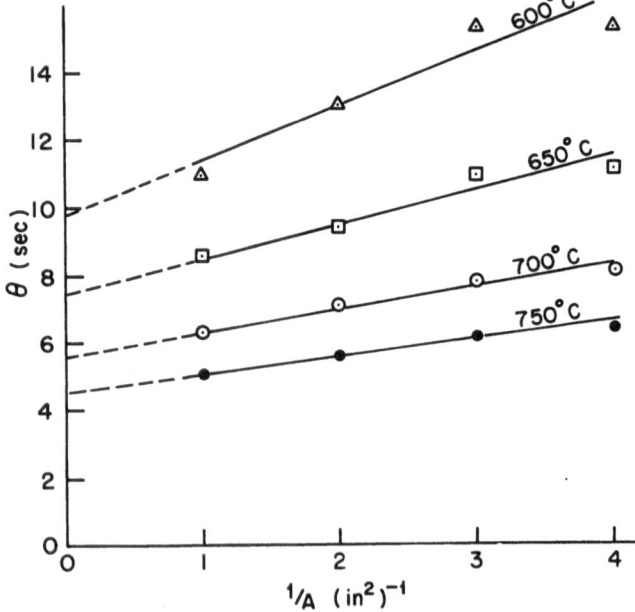

FIGURE 23. Autoignition time of nylon 66 as a function of bulk sample area; sample mass = 0.07 g.

arbitrarily, variations in these parameters have been studied not only to ascertain that the results are general and not peculiar to one specific set of circumstances but also to obtain further insight into the phenomenon of autoignition itself.

2.4.1. Sample Area

The influence of bulk sample surface area (as opposed to fiber surface area) on the autoignition time has been investigated for several materials by using samples of constant mass with varying area (i.e., one layer: 25×25 mm; two layers: 25×12.5 mm; three layers: 25×8.5 mm; and four layers: 12.5×12.5 mm). The samples were prepared by folding a 25×25 mm piece of fabric to obtain the desired bulk surface area and stapling it to a wire screen 25×25 mm. The results for cotton, nylon, and polyester fabrics are given in Figs. 22 through 24 as plots of observed ignition time, θ, versus the reciprocal of surface area, $1/A$, at several air temperatures. The observed linear relationships between θ and $1/A$ are not unexpected, since the rate of heat transfer should be inversely proportional to the surface area available for transfer of heat. For each temperature, the linear relationships have been extrapolated to infinite area ($1/A = 0$) to obtain an intrinsic ignition time, $(\theta_0)_A$, at the

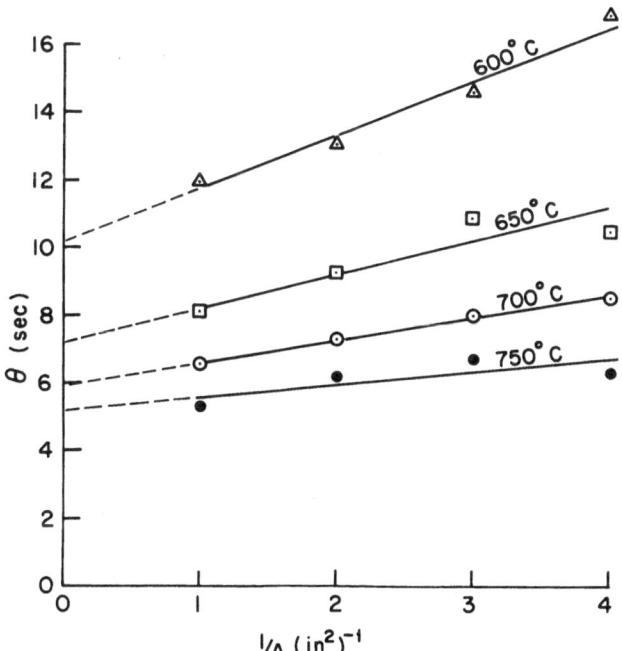

FIGURE 24. Autoignition time of PET as a function of bulk sample area; sample mass = 0.07 g.

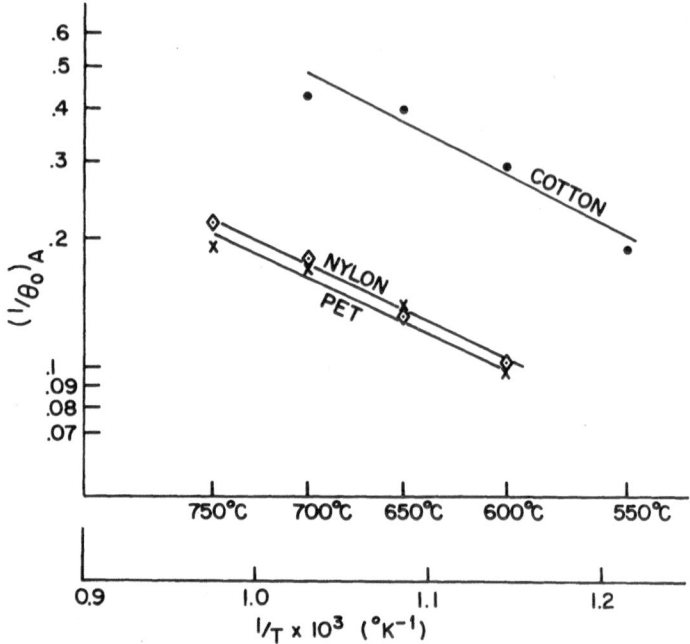

FIGURE 25. Arrhenius plots of intrinsic ignition time $(\theta_0)_A$ as a function of furnace air temperature.

mass level investigated. The temperature dependence of these intrinsic ignition times are shown in Fig. 25 as Arrhenius plots of log $(1/\theta_0)$ vs. $1/T$. The slopes of these lines correspond to activation energies of 8–10 kcal/mol, the same values as were obtained using the intrinsic ignition times from mass dependence (Fig. 21).

A further study of the combined effect of sample mass and area has produced the results shown in Fig. 26 for a cotton fabric at one air temperature. The relationships between ignition time and sample mass for three bulk sample areas 25 × 25 mm (1 × 1 in.), 25 × 12.5 mm (1 × ½ in.) and 12.5 × 12.5 mm (½ × ½ in.) are given. These data were obtained without the use of wire screens (multiple layers of fabric were sewn together with cotton thread) in order to eliminate any complications from the additional heat sink. The results show clearly that θ_0 is independent of the bulk sample area. Therefore, the meaning of intrinsic ignition time can be made more definitive; it is the ignition time for an infinitely small amount of material (one polymer molecule as the limiting case) independent of both mass and surface area factors. Figure 27 shows the data from Fig. 26 plotted as ignition time versus

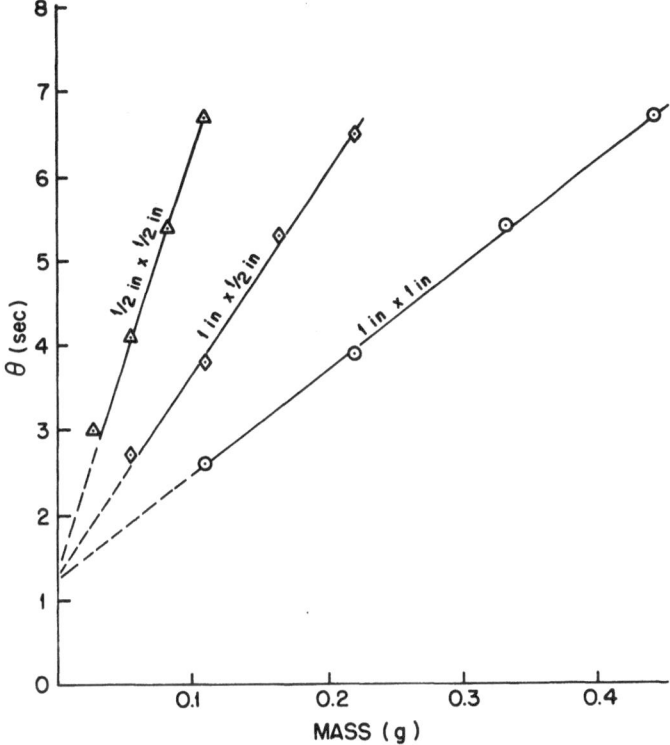

FIGURE 26. Autoignition times at 600°C for cotton samples with different areas.

mass to area ratio (m/A); the data all fall along one line, indicating that, for finite masses, ignition time is a unique function of m/A.

2.4.2. Fiber Surface Characteristics

The possible effect of fiber surface characteristics was investigated by comparing the autoignition behavior of two different forms of regenerated cellulose, rayon fabric and cellophane. The autoignition times were obtained for multilayered samples stapled together; no screens were used in order to avoid the possibility that they would mask any small differences that might exist between the two materials. The results in Fig. 28 show identical ignition times for both materials, indicating that autoignition is controlled by the bulk rather than by the surface characteristics of the sample. If this were not so, it would be unlikely that thermoplastic and nonthermoplastic materials would show such comparable behaviors.

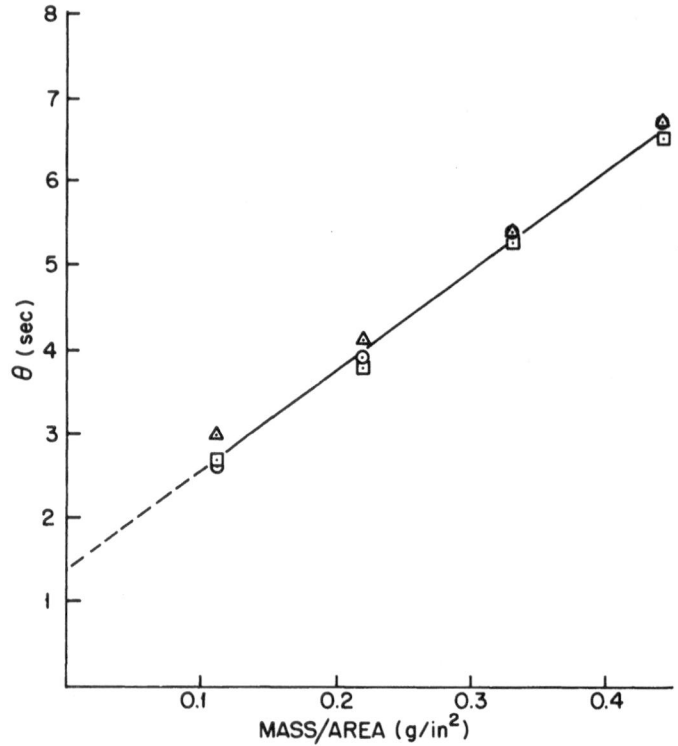

FIGURE 27. Autoignition time for cotton samples as a function of mass-to-area ratio at 600°C.
△ ¼ in. × ¼ in.; ▣ ¼ in. × 1 in.; ⊙ 1 in. × 1 in.

2.4.3. Oxygen Concentration

The possible influence of decreased oxygen concentration (i.e., less than 21%) on ignition time has been investigated for a representative group of polymers. The results, shown in Table 1, indicate no significant changes in ignition time with change in oxygen concentration. This is consistent with the results reported by Collins and Wendlandt for a similar experimental arrangement[17] and would seem to indicate that the rate-determining step in the ignition process is the diffusion of combustible gases from the decomposing sample rather than the diffusion of oxygen to it. If the latter were true, one would expect oxygen concentration to influence ignition time, since the diffusion rate of oxygen would be dependent on the partial pressure of the gas. As might be expected, it was observed that the intensity of the ignition explosion was reduced at low oxygen concentrations.

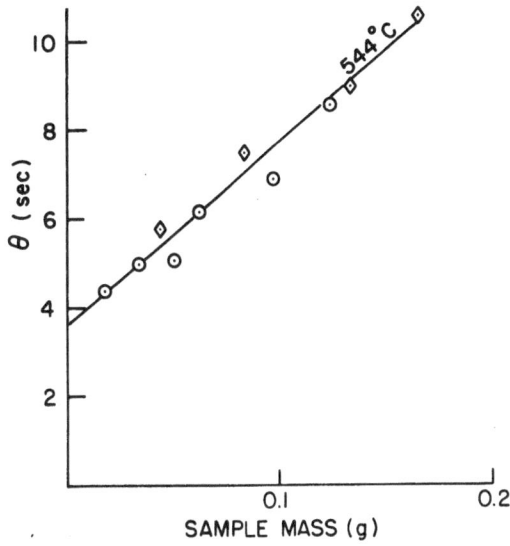

FIGURE 28. Autoignition of regenerated cellulose; comparative behavior of a film and fabric; air temperature, 544°C. ⟨⊙⟩ Rayon fabric; ⊙ cellophane.

TABLE 1

Ignition Times at Various Oxygen Concentrations
(one layer of fabric stapled to wire screen)

Oxygen concentration (%)	Rayon[a]	Cotton[a]	Acrylic[a]	PET[a]	Polypropylene[b]
21.0	5.6	7.2	15.3	17.7	11.0
16.4	5.5	6.7	16.2	17.2	11.2
11.1	5.7	6.5	16.1	17.3	11.7
4.7	6.4	6.7	16.9	16.4	12.6

[a]Air temperature = 550°C.
[b]Air temperature = 650°C.

Table 2 shows ignition data for rayon fabric at oxygen concentrations greater than 21%, indicating no dependence on oxygen concentration. That ignition time is apparently unaltered by oxygen concentrations greater or less than 21% provides additional support for the concept that observed ignition times are controlled by precombustion events and not by the final reaction with oxygen.

TABLE 2
Influence of Oxygen Concentration on the
Autoignition Times for Rayon
(furnace temperature = 500°C)

Oxygen concentration (%)	Ignition time (sec)
17.7	5.5
21	4.9
21 (nonflowing atmosphere)	5.0
23	5.4
26.3	5.2
28.2	5.1

3. Autoignition of Multicomponent Systems

3.1. Polymers with Flame-Retardant Additives

Autoignition behavior can be an important key to understanding the function of flame retardants. In particular, if a flame retardant has no effect on a polymer's intrinsic ignition time, that retardant must be active only in the postignition phase of the combustion sequence. Conversely, a flame retardant that changes the intrinsic ignition time has affected some preignition decomposition mechanism of the polymer; this, of course, does not preclude the possibility of it also having postignition effects once burning has begun.

Figure 29 compares the autoignition behavior of cotton sheeting treated with tetrakis(hydroxymethyl)phosphonium chloride (THPC), a commercial cellulose flame retardant, with an untreated cotton control. The presence of the phosphorus-based flame retardant lowers the intrinsic ignition times of cotton, indicating that this additive alters the cellulose decomposition mechanism.

The effect of diammonium phosphate (DAP), another phosphorus-based flame retardant, has also been studied. Figure 30 details the changes in intrinsic ignition times as a function of add-on level for cotton at four air temperatures. These results show that θ_0 decreases up to a 2% DAP level and is essentially constant for add-on levels greater than 2%. The effect of the additive decreases noticeably as the air temperature is raised.

Figure 31 shows that tetrabromophthalic anhydride (TBPA), a bromine-based flame retardant, does reduce cotton ignition times at finite masses (indicating a decreased value of θ_1, as defined by Fig. 4) but has no effect on the intrinsic ignition time. This is what would be expected as the result of the presence of an inert additive, so that this flame retardant is understood to be active only after ignition has occurred.

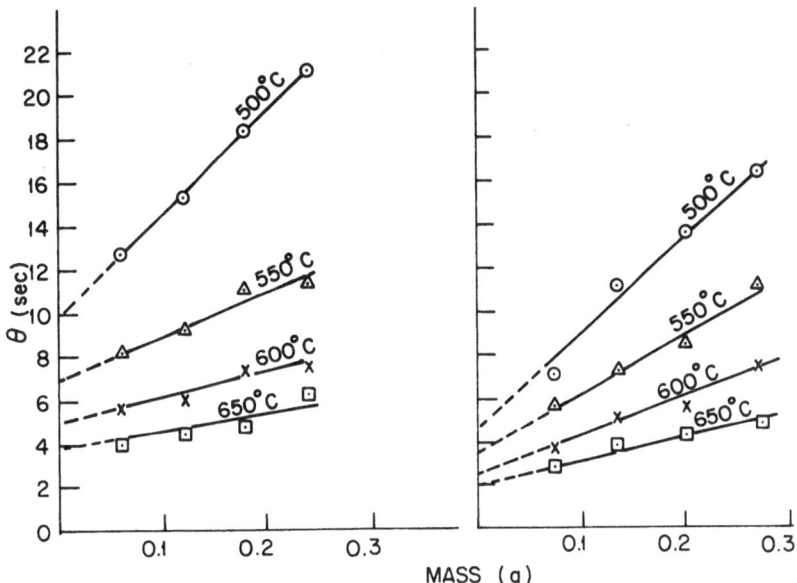

FIGURE 29. Autoignition of untreated cotton sheeting (left) and cotton sheeting + THPC (right).

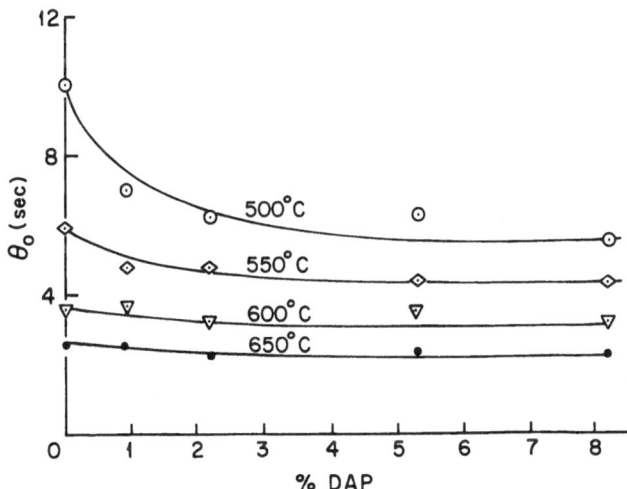

FIGURE 30. Effect of DAP add-on level on the intrinsic ignition time of cotton.

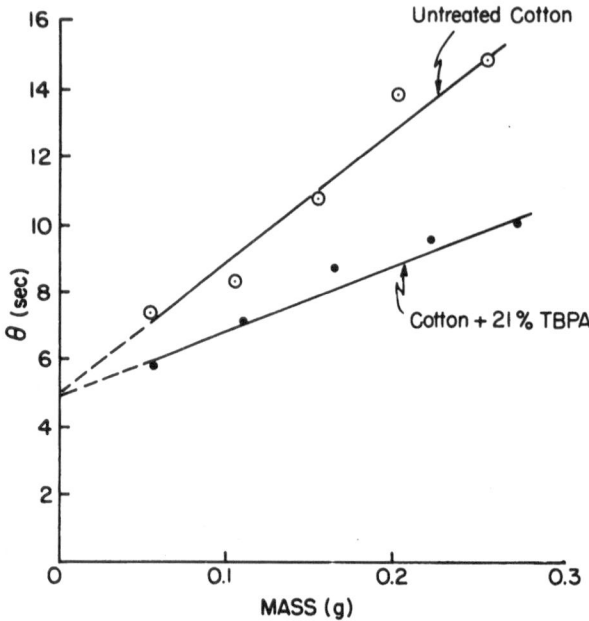

F<small>IGURE</small> 31. Autoignition of untreated cotton and flame-retardant cotton (21% tetrabromo-phthalic anhydride); control treated with benzyl alcohol and dried at elevated temperature. Air temperature = 550°C.

Tris-2,3-dibromopropyl phosphate (T23P), which contains both bromine and phosphorus, does reduce the intrinsic ignition time of cotton (Fig. 32). This compound very likely has two functions; it changes the pre-ignition decomposition mechanism and, since it contains bromine, also is active in the vapor phase after ignition.

Figure 33 demonstrates that the time-honored 70:30 borax–boric acid (B–BA) flame-retardant receipt for cotton has no apparent effect on the ignition behavior of cotton. Since this additive has a demonstrated effectiveness as a flame retardant, we must look elsewhere in the burning process for its function.

The effects of four flame retardants on the autoignition behavior of polyester are shown in Figs. 34 through 37. Figure 34 indicates that T23P, the most common flame retardant for polyester, has no effect on the latter's intrinsic ignition time. However, the slope of the ignition time–sample mass relationship is reduced by the additive. It is interesting to note that even though T23P effects a significant decrease in the intrinsic ignition time of cotton (Fig. 32), it has no such effect on polyester. The intrinsic ignition time of polyester is, however, reduced by DAP (Fig. 35).

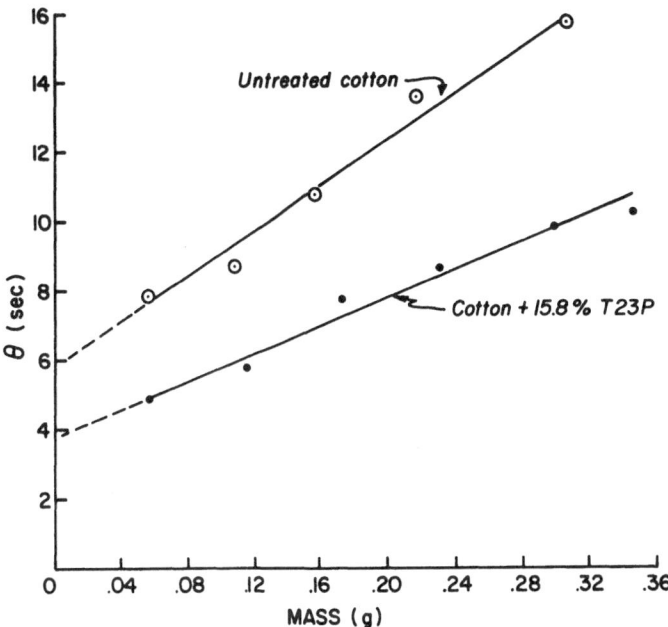

FIGURE 32. Autoignition of untreated and flame-retardant cotton (15.8% tris-2,3-dibromo-propyl phosphate). Air temperature = 550°C.

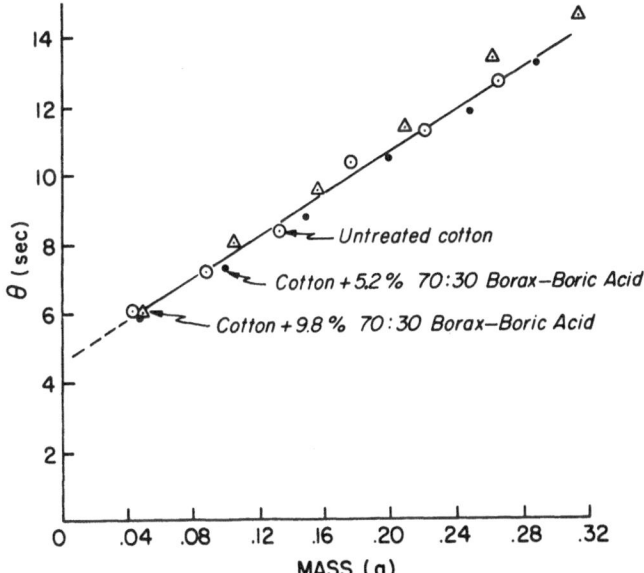

FIGURE 33. Autoignition of untreated and flame-retadrant cotton (5.2% and 9.8% 70:30 borax-boric acid). Air temperature = 600°C.

Figures 36 and 37 show the effects of two bromine-based flame retardants, octobromodiphenyl (OBDP) and TBPA, on polyester. Both these
materials have no effect on the intrinsic ignition time, indicating, as expected
for bromine-based flame retardants, that their activity is restricted to postignition processes. It is interesting to note that the OBDP additive increases
the ignition time for a finite sample mass, although the intrinsic ignition
time at zero mass is unaltered. This indicates an increase in the time required
to heat the material to its decomposition temperature.

Figures 38 and 39 compare the autoignition behavior of two other flame-
retardant materials (wool treated with zirconium–tungsten flame retardant
and nylon treated with thiourea) with their untreated controls. The
zirconium–tungsten flame retardant has no effect on the intrinsic ignition
time of the wool. Thiourea does cause a decrease in the intrinsic ignition

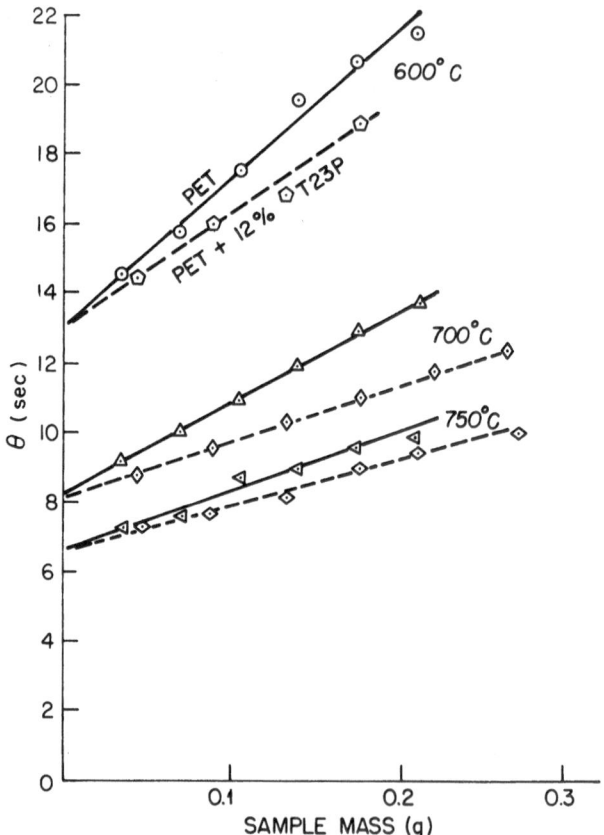

FIGURE 34. Autoignition of untreated and flame-retardant polyethylene terephthalate; solid
lines = untreated PET; dashed lines = PET + 12% T23P.

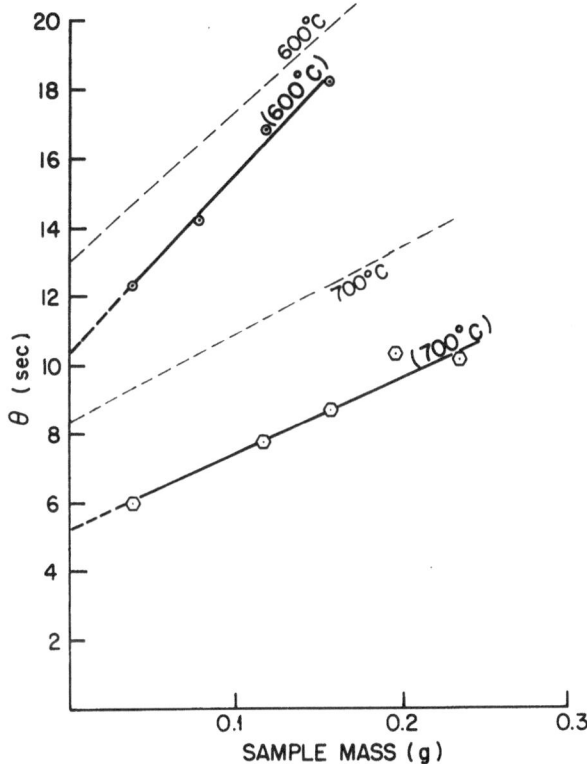

FIGURE 35. Autoignition of untreated and flame-retardant polyethylene terephthalate; solid lines = PET + 4.8% DAP; dashed lines = untreated PET.

time of the nylon, indicating its effect on the polymer decomposition mechanism.

All the recognized flame-retardant additives investigated to date either reduced intrinsic ignition times or had no effect on this property, which reflects fundamental decomposition characteristics. No additive was able to effect an increase in θ_0. While this may at first glance seem inexplicable, it should be realized that to achieve an increase in ignition delay an additive would most likely have to make the polymer backbone intrinsically more stable, an effect which could not be expected without some sort of chemical modification.

3.2. Multipolymer Systems

One can reasonably expect that, in the absence of any interactions, the ignition behavior of a mixed system, once corrected for mass effects, would

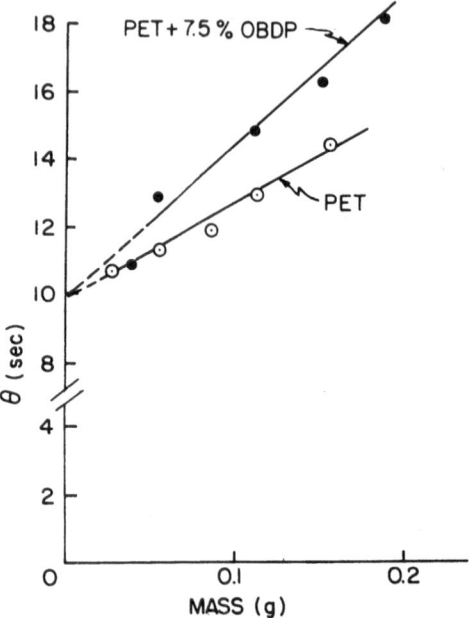

FIGURE 36. Autoignition of untreated and flame-retardant polyester (7.5% octobromodi-phenyl). Air temperature = 650°C.

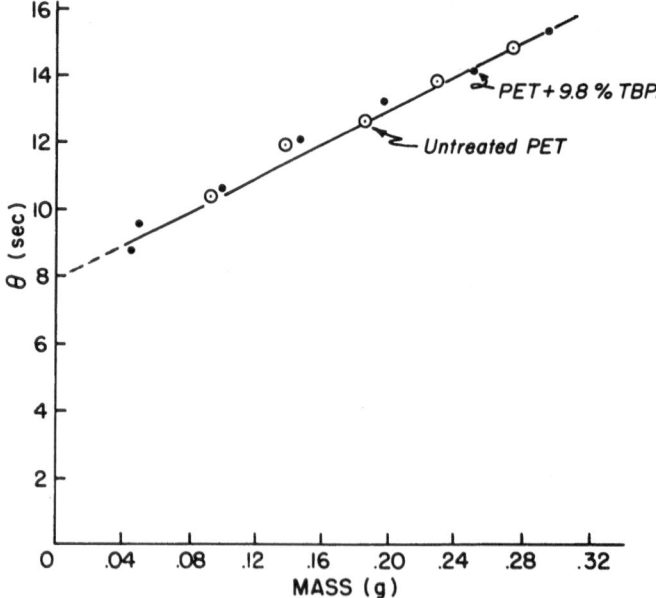

FIGURE 37. Autoignition of untreated and flame-retardant polyester [9.8% tetrabromophthalic anhydride (TBPA)]; control treated with benzyl alcohol and dried at elevated temperature. Oven temperature = 650°C.

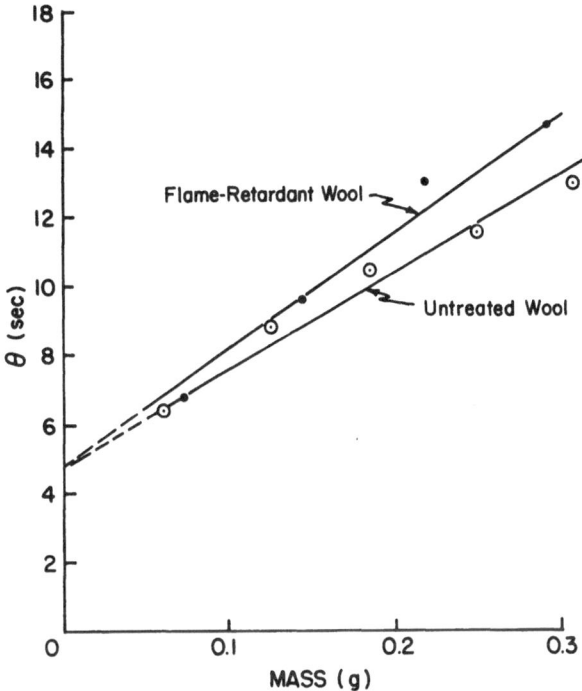

FIGURE 38. Autoignition of untreated and flame-retardant wool (zirconium–tungsten treated). Air temperature = 600°C.

be controlled by the faster igniting component. It should be kept in mind that the reverse of this statement cannot be used as a criterion; all non-interacting mixtures behave in this manner, but some interacting ones may also exhibit such behavior. This idea is supported by data collected for a large number of combinations; examples of such systems (rayon–acrylic, rayon–Nomex®, and nylon–wool) are shown in Figs. 40 through 42. In all these cases, the points for the blend lie along the line for the faster-igniting component. Additional vertification for the proposed noninteracting criterion has been obtained by studying the behavior of polymer–glass blends where the glass component functions only as a competing heat sink. Figures 43 and 44 show the ignition behavior of two such systems: Nomex®–glass and nylon–glass. For the Nomex®–glass combination, the ignition times for the polymer–glass blends fall along the same line as for the polymer alone. This occurs because the thermal absorptivity of Nomex® is approximately equal to that of the glass yarn. For the nylon–glass blend, the ignition times at finite masses are less than those of the polymer alone, but the lines for the blend and the pure polymer converge to the same intrinsic ignition time. This decrease in the slope of the ignition time–sample mass relationship

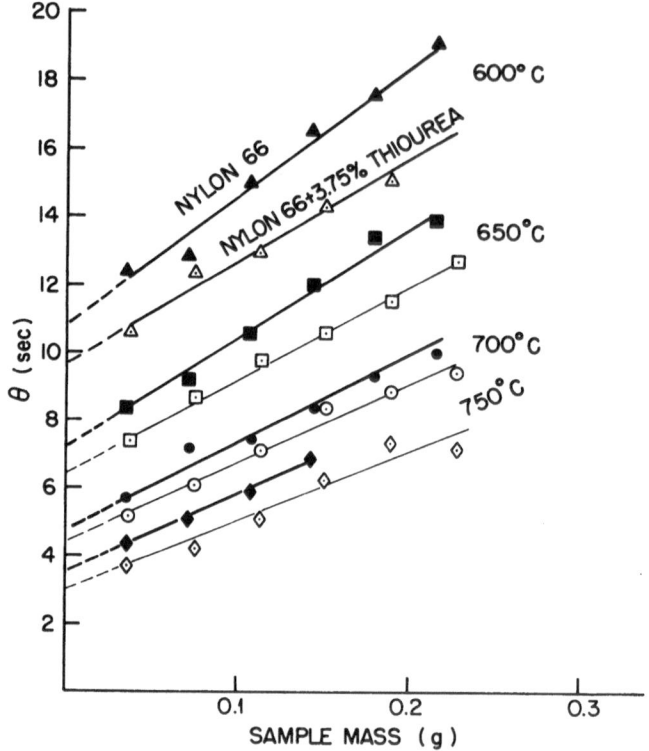

FIGURE 39. Autoignition of untreated and flame-retardant nylon 66; closed points = untreated nylon; open points = nylon + 3.75% thiourea.

is indicative of a reduction in the time required to heat a sample of finite mass to its decomposition temperature (i.e., reduction in θ_1, as defined in Fig. 4). For thermoplastic–glass blends this behavior is not unexpected, since, for a given weight of sample, the presence of the glass means that only half the sample undergoes melting, thus lessening the time required to heat the system to decomposition. Such data support the argument that the effect of an inert additive, if any, is to alter the slope of the ignition time–sample mass relationship without changing the intrinsic ignition time.

Of greater technical interest are considerations of interacting systems where the intrinsic ignition time is different from that of the faster-igniting component. Figure 45 shows the ignition characteristics of a polyester, a nylon, and a 50:50 mixture of the two which ignites faster than either of the two parent components. This behavior is a manifestation of chemical and/or physical interaction(s) which affect preignition decomposition at high heating rates.

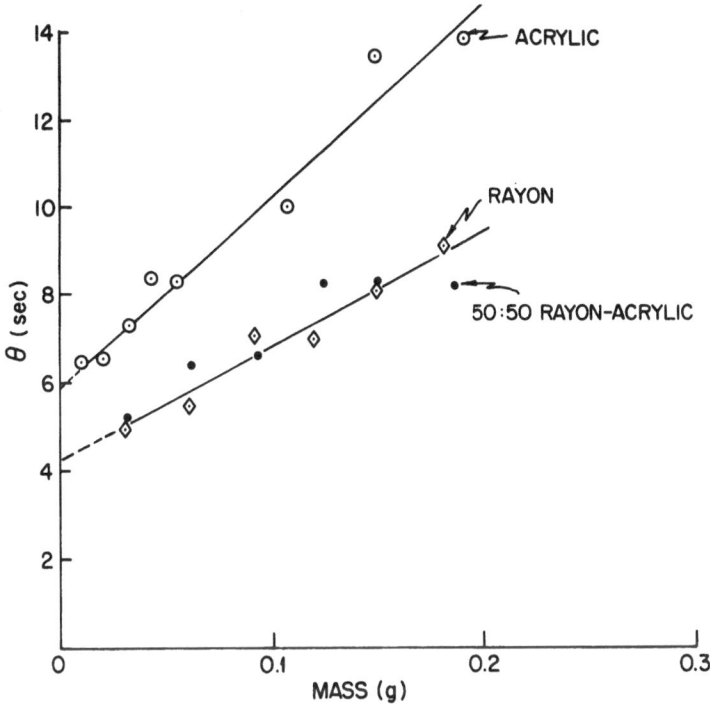

FIGURE 40. Autoignition of rayon, acrylic, and 50:50 rayon–acrylic yarns. Oven temperature = 600°C.

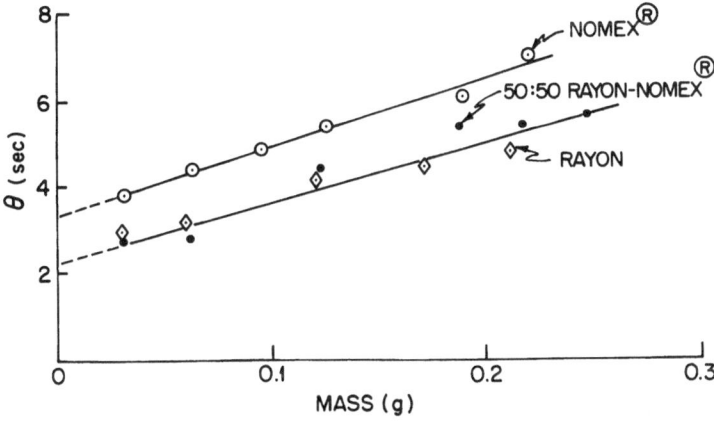

FIGURE 41. Autoignition of rayon, Nomex®, and 50:50 rayon–Nomex® yarns. Oven temperature = 700°C.

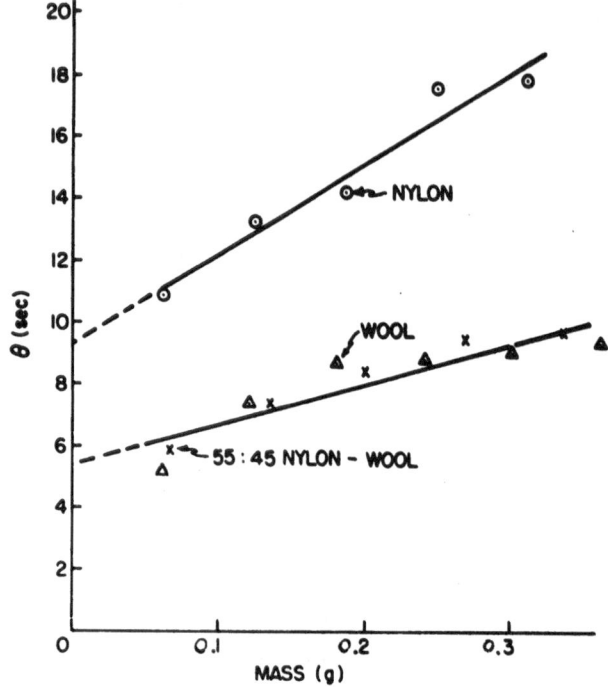

FIGURE 42. Autoignition of nylon, wool, and a 55:45 nylon–wool blend. Air temperature = 650°C.

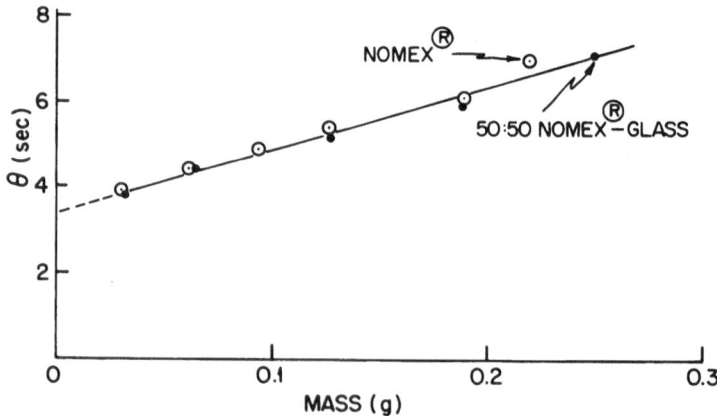

FIGURE 43. Autoignition of Nomex® and 50:50 Nomex®-glass yarns. Oven temperature = 700°C.

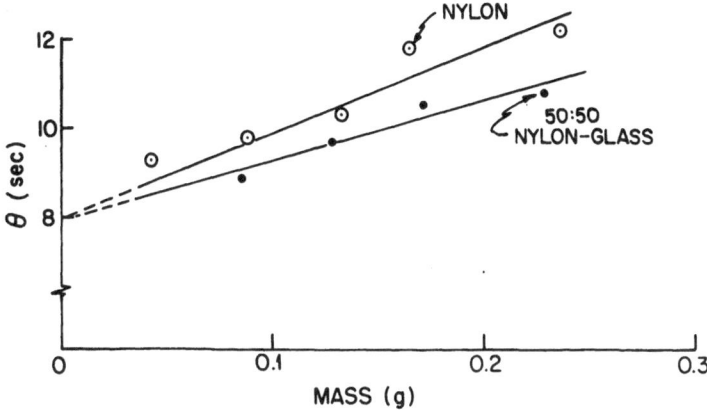

FIGURE 44. Autoignition of nylon and 50:50 nylon–glass yarns. Oven temperature = 650°C.

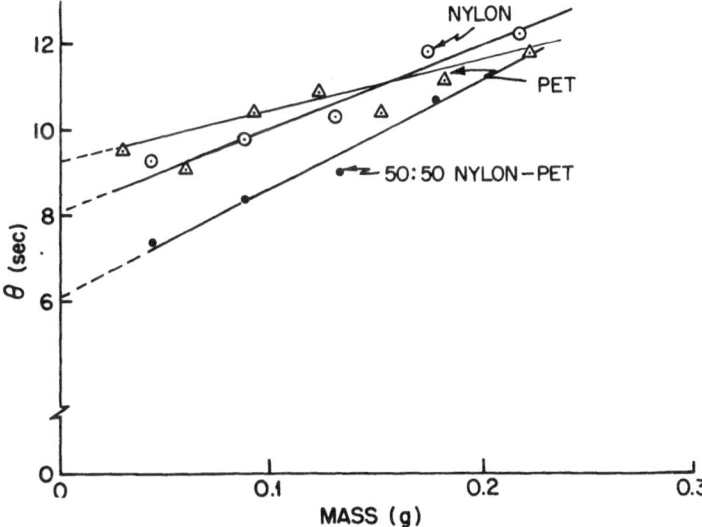

FIGURE 45. Autoignition of nylon, polyester, and 50:50 nylon–polyester yarns. Oven temperature = 650°C.

Table 3 lists the intrinsic ignition times for combinations of two other interacting systems, polyester–cotton and polyvinyl chloride–rayon. For both these systems, the intrinsic ignition times of the blends are less than for either component. The effect of blend composition is not significant over the range investigated.

TABLE 3
Intrinsic Ignition Times (sec) for
Multicomponent Systems
(oven temperature = 550° C)

100% polyester (Fortrel®)	11.4
65:35 polyester–cotton	5.7
50:50 polyester–cotton	5.7
35:65 polyester–cotton	5.4
100% cotton	10.0
100% PVC	$(5.3)^a$
75:25 PVC–rayon	3.0
30:70 PVC–rayon	3.5
100% rayon	4.4

^{a}Extrapolated from higher-temperature results.

A more detailed discussion of the flammability behavior of multipolymer systems can be found elsewhere,[18] including a discussion of other flammability characteristics in addition to ignition behavior.

4. *The State of the Art*

The technical importance of polymer ignition phenomena has grown significantly in recent years, and our need of better methodology for studying it has become more critical. It would be of appreciable consequence if any of the following advances could be accomplished:

1. Definitive evidence showing that the mode of heat supply (i.e., radiative, conductive, etc.) to a polymer does not alter the basic process by which it ignites.
2. Establishment of the fundamental criteria which must be met in order for a polymer to ignite.
3. Determination of whether or not it is possible to reduce the ignitability of a polymer by any method other than bulk dilution with a less ignitable species.
4. Resolution and/or explanation of why radiative ignition studies show a strong dependence on ambient oxygen concentration, while autoignition seems to be independent of this parameter.
5. Accumulation of reliable experimental data on the self-ignition of oxygen-containing polymers in the absence of ambient oxygen.

5. References

1. T. Kashiwaga, B. W. MacDonald, H. Isuda, and M. Summerfield, in *13th International Symposium on Combustion*, The Combustion Institute, Salt Lake City (1971), p. 1073.
2. J. R. Welker, *J. Fire Flammability* **1**, 12 (1970).
3. N. J. Alvares and S. B. Martin, *13th International Symposium on Combustion*, The Combustion Institute, Salt Lake City (1971), 905.
4. U. K. Shivadev and H. W. Emmons, *Combust. Flame* **22**, 223 (1974).
5. N. Rangaprasad, C. M. Sliepcevich, and J. R. Welker, *J. Fire Flammability* **5**, 107 (1974).
6. J. R. Hallman, J. R. Welker, and C. M. Sliepcevich, *Polym. Plast. Technol. Eng.* **6**, 1 (1976).
7. T. J. Ohlemiller and M. Summerfield, *13th International Symposium on Combustion*, The Combustion Institute, Salt Lake City (1971), p. 1087.
8. N. P. Setchkin, *J. Res. Nat. Bur. Stand.* **43**, 591 (1949).
9. ASTM Standard D-1929, American Society for Testing & Materials, Philadelphia (1964).
10. W. Wulff, A. Alkidas, R. W. Hess, and N. Zuber, *Text. Res. J.* **43**, 577 (1973).
11. L. W. Sayers, *Text. Inst. Ind.* **3**, 168 (1965).
12. P. Lennox-Kerr, *Text. Ind. Atlanta* **133** (4), 156 (1969).
13. W. K. Smith and J. B. King, *J. Fire Flammability* **1**, 272 (1970).
14. B. Miller, J. R. Martin, and C. H. Meiser, Jr., *J. Appl. Polym. Sci.* **17**, 629 (1973).
15. R. F. Schwenker, L. R. Beck, and R. K. Zuccarello, *Am. Dyest. Rep.* **53** (19), 30 (1964).
16. A. Alkidas, R. W. Hess, W. Wulff, and N. Zuber, Final Report for Government Industry Research Committee on Fabric Flammability, Office of Flammable Fabrics, National Bureau of Standards, Washington, D.C. (Dec. 1971).
17. L. W. Collins and W. W. Wendlandt, *Thermochim. Acta* **7**, 201 (1973).
18. B. Miller and J. R. Martin, *J. Fire Flammability* **6**, 105 (1975).

4

Phosphorus-Based Flame Retardants

Edward D. Weil

1. Introduction

Compounds of phosphorus were recognized as flame retardants at the beginning of the nineteenth century when the French chemist Gay-Lussac recommended ammonium phosphate to prevent the burning of theater curtains. The advent of organic phosphorus compounds as important flame retardants for plastics dates back to the 1910–1920 era, when the extreme flammability of cellulose nitrate was brought under some degree of control by the use of tricresyl phosphate. Tricresyl phosphate was the first major commercial organophosphorus compound. Today, phosphorus flame retardants are among the "workhorse" products in the flame-retardant field, and research is proceeding at an accelerating pace in this area in view of increasingly stringent flame-retardant requirements for plastics and textiles. This chapter encompasses the phosphorus flame retardants which are actually in commercial use or which appear to be at a serious stage of commercial development for plastics and textiles. In addition to these compounds, there have been many thousands of phosphorus compounds suggested in patents and publications as having flame-retardant utility; a survey has been published by Lyons.[1] A particular aspect of phosphorus flame retardancy on which a great deal of work has been done (with only a few

Edward D. Weil • Stauffer Chemical Company, Dobbs Ferry, New York.

commercial products being arrived at) has been the synthesis of addition and condensation polymers from phosphorus monomers. An excellent review of this area was published by Sander.[2]

2. Inorganic Phosphorus Compounds

2.1. Red Phosphorus

Red phosphorus is relatively nontoxic and, unlike white phosphorus, is not spontaneously flammable (although easily ignited). In finely divided form, it has been found to be effective as a flame-retardant additive, especially when accompanied by a bromine-containing additive, in polyolefins and styrenics.[3] Commercial usage in Europe has been reported in molded nylon electrical parts and in polyurethanes. Handling hazards, such as flammability and partial reversion to white phosphorus, may hold back major usage of red phosphorus despite its attractive cost–effectiveness relationship.

2.2. Ammonium Phosphates

Where water solubility is desirable or not harmful, monoammonium and diammonium phosphates, or mixtures of the two (which are more soluble in water and nearly neutral), are very effective and inexpensive flame-retardant additives. For example, to flame-retard paper, canvas, cotton batting, draperies, fiberboard, or other cellulosic materials, a solution of 1 to $1\frac{1}{2}$ lb/gal of MAP–DAP can be padded or sprayed onto the substrate. A dry add-on of 7–20% on cotton will generally pass a vertical flammability test. Ammonium phosphate finishes are not resistant to laundering but are somewhat resistant to dry cleaning. Applications for treatments of this sort are disposable nonwoven textiles, wallboard, corrugated cardboard, packaging, and cotton mats. One particularly important advantage of these inorganic phosphorus flame retardants (and phosphorus flame retardants in general) is their effectiveness in preventing afterglow.

The crystalline character of ammonium phosphates may cause a harsh texture on the surface; to alleviate this problem, commercial ammonium phosphates are available containing softening and penetrating agents.

Self-cross-linking acrylic latices have been formulated with diammonium phosphate plus organic phosphates to obtain flame-retardant textile back-coatings and nonwoven binders with a useful degree of durability to laundering and dry cleaning.

2.3. Insoluble Ammonium Polyphosphate (*Monsanto's Phoschek P/30®*)

By heating ammonium phosphate with urea, NH_3 and CO_2 are given off and a long-chain relatively water-insoluble ammonium polyphosphate is produced. The material is available in finely divided solid form. Its principal use is as a flame retardant for intumescent paints and mastics.[4] In such formulations, ammonium polyphosphate is used at high loadings and is considered to function as a "catalyst." When the intumescent coating is exposed to a flame, the ammonium polyphosphate yields a phosphorus acid which then interacts with an organic component such as dipentaerythritol to form a carbonaceous char. A blowing agent such as melamine or chlorowax is also present to give the char the required porosity. These ingredients plus binder, pigment, and various fillers comprise the typical intumescent formulation. Mastics are intended to be applied in much thicker layers to girders, trusses, and decking; they generally contain asbestos fibers for coherence.

2.4. Ammonia–P_2O_5 Products (*Stauffer's Victamide®*)

The reaction of ammonia gas with phosphorus pentoxide affords a white amorphous finely divided solid which may be described as an ammonium salt of a metaphosphorimidic acid.[5] Analysis shows about two ammonium nitrogen atoms, one imide nitrogen atom, and two phosphorus atoms. Victamide® is known to be a complex mixture of salts, but a typical component is believed to have the following structure[6]:

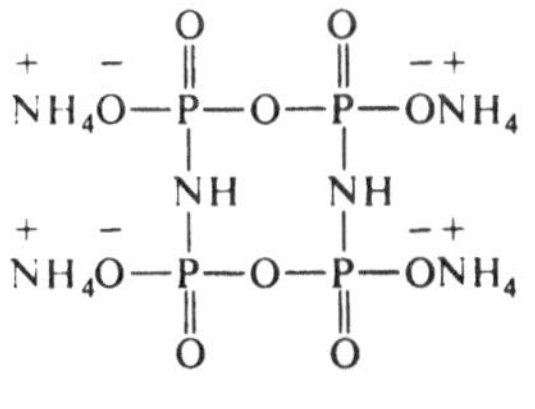

$$
\begin{array}{ccc}
& O & O \\
+ \quad - & \parallel & \parallel \quad - + \\
NH_4O-P-O-P-ONH_4 \\
& \mid & \mid \\
& NH & NH \\
+ \quad - & \mid & \mid \quad - + \\
NH_4O-P-O-P-ONH_4 \\
& \parallel & \parallel \\
& O & O
\end{array}
$$

The substance is slowly soluble in water. As in the case of ammonium phosphates, its aqueous solution can be applied to paper, cloth, batting, and nonwovens. When dry, it produces a less harsh feel than that caused by the crystalline salts, since it forms a coherent smooth film.

Ammoniation of Victamide® by concentrated ammonia affords a more nitrogenous product which when applied to a cellulosic substrate and heated yields a semidurable f.r. (flame-retardant) finish which will withstand several aqueous washes.[7,8]

2.5. Phosphoric-Acid-Based Systems for Cellulosics

Going one step beyond the physical deposition of soluble phosphates on cellulose, it is possible to actually attach phosphoric acid chemically to give a cellulose phosphate ester. This was originally accomplished by heating cotton or paper with phosphoric acid in the presence of basic compounds such as urea at about 300–350° F. The most successful applications of this approach are a series of systems developed by American Cyanamid Company which are based on phosphoric acid and cyanamide or dicyandiamide–formaldehyde resins.[9] The nitrogenous component catalyzes the phosphorylation of the cellulose, has a buffering action which retards the acid degradation of the cellulose, and has a synergistic action on the phosphorus in regard to flame retardancy (more will be said later about N–P synergism. A fair degree of durability to laundering is achieved; typically, five to ten home launderings can be tolerated, and dry-cleaning resistance is good. A good part of the decline in flame retardancy is caused by ion exchange wherein the phosphoric acid groups bind sodium, calcium, and magnesium cations. It is well established that phosphoric acid salts of strong nonvolatile bases do not act as flame retardants in most systems. These finishes have had limited utility because of fabric damage during cure and because of the limited durability. Some usage has been achieved on draperies.

Phosphoric acid has been formulated with curable aminoplast resins to prepare leach-resistant clear flame-retardant coatings for wood.[10] Commercialization of such coatings has apparently taken place in Canada.

3. Organic Phosphorus Flame Retardants—Additive Types

3.1. Alkyl Acid Phosphates

Alkyl acid phosphates such as methyl acid phosphate and butyl acid phosphate, in view of their high percentage of phosphorus and low cost, have been employed as additive flame retardants in plastics, especially those which are fabricated at low temperatures and wherein these relatively hydrophilic compounds are protected from moisture by a water-resistant polymer matrix. Thus, about 1% phosphorus as butyl acid phosphate gives a chlorendic-acid-derived polyester a significant increase in flame retardancy. In cast polymethyl methacrylate, alkyl and haloalkyl acid phosphates have been found effective.[11-13] To pass practical flame retardancy tests in PMMA and polyester laminates, compounders generally find that inclusion of chlorine or bromine is also desirable. A dibromopropyl methyl acid phosphate, Fyrol® 72, containing 10% P and 52.5% Br and having good light stability, has been introduced for these application by Stauffer.[14] This product and related mono- and bis(dibromopropyl) acid phosphates have also been found commercially useful, in the form of their water-soluble ammonium salts, for producing semidurable flame-retardant finishes on textiles.[15,16]

3.2. Trialkyl Phosphates

Triethyl phosphate is a colorless liquid boiling at 209–218°C. It contains 17% phosphorus and is rather low in cost. Triethyl phosphate has been commercially used as an additive for polyester laminates and in cellulosics. In polyester resins, it also serves the function of a viscosity depressant. As shown in Fig. 1, even at 5% loading, a highly significant improvement in the flammability rating of a halogenated polyester is observed.[17] Above about 10% loading (1.7% P), further addition of triethyl phosphate produced no further improvement. The viscosity-depressant effect of triethyl phosphate in polyester resin permits high loadings of alumina trihydrate (a flame-retardant smoke-suppressant filler).[18] Because of its water solubility, triethyl phosphate is limited in utility to situations where weathering resistance is unimportant. The halogenated alkyl phosphates are generally used for applications where lower volatility and greater resistance to leaching are required. Trioctyl phosphate is a specialty flame-retardant plasticizer for vinyl compositions where low-temperature flexibility is critical, for example, in military tarpaulins. It can be included in blends with general-purpose plasticizers such as DOP to improve low-temperature flexibility.

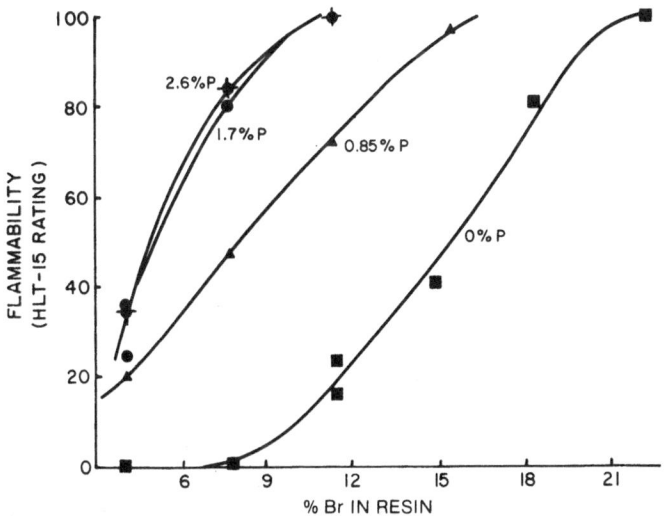

FIGURE 1. Effect of triethyl phosphate (5, 10, 15%) in a tetrabromophthalic polyester.[17]

3.3. Dimethyl Methylphosphonate (DMMP)

This phosphonate ester, made by rearrangement of trimethyl phosphite, contains 25% phosphorus and is therefore highly efficient as a flame retardant. It is a low-viscosity colorless liquid, boiling at 185°C. Because of volatility, it has been useful mainly in thermoset systems. DMMP functions in polyester resin as a viscosity depressant and flame retardant of somewhat greater efficiency than triethyl phosphate. Significant commercial usage has developed for sheet molding and bulk molding polyester compounds where incorporation of DMMP permits use of a wider range of loadings of alumina trihydrate filler. In this way low-smoke- and low-flame-spread reinforced polyester resin formulations suitable for sanitary ware applications can be achieved.[19] DMMP has also been found commercially useful for increasing the phosphorus content of flame retardants used in rigid foams. In such applications, it also is said to aid dimensional stability.[20] It is too volatile for use in flexible foams. DMMP has also been found useful as a chemical intermediate for making other flame retardants (see Secs. 3.5, 3.6, and 4.3.4).

3.4. Halogenated Alkyl Phosphates and Phosphonates

This class encompasses several of the "workhorse" flame-retardant additives, each of which has multiple uses. The presence of the halogen accomplishes several desirable objectives; the halogen (1) contributes flame retardancy, (2) depresses vapor pressure, and (3) depresses water solubility, thus improving greatly the permanency of these additives. Good manufacturing routes are available which permit these compounds to be priced favorably. Individual compounds of commercial importance are discussed below.

3.4.1. Tris(2-chloroethyl) Phosphate (Stauffer's Fyrol® CEF)

This is a colorless and practically odorless liquid containing 10.8% P and 36.7% Cl. It can be made from ethylene oxide and phosphorus oxychloride as shown[21]:

$$POCl_3 + 3CH_2{-}CH_2 \overset{O}{\longrightarrow} O{=}P(OCH_2CH_2Cl)_3$$

This ester has proved to be versatile because of its wide compatibility with most important polymers except polyolefins and polystyrene. It is more volatile than tricresyl phosphate or dioctyl phthalate but less volatile than dibutyl phthalate and is fairly stable to hydrolysis. Tris(2-chloroethyl) phosphate affords good retention except under very hot and humid conditions.

Commercial uses are quite numerous, including urethane foam (especially rigid foam), carpet backing, flame-laminated foam, flame-retardant paints and lacquers, epoxy, phenolic resins, amino resins, polyvinyl acetate coatings and adhesives, urethane coatings, cast acrylic sheet, polyester resins, and polyisocyanurate foams.

Emulsions of tris(2-chloroethyl) phosphate blended with a binder such as a vinyl or acrylic emulsion can be used for applications such as the back-coating of upholstery. Tris(2-chloroethyl) phosphate is an excellent secondary plasticizer for use in polyvinyl chloride, as a partial replacement for flammable plasticizers such as phthalates, or, where a particularly high degree of flame retardancy is desired, in combination with TCP or cresyl diphenyl phosphate. It can be substituted for antimony oxide where opacity is undesirable.

3.4.2. Tris(dichloropropyl) Phosphate (Stauffer Fyrol® FR-2)

This liquid additive containing 49% chlorine and 7.2% phosphorus is made from epichlorohydrin and phosphorus oxychloride. The resultant dichloropropyl groups are mainly β,β'-dichloroisopropyl groups; therefore the principal component is the following:

$$O{=}P\left(-OCH\underset{CH_2Cl}{\overset{CH_2Cl}{\diagdown}}\right)_3$$

In comparison to tris(2-chloroethyl) phosphate, this product shows lower volatility, lower water solubility, a practically negligible rate of hydrolysis, and resistance to attack by bases. This is probably the result of its sterically hindered ester structure. In urethane foams, it acts as an additive flame retardant without adverse effect on the catalyst system and without causing scorch or adverse effects on tensile strength. Levels of 10–20% are generally used.[22] Many producers of flexible urethane foams find they can successfully meet the Motor Vehicle Safety Standard 302 requirements for automotive seating by use of this additive.

Tris(dichloropropyl) phosphate also has been found useful in acrylic latices for textile backcoating and binding of nonwoven fabrics.

3.4.3. Tris(β-chloropropyl) Phosphate (Stauffer Fyrol® PCF)

This phosphate ester containing 33% chlorine and 9.5% phosphorus is made by the reaction of phosphorus oxychloride with propylene oxide. The product has both β-chloroisopropyl and β-chloro-n-propyl groups in the molecule, but the former group predominates. This compound has been sold commercially by ICI as Daltogard F® and by Stauffer as Fyrol® PCF. Properties are rather close to Fyrol® CEF (the 2-chloroethyl analog). This compound has less reactivity to water and bases and thus has been found advantageous in some urethane foam, especially flexible foam, recipes where scorch might be encountered,[23] or in rigid foam systems where improved storage stability in the isocyanate or in the polyol–catalyst mixture is required. This compound has also been found effective for reducing brittleness in isocyanurate foams.

3.4.4. Tris(2,3-dibromopropyl) Phosphate

This compound is widely used as an additive in a variety of plastics and textiles. It is made by addition of bromine to allyl alcohol followed by reaction of the resultant 2,3-dibromopropanol with phosphorus oxychloride.[24] This product, available as Michigan Chemical's Firemaster® T23 or Stauffer's Fyrol® HB-32, is a dense, nearly colorless, fairly viscous liquid, insoluble in water, very stable to hydrolysis, and stable up to about 200°C. The compound itself has negligible vapor pressure, but the commercial grade contains small amounts of volatile components such as the parent dibromopropanol.

Despite generalizations which are sometimes made about poor light stability of bromine compounds, the light stability of this compound has been adequate for use in clear cast acrylic sheet and in cellulose acetate or acrylic fibers where serious discoloration cannot be tolerated. Some of the other applications of this compound include polystyrene (especially expanded foam), urethane foams, lacquers, SBR latices, cured unsaturated polyesters, and epoxies. Textile finishes having good durability to laundering have been developed. For example, tris(dibromopropyl) phosphate is applied as an emulsion to 100% polyester fabric and the fabric then heated to "thermofix" the phosphate.[25] This flame retardant can also be "exhausted" onto the fabric in an operation similar to dyeing. The tris(dibromopropyl) phosphate enters the fibers and becomes tightly locked in, that is, resistant to multiple launderings. This finishing technique has been successfully used on 100% polyester to pass the federal requirements for childrens' nightwear. Emulsions and emulsifiable concentrates of tris(dibromopropyl) phosphate have been marketed.[26,27] Recently, toxicological concerns have resulted in substantial discontinuance of this application.

3.4.5. Oligomeric Chloroalkyl Phosphonate (Monsanto Phosgard® C22R)

Where reduced vapor pressure and thus increased permanence of the flame-retardant additive is required, phosphorus additives having several phosphorus atoms per molecule have been found commercially useful.

The first of this class to become commercial is Monsanto's Phosgard® C22R, a chloroethyl phosphonate which can be made from PCl_3, ethylene oxide, and acetaldehyde. Its structure is the following[28]:

$$\underset{\underset{ClCH_2CH_2O}{|}}{\overset{\overset{O}{\|}}{ClCH_2CH_2P}} - OCH \underset{\underset{CH_3}{|}}{\overset{\overset{CH_3}{|}}{}} \left(\underset{\underset{OCH_2CH_2Cl}{|}}{\overset{\overset{O}{\|}}{-P}} - O - \underset{\underset{|}{|}}{\overset{\overset{CH_3}{|}}{CH}} - \right)_n \overset{O}{\overset{\|}{P}}(OCH_2CH_2Cl)_2$$

Phosgard® C22R has been recommended for use in rigid and flexible urethane foams to obtain self-extinguishing properties. Its use in cast PMMA at levels of about 17.5%–25% has been suggested. Phosgard® C22R has also been found useful in thermoset resins of the aminoplast, phenolic, and epoxy types.

3.4.6. Bis(2-chloroethyl) 2-Chloroethylphosphonate (Mobil Antiblaze® 78)

This compound is an inexpensive phosphorus–chlorine-containing phosphonate ester made by the thermal rearrangement of tris(2-chloroethyl) phosphite. One commercial application is as an intermediate to Stauffer Chemical's Fyrol® Bis-Beta, bis(2-chloroethyl) vinylphosphonate, and Fyrol® 76. It also is an intermediate for hydrolysis to 2-chloroethylphosphonic acid, a plant growth regulator.

Antiblaze® 78 has volatility comparable to that of tris(2-chloroethyl) phosphate, and it therefore has been found most suitable for closed-cell rigid urethane foams. This compound is readily dehydrochlorinated by bases such as amines, which may impose some limitation on the formulations in which it can be used.

3.4.7. Diphosphates

Three diphosphates having a rather similar pattern of utility as low-volatility flame-retardant additives are available at the present time.

The first of these to be introduced was Monsanto's Phosgard® 2XC20. From information in patents, its synthesis appears to be as follows[29]:

Phosgard® 2XC20 is recommended by its manufacturer as a nonvolatile flame-retardant additive for urethane foam, acrylics, and epoxies. The neopentyl structure in the molecule may be expected to enhance its stability. The principal use for Phosgard® 2XC20 may be in flexible urethane foams.

A more recently introduced diphosphate is Olin's Thermolin® 101, having the following structure:

$$\begin{array}{c} ClCH_2CH_2O \\ \\ ClCH_2CH_2O \end{array} \overset{\overset{O}{\|}}{P}-O-CH_2CH_2-O-\overset{\overset{O}{\|}}{P} \begin{array}{c} OCH_2CH_2Cl \\ \\ OCH_2CH_2Cl \end{array}$$

It is a liquid of rather low viscosity (242 cps/25°C) containing 13% phosphorus and 31% chlorine.

The synthesis described in the patent literature proceeds from ethylene oxide and phosphorus trichloride to tris(2-chloroethyl) phosphite, which is then chlorinated to bis(2-chloroethyl) phosphorochloridate. This acid chloride is then reacted with ethylene glycol, possibly in the presence of a base.[30]

Another recently introduced diphosphate of related structure is Monsanto's Phosgard® 1227, having the following structure:

$$\begin{array}{c} ClCH_2CH_2O \\ \\ ClCH_2CH_2O \end{array} \overset{\overset{O}{\|}}{P}-OCH_2CH_2OCH_2CH_2O-\overset{\overset{O}{\|}}{P} \begin{array}{c} OCH_2CH_2Cl \\ \\ OCH_2CH_2Cl \end{array}$$

The compound was described in an earlier Russian publication.[31] The synthesis could be similar to that of Thermolin® 101, but using diethylene glycol instead of ethylene glycol. A comparison of these diphosphates to related monophosphates has been published, showing the diphosphates to have better permanency in flexible foams subjected to accelerated aging.[40]

3.4.8. Oligomeric Phosphate (Fyrol® 99)

This product contains the ethylene glycol diphosphate described above as well as higher oligomers. Its synthesis may be represented by the following equation[32]:

$$(ClCH_2CH_2O)_3PO \xrightarrow[\substack{heat;\ basic \\ catalyst}]{-ClCH_2CH_2Cl} \begin{array}{c} ClCH_2CH_2O \\ \\ ClCH_2CH_2O \end{array} \overset{\overset{O}{\|}}{P}-O-\left[CH_2CH_2-O-\overset{\overset{O}{\underset{\underset{ClCH_2CH_2O}{|}}{\|}}}{P}-O\right]_n CH_2CH_2Cl \quad n=2\text{-}4$$

Because of its higher oligomer content, Fyrol® 99 is found to be low in volatility and has advantageous resistance to solvents. It has found utility in amino resin-impregnated air filters, in flexible urethane foam, in rebonded foam, and in structural foam. It also shows promise for use in phenolics, epoxy resins, and other thermoset polymer systems.

3.5. Oligomeric Halogen-Free Phosphorus Esters

It is generally found that the intrinsic efficacy of a flame retardant depends more on its phosphorus content than on its halogen content. Therefore, a strategy for achieving highly efficient flame-retardant compositions is to build in a high percentage of phosphorus while retaining nonvolatile physical characteristics. It has been found that substantially halogen-free oligomers of high phosphorus content can be prepared by the following reaction[33]:

$$2CH_3PO(OCH_3)_2 + (ClCH_2CH_2O)_3PO \xrightarrow{-3CH_3Cl} (OCH_2CH_2O-\overset{\overset{\displaystyle O}{\|}}{\underset{\underset{\displaystyle \underset{\displaystyle CH_3O}{|}}{\overset{\displaystyle |}{\underset{\displaystyle \|}{POCH_3CH_2}}}}{P}}-OCH_2CH_2O\overset{\overset{\displaystyle O}{\|}}{\underset{\underset{\displaystyle CH_3}{|}}{P}}-)_x$$

This product is water-soluble, low in volatility, and compatible with a variety of thermosetting resins. Diluted with a small percentage of alcohol to a convenient viscosity, it is sold as Fyrol® 58 for use in amino resin-impregnated automotive air filters. The absence of halogen from this product avoids the release of "smoke" during the curing operation.

3.6. Oligomeric Cyclic Phosphonate (Antiblaze® 19)

This flame-retardant additive is a highly viscous, colorless, water-soluble syrup having a high phosphorus content (21%) and no halogen. From the published data and patent, it may be deduced to be a mixed cyclic phosphonate ester, made by the following chemistry[34]:

$$C_2H_5C(CH_2OH)_3 + PCl_3 \longrightarrow C_2H_5C\underset{\displaystyle CH_2-O}{\overset{\displaystyle CH_2-O}{\big\langle}CH_2-O\big\rangle}P \xrightarrow{CH_3P(OCH_3)_2}$$

$$\left(CH_3\overset{\displaystyle O}{\overset{\displaystyle \|}{P}}\underset{\displaystyle O-CH_2}{\overset{\displaystyle O-CH_2}{\big\langle}} \underset{\displaystyle CH_2-O}{\overset{\displaystyle C_2H_5}{C}} \right)_n \overset{\displaystyle O}{\overset{\displaystyle \|}{P}}\underset{\displaystyle (OCH_3)_{2-n}}{\overset{\displaystyle OCH_3}{\big\langle}}$$

where n is between 1 and 2. Although the bicyclic phosphite intermediate is very toxic,[35] the end product is low in toxicity and has good thermal and hydrolytic stability. Antiblaze® 19 has been introduced for thermofixation into polyester fabric, an application for which tris(2,3-dibromopropyl) phosphate was previously the dominant product. Antiblaze® 19 has also found use in urethane foams and thermoset resins.

3.7. Oligomeric Phenylphosphonates

Oligomeric phenylphosphonates of dihydric phenols have been found to have sufficient thermal stability and freedom from volatility to permit their use in thermoplastics processed above 250°C. In particular, the phenylphosphonate oligomer from sulfonylbisphenol, made by the following equation,

has been used commercially in a polyester fiber "Heim" developed in Japan.[36]

3.8. Tricresyl Phosphates and Related Phosphates

As mentioned earlier, tricresyl phosphate has a half-century of commercial use history, having been originally introduced for cellulose nitrate and other cellulosics. It is now mainly used in polyvinyl chloride, particularly in electrical insulation, automotive and furniture upholstery, and conveyor belts. The use of aryl phosphate plasticizers in vinyl has been accelerated by the Motor Vehicle Safety Standard 302 flammability requirement for auto interiors. Increased flame-retardancy requirements for electrical insulation and for vinyl film and sheet used in construction also have stimulated the growth of these flame-retardant plasticizers.

Tricresyl phosphates are produced by the reaction of mixtures of phenol, cresols, and xylenols (cresylic acids) in various proportions with phosphorus oxychloride, generally in the presence of a catalyst. The esters are subsequently distilled and washed to a high degree of purity. Cresylic acids from petroleum and from coal tar are both used; to avoid the toxic *o*-cresyl phosphates, cresylic acids containing very little of the ortho isomer are used.

Typical tricresyl phosphates are Stauffer's Lindol® and Phosflex® 179C, which are widely used as flame-retardant plasticizers in vinyls and cellulosics. A lower-volatility trixylenyl phosphate is represented by Phosflex® 179A. A special grade is also made with extremely high electrical resistivity for vinyl-covered wire.

A more recent development is the use of synthetic cresylic acid substitutes, which has occurred largely in response to the upward trend in prices of coal tar and petroleum-derived cresylic acids, especially *m/p*-cresol. The first synthetic substitutes which have appeared on the market are isopropylphenyl phosphates, the isopropylphenol having been made by alkylation of phenol by propylene. The use pattern for these materials is similar to TCP.[37,38] Tertiary-butylphenyl phosphates have also reappeared in the marketplace, having been sold earlier (in the 1950 era) by Dow.

Two new chlorine-containing blended aliphatic–aryl phosphates, Phosflex® 300 and 400, have recently been introduced by Stauffer for use in vinyls. These materials have a use pattern similar to TCP and cresyl diphenyl phosphate but appear to have advantages in regard to smoke- and flame-retardant effectiveness.

A newer member of this chlorine-containing aryl phosphate group, Phosflex® 500, appears effective for flame-retarding the adhesive in rebonded urethane foam.

Another variation on the tricresyl phosphate theme, cresyl diphenyl phosphate (CDP), is a versatile plasticizer and flame-retardant additive. CDP can be used in the same plastics as TCP and gives comparable flame retardancy but is found to have advantages with respect to flex temperature, heat and light stability, and plasticizer efficiency. Its volatility is of course somewhat greater than that of TCP but for many applications is adequately low. The pricing of CDP is relatively favorable since phenol is available from many sources.

The use of tricresyl phosphate in vinyls tends to limit their low-temperature flexibility. Related phosphate plasticizers having two phenyl groups and one long-chain alkyl group were designed to avoid this difficulty. The two principal commercial examples of this group are Monsanto's isodecyl diphenyl phosphate and 2-ethylhexyl diphenyl phosphate.[39] The latter obtained FDA approval for use in food packaging.

When antimony oxide prices are high, vinyl compounders concerned with cost minimization turn to the alkyl diphenyl phosphates and to cresyl diphenyl phosphate as alternatives to antimony oxide. A comparison of the flame-retardant effectiveness of these ingredients, as determined by the LOI method in a vinyl containing 40 phr of total plasticizer (DOP used to make up the difference), is shown in Fig. 2.

3.9. Triphenyl Phosphate

Triphenyl phosphate is a solid, melting at 48°C, having low odor, low color, and low water solubility. It has been used as a flame retardant in cellulosic polymers such as cellulose acetate. More recently the value of

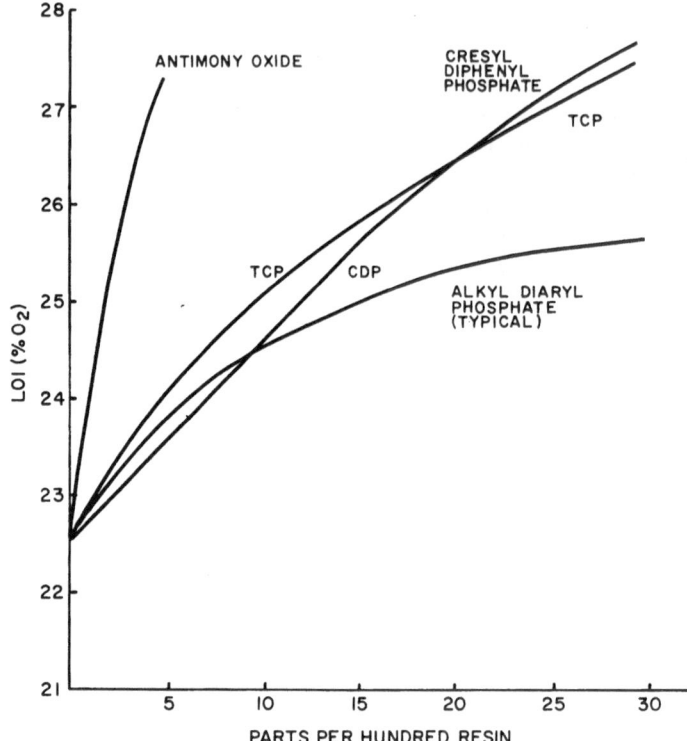

FIGURE 2. Flammability of polyvinyl chloride films at 60 phr total plasticizer (DOP or DOP + phosphate), as determined by limiting oxygen index (LOI).

triphenyl phosphate as a flame retardant for engineering thermoplastics has been recognized, especially in view of the high thermal stability of the compound (boiling undecomposed at 370°C) and satisfactory cost-effectiveness relationship. It has also been found useful in systems where high stability is needed toward bases, such as in urethanes where the flame retardant, catalyst, and some of the other ingredients may be premixed and held for long periods before use.

3.10. Phosphonitrilics

A phosphonitrilic additive for rayon represents the first serious commercial application of the much-investigated phosphonitrilic chlorides, which, as chemicals, have been known for 135 years. The chemistry of the flame-retardant additive, introduced by Avisco Division of FMC,[40] is as follows:

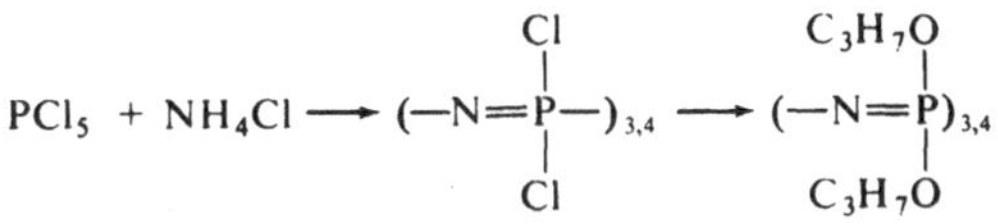

$$PCl_5 + NH_4Cl \longrightarrow (-N{=}\overset{\displaystyle Cl}{\underset{\displaystyle Cl}{P}}-)_{3,4} \longrightarrow (-N{=}\overset{\displaystyle C_3H_7O}{\underset{\displaystyle C_3H_7O}{P}})_{3,4}$$

The additive is an odorless liquid of low toxicity, insoluble in water and stable to acid and base. It thus can be admixed into the highly alkaline viscose and spun into acid to obtain rayon fibers where the additive remains locked in as liquid droplets. Economics appear to have deterred large-volume usage of this product, but it has been produced in developmental quantities.

3.11. Phosphonium Bromides (American Cyanamid's Cyagards®)

These are very high-melting thermally stable solids which can be milled into thermoplastics such as polypropylene, impact polystyrene, and ABS. They are effective at low concentrations (one half or less of the concentrations required with chlorine–antimony systems), and being solids, they do not cause plasticization. The synthesis of a representative compound of this series is as follows[41]:

$$PH_3 + 3CH_2{=}CHCN \longrightarrow P(CH_2CH_2CN)_3 \xrightarrow{BrCH_2CH_2Br}$$

$$(NCCH_2CH_2)_3\overset{+}{P}CH_2CH_2\overset{+}{P}(CH_2CH_2CN)_3$$

$$Br^- \qquad Br^-$$

Despite its ionic structure, the bis-phosphonium compound has very low water solubility. These compounds are rather expensive but can be used in combinations with cheaper halogenated hydrocarbons[42] or in combination with ammonium polyphosphate.

3.12. Phosphine Oxides

A phosphine oxide of the following structure

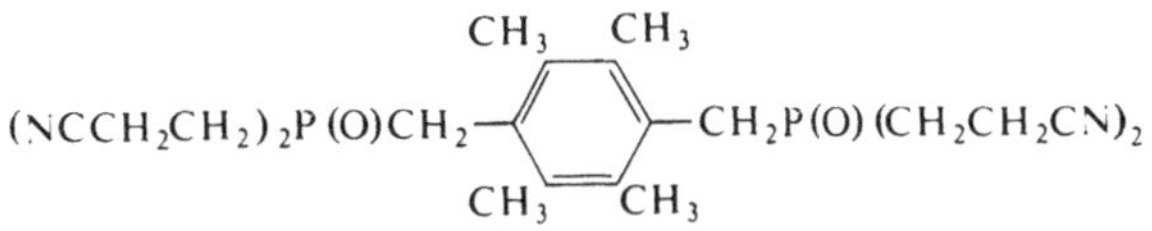

$$(NCCH_2CH_2)_2P(O)CH_2{-}\!\!\left\langle\!\!\begin{array}{c}CH_3 \quad CH_3\\ \\ CH_3 \quad CH_3\end{array}\!\!\right\rangle\!\!{-}CH_2P(O)(CH_2CH_2CN)_2$$

has been found highly effective in halogen-free polypropylene formulations containing TiO_2 and ammonium polyphosphate, giving flame retardancy with low smoke and freedom from dripping. The mechanism of action appears to involve formation of a char coating.[43]

4. Organic Phosphorus Compounds—Reactive Types

4.1. Organophosphorus Monomers

Many hundreds of vinyl monomers containing phosphorus have been described in the literature,[2] but few of these have appeared commercially. Bis(2-chloroethyl) vinylphosphonate (Stauffer's Fyrol® Bis-Beta) is a relatively inexpensive phosphorus-containing monomer made from ethylene oxide and phosphorus trichloride:[44]

$$PCl_3 + CH_2\!-\!CH_2 \xrightarrow{\quad\quad} P(OCH_2CH_2Cl)_3 \xrightarrow[heat]{} ClCH_2CH_2PO(OCH_2CH_2Cl)_2 \xrightarrow[base]{}$$

$$CH_2{=}CHPO(OCH_2CH_2Cl)_2$$

Several fields of application have been found for bis(2-chloroethyl) vinylphosphonate. Emulsion copolymers with other vinyl monomers have been developed by B. F. Goodrich and by Stauffer. Several such products are available as flame-retardant latices for textile and paper applications.[45-47] Fyrol® Bis-Beta has also been found effective as a flame-retardant cross-linking agent for cured unsaturated polyesters, especially those containing tetrachlorophthalic, tetrabromophthalic, chlorendic acid, or UOP's Chloran® in the backbone,[48-51] and at least one such product has been on the market.

The vinylphosphonates have been found to copolymerize particularly well with vinyl halides and vinylidene halides. Textile fibers having the flame retardancy of the modacrylics with superior physical properties resembling the acrylics have been made by incorporating small percentages of the vinylphosphonate monomer.[52]

Vinyl chloride–vinylphosphonate copolymers which afford low-smoke film and sheet have also been described recently.[53]

4.1.1. Allyl Phosphorus Monomers

A great deal of research work has been reported on allyl phosphates and phosphonates (M. Sander[2]), but they have been slow to find commercial utility. Triallyl phosphate and several other allyl phosphorus compounds have been produced in commercial or developmental quantities. One plausible use for triallyl phosphate is as a cross-linking monomer in polyolefins for electrical insulation, in somewhat the manner in which triallyl cyanurate is now used, with the additional advantage of imparting flame retardancy.

4.2. Phosphorus-Containing Diols and Polyols

The commercial development of several phosphorus-containing diols occurred in response to the need to flame-retard rigid urethane foams used for insulation in the transportation and construction industry. A 1970 review[54] shows 337 references, mostly to patents on phosphorus polyols, but few have been used commercially. Perhaps the earliest product introduced (and still used) for this application was Mobil's Vircol® 82, a diol obtained by reacting propylene oxide and dibutyl acid pyrophosphate. The substance has an OH number of 205 and contains 11.3% P.[55,56]

This phosphorus content and OH number fit the following average structure:

$$C_4H_9O-\overset{\overset{O}{\|}}{\underset{\underset{O(C_3H_6O)_x H}{|}}{P}}-O-C_3H_6-O-\overset{\overset{O}{\|}}{\underset{\underset{O(C_3H_6O)_y H}{|}}{P}}-OC_4H_9$$

where $x + y = 3.4$.

A related product, Vircol® 611, has a lower viscosity, a higher percentage of phosphorus (16%), and a lower OH number (150). It is believed to be a mixture of Vircol® 82 and dimethyl methylphosphonate.

Somewhat later, a phosphorus-containing diol was developed that also contains one atom of nitrogen per molecule. The nitrogen appears to have a synergistic effect on the flame-retardant action of the phosphorus. This diol, Stauffer's Fyrol® 6, can be synthesized as shown[57]:

$$(HOCH_2CH_2)_2NH + HCHO + HP(O)(OC_2H_5)_2 \longrightarrow (HOCH_2CH_2)_2NCH_2P(O)(OC_2H_5)_2$$

The product contains 12.6% phosphorus and has an OH number in the 450 mg KOH/g range.

A typical formulation of a rigid urethane foam containing this phosphonate diol is shown in Table 1.

A particular advantage of this diol phosphonate is the stability of polyol–catalyst mixtures containing it and the resistance of foams containing it to humid aging.

The need for flame retardants for flexible foams has until now been largely met by additives such as tris(dichloropropyl) phosphate. Recently there has been active research in this field on reactive diols and triols which could be of value in increasing the permanence of flame retardancy. The requirements for a flame-retardant polyol for flexible foam have been gen-

TABLE 1

Typical Rigid Foam Formulation Using Fyrol® 6[diethyl
N,*N*-bis(2-hydroxyethyl)aminomethylphosphonate][a]

	% by weight
Polymethylene polyphenylisocyanate ("PAPI")	43.4
Propylated sucrose polyol (OH no., 450)	28.9
FYROL 6	9.5
Silicone surfactant	0.7
Fluorotrichloromethane	15.8
Dimethylaminoethanol catalyst	0.46
Dibutyltin dilaurate catalyst	0.23
Catalyst diluent	1.09
NCO index	105.7
% Phosphorus in formulation	1.22
Foam density (lb/ft^3)	1.96
Compressive strength (psi, parallel to rise)	31–33
Tensile strength (psi)	30
% Closed cells	>90
K factor (Btu/hr/ft^2/°F/in.)	0.12
Flammability (ASTM D-1692)	S.E.

[a] Data from J. Biranowski, Stauffer.

erally more difficult to meet because of the open-cell nature of these foams,
their lower cross-link density (more combustible pyrolysis products, less
char), and the stringent specifications and economic considerations which
must be met.

In Europe, some commercial use was made of an oligomeric hydroxy-
alkyl chloroalkyl phosphate.[58]

Three related polyols containing 24–36% bromine and 2.3–2.7% phosphorus have been recommended for use in semirigid foam; one of them,
Brominex® 160P, has an OH number of 47 and is suggested for flexible
foam. Some suggestion of the chemistry which may be involved in the synthesis of these products is given by a patent,[59] which describes addition of
bromine to castor oil and subsequent introduction of phosphonate groups
by reaction with trialkyl phosphite.

4.3. Reactive Organophosphorus Compounds in Textile Finishing

Although synthetic fibers can be flame-retarded with additives, the
durable flame retarding of cotton is best accomplished by a reactive finish.
Before 1972 the principal market for such finishes was military goods and

industrial protective clothing. However, a much larger yardage of cotton goods now must be treated to pass federal requirements established for children's nightwear under the Flammable Fabrics Act Amendment of 1967. Several durable f.r. reactive cotton finishes have reached the market.

4.3.1. Tetrakis(hydroxymethyl) Phosphonium Salts

The reaction of formaldehyde with phosphine in hydrochloric acid yields tetrakis(hydroxymethyl) phosphonium chloride. This is a water-soluble crystalline compound, generally sold in concentrated aqueous solution, and is capable of being cured by reaction with amino compounds (ammonia, urea, melamine, etc.) to form a durable thermoset resin.[60]

More recently, the chloride has been replaced in commercial usage by the corresponding sulfate in order to avoid the possibility of formation of bis(chloromethyl) ether during textile processing. Tetrakis(hydroxymethyl) phosphonium oxalate and a phosphate–acetate mixture are also commercially available.

Finishes for cotton based on tetrakis(hydroxymethyl) phosphonium salts plus melamines, ureas, or other co-reactants have been developed by Albright and Wilson in England, the USDA Southern Regional Laboratory, and Hooker Chemical Company. Hooker's Roxel® finishes have been used extensively on industrial garments and military cotton goods.

Considerable effort has been expended in recent years on developing modifications of these finishes for lighter-weight cotton goods to try to overcome the excessive fabric stiffness caused by the earlier finishes. One improved finish developed by the USDA and Hooker Chemical Company is the so-called THPOH–ammonia finish, wherein the tetrakis(hydroxymethyl) phosphonium salt is first converted to a free organic base by careful reaction with alkali and then cured on the fabric by reaction with ammonia gas. The cloth is then treated with hydrogen peroxide or perborate to convert phosphine groups to more stable phosphine oxide groups. Durable flame retardancy with good hand and strength retention has been obtained. Perhaps because of the ammonia gas requirement, which necessitates special equipment, the commercialization of the process has proceeded slowly. Recent efforts by Hooker, Cotton Inc., and United Merchants appear to have made progress in ammoniator design, and the finish is now in commercial production.

A prepolymer, obtained by heating a tetrakis(hydroxymethyl) phosphonium salt with elimination of water, was said by Ciba-Geigy to show promise for the flame retarding of cotton–polyester blends.[61] Odor problems are reported to have retarded this development. Other types of prepolymers are currently in development for use on cotton and blends.

4.3.2. Pyrovatex CP®

This finish, developed by Ciba-Geigy, is based on the following chemistry[62,63]:

$$(CH_3O)_2PHO + CH_2=CHCONH_2 \longrightarrow (CH_3O)_2P(O)CH_2CH_2CONH_2 \xrightarrow{HCHO}$$

$$(CH_3O)_2P(O)CH_2CH_2CONHCH_2OH$$

Properly applied finishes based on Pyrovatex CP® permit cotton to pass the vertical flammability test after 50 detergent washes; the initial add-ons are in the 20–40% range depending on the weight of the fabric. Children's nightwear with this finish has been on the market for several years.

4.3.3. MCC100/200/300 (Monsanto)

This is a three-package system comprising trimethylphosphoric triamide, a methylolmelamine resin, and an amine hydrochloride catalyst.[64] This system appears to be capable of giving an f.r. finish durable to 50 launderings. This process achieved small-scale commercial use but had disadvantages in regard to fabric "hand" and has been withdrawn.

4.3.4. Oligomeric Vinylphosphonate (Stauffer Fyrol® 76)

The reactive textile flame retardant differs from the others described above in that curing is accomplished by a free-radical mechanism. The finish therefore lends itself to a very rapid cure over a wide range of temperatures in conventional equipment.

Fyrol® 76 is a water-soluble vinylphosphonate monomer containing 22.5% P. It is most commonly used in conjunction with methylolacrylamide using a free-radical catalyst such as persulfate.[65,66] Fyrol® 76 can be applied to cellulosics by conventional pad-dry cure or may be simultaneously dried and cured. Little or no odor is present during formulation or throughout the plant area during processing. Conventional cure temperatures of 300–350° F are most commonly employed.

With a lightweight cotton flannel approximately 25–30% total applied solids is required to pass DOC FF 3-71. At this add-on the fabric exhibits a soft hand, good strength, stabilization, wrinkle recovery, and permanent-press properties. Tensile strength retention of 80–100% and tearing strength retention of 65% have been obtainable. On cotton knits, burst strength retention of 60–80% has been achieved. In addition, Fyrol® 76-treated fabrics exhibit durability to dry cleaning and commercial laundering.

Laundering the finish with low-phosphate or phosphonate-free liquid detergents in very hard water does not adversely affect durable flame-retardant properties, but highly alkaline detergents are deleterious, especially in hard water. In soft water, soap and carbonate detergents can be used without affecting the fire-retardant properties. Chlorine-containing bleaches limit durability to 25–30 cycles. This durability is comparable to the best achieved so far with other flame-retardant finishes for cotton. Perborate bleach can be used without adversely affecting flame-retardant properties after 50 wash-and-dry cycles.

Good results have been demonstrated on cotton flannel, knits, print cloth, sheeting, terry cloth, corduroy and heavyweight fabrics, cotton–rayon blends, 100% rayon, and certain polyester–cotton blends and acetate–cotton blends.

The use of Fyrol® 76 essentially as "homopolymer," with little or no nitrogen co-reactant, has shown promise for flame-retardant finishing of polyester–cotton blends.[66a] Combinations of such finishes with bromine compounds, such as tris(dibromopropyl) phosphate, permit use of reduced add-ons of the vinylphosphonate, as does the use of a blend in which the polyester component contains built-in bromine.[67]

5. Mode of Action of Phosphorus Flame Retardants

5.1. Condensed Phase Mechanisms

The mechanism by which phosphorus acts in flame-retarding cellulose appears to be rather well understood, but in most other polymer systems not much more than reasonable hypotheses can be presented. In cellulose, two alternative routes of thermal breakdown are possible: One route leads by way of an intermediate known as levoglucosan to volatile combustible organic fragments; the other route (catalyzed by acids) leads to water and difficultly combustible carbonaceous char[68]:

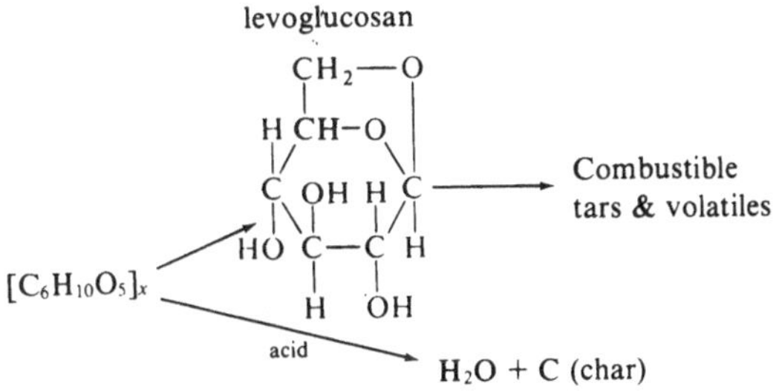

Although a variety of acids catalyzes the H_2O-and-char-forming reaction, phosphoric acid is obviously an ideal acid because it is nonvolatile and when strongly heated yields an even more powerfully dehydrating catalyst, poly-phosphoric acid.

The flame-retardant action of phosphorus compounds in a cellulosic system is believed to involve phosphorylation of the cellulose. The phosphorylated cellulose then breaks down to water and carbonaceous char. Nitrogenous compounds such as amides and amines appear to catalyze the cellulose phosphate-forming steps and are found to strongly synergize the action of phosphorus in cellulosic systems.[69,69a,70]

Synergism of phosphorus by nitrogen is by no means a general phenomenon in f.r. systems. It is observed in cellulose, and it is believed to occur in urethane foams. Reeves et al.[71] have shown that some nitrogen compounds do not show the synergistic effect. For example, they found no synergism of phosphorus f.r. action in cellulose by nitriles such as poly-acrylonitrile. It was shown by Willard and Wondra[72] that under conditions where melamine resin synergized phosphorus a uron resin had little effect and a triazone resin was actually somewhat antagonistic.

In polymethyl methacrylate, it has been shown[73] that the presence of phosphorus additives, especially nonvolatile ones, increases the yield of charred residue, and in this case the mechanism of action somewhat resembles that in cellulose.

In rigid polyurethane foams, the action of phosphorus flame retardants also appears to involve effects on char formation. The involvement of a "char coat" in the extinguishment mechanism was suggested by Anderson.[74]

The physical character of the char was also found to be affected in some studies. Einhorn showed that the presence of a built-in phosphorus flame retardant caused a rigid urethane foam to produce a more coherent barrier-like char which was believed to act as a physical barrier to the combustion process.[75]

Several workers have reported that the char is mechanically stronger when the foam contains a phosphorus flame retardant.[76,77] Although some of the phosphorus can be volatilized during burning, in the case of tris(halo-alkyl) phosphate additives[78] or even in the case of reactive diol-type phosphorus flame retardants,[79] there is also evidence that a substantial fraction of the phosphorus can remain behind in the char.[80,81] Obviously, the ratio of retained to released phosphorus will be dependent on the structure of the retardant, the substrate, and the combustion conditions.

A recent study by Kresta and Frisch affords further insight into the mechanism of action of phosphorus flame retardants in rigid urethane and urethane–isocyanurate foams.[82] They show that, to a rather good approximation, the oxygen index is a linear function of the logarithm of percent phosphorus in the foam and is not much affected by the structure of the flame retardant (within the group of retardants studied). They also show that the so-called synergism of chlorine and phosphorus is an artifact of the nonlinear response of flame retardancy (expressed by oxygen index) to phosphorus content. When the % Cl requirement is graphed against log(% P) for foams having a constant oxygen index, a straight-line relationship was found. These findings tend to substantiate the view that phosphorus-halogen synergism is generally an artifact of nonlinear f.r.–concentration responses and not a consequence of specific phosphorus–halogen chemical interactions.[70]

Kresta and Frisch concur with the view that phosphorus flame retardants function in rigid foams by inducing char; they explain the leveling off of oxygen index with increase in phosphorus on the basis that the oxygen index of carbon (or impure carbon) is the limiting value for oxygen index. In polymers such as polystyrene or polypropylene, containing no oxygen groups, the acid-catalyzed char-forming mode of action of phosphorus plays a less evident role, and, in fact, phosphorus is not so effective in such polymers as it is in more oxygenated polymers. It is, nonetheless, somewhat effective, and a basis for its action must be sought. The reasonable suggestion was made by Sherr and co-workers that phosphorus additives may become converted thermally to polyphosphoric acid, which forms a continuous coating over the burning polymer.[83] Some evidence that such a coating does form has been seen in the case of polystyrene,[84] and it is not unreasonable to believe that the coating acts as a barrier to passage of heat, oxygen, or fuel.

There is also some evidence that phosphorus-containing additives can work in some cases by catalyzing thermal depolymerization of the polymer melt, facilitating the flow or drip of the melt from the combustion zone.[84] In polystyrene, tris(2,3-dibromopropyl) phosphate acts to a significant degree by this mechanism.[78] A similar mechanism is believed to prevail in phosphorus flame-retarded flexible urethane foams, which are able to pass tests such as Motor Vehicle Safety Standard 302 by melting away from the flame zone.

Polyester fabrics flame-retarded with tris(dibromopropyl) phosphate or other phosphorus additives are generally seen to melt and drip more readily than the fabrics without the flame retardant. This mechanism can be defeated by the presence of nonthermoplastic threads, which can serve as wicks, or by silicone oils, which appear to impede the melt flow.[85,86]

Enhancement of the melt–drip phenomenon by incorporation of phosphorus compound is an important flame-retardant mechanism in polyester fabrics.

5.2. Vapor Phase Mechanisms

Considerable attention has been directed in recent years to a possible vapor phase mechanism of action of phosphorus flame retardants.[84]

It has been reported by Rosser *et al.* that trimethyl phosphate retards the velocity of a methane–oxygen flame with about the same molar efficiency as SbCl₃.[87,88] Both physical and chemical mechanisms have been adduced for the vapor phase action of a utilized phosphorus compound. In studies by Steutz *et al.*, tris(dibromopropyl) phosphate was found not to change the activation energy of thermooxidative degradation of polypropylene, although it raised the oxygen index. A vapor phase physical shielding effect was postulated for tris(dibromopropyl) phosphate in this polymer.[89]

Studies done recently at the National Bureau of Standards have utilized triphenylphosphine oxide as a "model" phosphorus flame retardant.[90] This compound was shown by mass spectroscopy to break down in the flame to give small molecular species such as PO·. The rate-controlling hydrogen atom concentration in the flame was shown spectroscopically to be greatly reduced when the phosphorus species were present. The mechanism proposed for flame retardancy in this situation is scavenging of H· by phosphorus radicals such as PO·.

In evaluating the importance of vapor phase versus condensed phase mechanisms of phosphorus flame retardency, it should be recalled that highly stable volatile phosphorus compounds such as triphenylphosphine oxide are not typical of the types of retardants actually found commercially useful. The Bureau of Standards study shows that there is a vapor phase mechanism possible but does not assess its importance in practical systems.

In a model system (PMMA) which offers the advantage of giving sharp, easily reproducible, oxygen indices and compatibility with a wide variety of additives, the present author has compared a variety of simple phosphorus additives at equivalent phosphorus loadings. Poor flame retardancy was found with trimethylphosphine oxide, a volatile stable species, whereas the maximum oxygen index elevation that was observed was with phosphoric acid.[91]

In a recent study where phosphine oxides were shown to be highly effective retardants for polypropylene, evidence was presented for a condensed phase mode of action involving char coating.[43]

Some evidence has been adduced for an important vapor phase mechanism of action of certain phosphorus-containing textile finishes. When a tetrakis(hydroxymethyl) phosphonium salt or hydroxide is polymerized on a cellulosic fabric by co-reaction with amino compounds, and especially when subsequently treated with hydrogen peroxide, the resultant cross-linked polymer contains phosphorus mainly `in the form of phosphine oxide structures.[92]

When polyester–cotton blends having such finishes are burned, part of the phosphorus is volatilized and part is retained in the char. When the same blend fabric bearing a phosphonate finish is burned, much more of the phosphorus is retained in the char and much less is volatilized.[93] These observations offer an explanation for the observation that the phosphine oxide (phosphonium-salt-derived) finish is more efficient than the phosphonate finish on a phosphorus basis on polyester–cotton blends but not on cotton. It has been postulated that the polyester–cotton blend fabric needs both a condensed phase mechanism of action and a vapor phase mechanism of action for efficient flame retardancy.[94] It seems reasonable that polyester–cotton blends would respond to a vapor phase flame retardant, since such blends have been shown to give off substantial quantities of ethylene upon pyrolysis (while cotton or polyester separately give little ethylene).[95]

Polyethylene terephthalate has been shown to afford a higher oxygen index with 5% phosphorus incorporated in the backbone of the polymer as phenylphosphinyl groups, as contrasted to 5% phosphorus incorporated as a relatively volatile additive, triphenylphosphine oxide.[96] This result suggests that a condensed phase mechanism may be more effective than a vapor phase mechanism in this polymer.

The problem of determining whether a retardant operates mainly by a condensed phase mechanism or mainly by a vapor phase mechanism is particularly complicated in the case of the haloalkyl phosphorus esters. A number of these compounds have been shown to be capable of thermal degradation to release volatile halogenated hydrocarbons (plausible flame inhibitors) while their phosphorus remains behind as relatively nonvolatile phosphorus acids (plausible condensed phase flame retardants).[97]

6. Trends and Future Developments

Generally, the trends in commercial development of phosphorus flame retardants have been in the direction of less volatile compounds and built-in or oligomeric additive types as permanence requirements in flammability regulations, specifications, and codes become stricter.

Increasing attention is being directed to questions of smoke and toxic fumes from burning materials, since statistics show that a major fraction of fire casualties arise from these causes. Standards are beginning to be developed by government agencies, model code organizations, and voluntary standards organizations for smoke evolution from materials of construction and interior furnishings. Toxicity standards in the United States must await further progress on methods of measurement and some consensus regarding the relation of these measurements to actual fire hazards. Although examples have been published where specific phosphorus flame retardants increased smoke relative to the non-flame-retarded polymer, there are also instances where the opposite was found[98]; in particular, such favorable results were seen in polymer systems where char coating resulted from the presence of the phosphorus compound.

In regard to toxic fumes, no documented instances of phosphorus flame retardants contributing to human fire casualties were located in the literature. Laboratory experiments using analytical means or experimental animals have shown no universal trend. In pyrolysis or in combustion of cellulosics treated with phosphorus retardants, the CO/CO_2 ratio can be increased or decreased depending on experimental conditions.[99,100] In the case of polyurethanes where HCN is a major contributor to smoke toxicity, the HCN level was found to decrease in one study when typical phosphorus flame retardants were added,[101] and there is a recent animal study which showed reduced mortality with phosphorus retardants present.[102] One widely publicized instance of a specific toxic effect was found at the University of Utah with rigid urethane foams based on trimethylolpropane polyol and containing phosphorus flame retardants. The active toxic species was identified as a bicyclic trimethylolpropane phosphate formed during the combustion process from the breakdown products of both the polyol and the flame retardant.[103] The toxic bicyclic phosphate is avoided when other types of polyols are substituted for the trimethylolpropane polyol.[98]

It has been shown that flame retardants operating by flame inhibition can be defeated by even moderately elevated temperatures.[104] On the other hand, a coherent char layer is a powerful means for the protection of a combustible material from heat and flame[104] and for reducing the emission of smoke and fumes. Phosphorus-containing flame retardants which enhance char formation can therefore be expected to attain increased importance as flammability, smoke, and toxic fume standards become more stringent.

7. References

1. J. W. Lyons, *The Chemistry and Uses of Flame Retardants*, Wiley-Interscience, New York (1970).
2. M. Sander, *J. Macromol. Sci. Rev. Macromol. Chem.* C1, 1 (1967); C2, 1 (1968).
3. C. F. Raley, paper given at Polymer Conference Series, Recent Advances in Flame- and Smoke-Retardants of Polymers, University of Detroit, Detroit, May 12, 1975.
4. H. L. Vandersall, *J. Fire Flammability* 2, 97 (1971).
5. W. H. Woodstock, U.S. Pat. 2,122,122 (1938) (to Victor Chemical Works).
6. M. Becke-Goehring and J. Sambeth, *Z. Anorg. Allg. Chem.* 297, 287 (1958).
7. J. R. Costello, U.S. Pat. 2,964,377 (1960) (to Stauffer Chem. Co.).
8. J. C. Chapin, U.S. Pat. 3,684,559 (1972) (to Ventron).
9. S. J. O'Brien, *Text. Res. J.* 38, 256 (1968).
10. S. C. Juneja, U.S. Pat. 3,887,511 (1975) (to Canadian Patents & Development Ltd.).

11. Rohm and Haas GmbH, Br. Pat. 996,914 (1965).
12. W. L. Howe, U.S. Pat. 3,347,818 (1967) (to duPont).
13. British Celanese, Br. Pat. 1,100,283 (1968).
14. E. D. Weil, U.S. Pat. 3,666,712 (1972) (to Stauffer Chemical Co.).
15. J. W. Weaver, J. G. Frick, Jr., and J. D. Reid, U.S. Pat. 2,711,998 (1955) (to U.S. Secretary of Agriculture).
16. R. C. Putnam and J. W. Young, U.S. Pats. 3,676,389 (1972) and 3,762,942 (1973) (to Polaris Chemical Corp.).
17. R. Nametz, *Ind. Eng. Chem.* **59**, 107 (1967).
18. A. N. Beavon, U.S. Pat. 3,909,484 (1975) (to Universal-Rundle Corp.).
19. P. V. Bonsignore and J. H. Manhart, *Proc. Annu. Conf. Reinf. Plast. Compos. Inst. Soc. Plast. Ind.* **29**, 23C (1974).
20. J. J. Anderson, Belg. Pat. 674,447 (1966) (to Socony Mobil Oil Co.).
21. A. L. Smith, U.S. Pat. 3,100,220 (1963) (to Celanese Corp.).
22. R. Polacek, U.S. Pat. 3,041,293 (1962) (to Celanese Corp.).
23. J. W. Crook and G. A. Haggis, *J. Cell. Plast.* **5** (2), 119 (1969).
24. D. E. Overbeek and R. C. Nametz, U.S. Pat. 3,046,297 (1962) (to Michigan Chemical).
25. Eastman Kodak, Br. Pat. 1,292,878 (1972).
26. R. E. Smith and R. C. Nametz, in *Book of Papers, 1974 National Technical Conference of the AATCC*, American Association of Textile Chemists and Colorists, Research Triangle Park, North Carolina (1974) pp. 452–461.
27. E. Baer, in *Proceedings of the 1973 Symposium on Textile Flammability*, LeBlanc Research Corp., E. Greenwich, R.I. (1973), p. 117.
28. G. H. Birum, U.S. Pats. 3,014,956 (1961) and 3,042,701 (1962) (to Monsanto).
29. G. H. Birum, U.S. Pat. 3,192,242 (1965) (to Monsanto).
30. J. S. Babiec, J. J. Pitts, W. L. Ridenour, and R. J. Turley, *Plast. Technol.* **1975**, 47 (June 1975).
31. T. M. Moshkina and A. N. Pudovik, *Zh. Obshch. Khim.* **32** (5), 1671 (1962).
32. E. D. Weil, U.S. Pats. 3,896,187 (1975) and 3,513,644 (1970) (to Stauffer Chemical Co.).
33. E. D. Weil, U.S. Pat. 3,891,727 (1975) (to Stauffer Chemical Co.).
34. J. J. Anderson, J. G. Camacho, and R. E. Kinney, U.S. Pat. 3,789,091 (1974) (to Mobil Oil Corp.).
35. E. M. Bellet and J. Casida, *Science* **182**, 1135 (1973).
36. Y. Masai, Y. Kato, and N. Fukui, U.S. Pat. 3,719,727 (1973) (to Toyo Spinning Co.).
37. J. M. Heaps, *Plastics London* **1969**, 410 (April 1969).
38. D. R. Randall and W. Pickles, U.S. Pat. 3,919,158 (1975) (to Ciba-Geigy AG).
39. J. R. Darby and J. K. Sears, in *Encyclopedia of Polymer Science and Technology* (H. F. Mark and N. G. Gaylord, eds.), Wiley-Interscience, New York, pp. 229–306 (1969).
40. L. E. A. Godfrey and J. W. Schappel, *Ind. Eng. Chem. Prod. Res. Dev.* **9**, 426 (1970).
41. L. Cipriani, *Tech. Pap. Reg. Tech. Conf. Soc. Plast. Eng. Palisades Sect.* (Oct. 27–28, 1970).
42. J. F. Cannelongo, U.S. Pats. 3,422,048 (1969) and 3,513,119 (1970) (to American Cyanamid Co.).
43. C. Savides, A. Granzow, and J. F. Cannelongo, *Tech. Pap. Reg. Tech. Conf. Soc. Plast. Eng.* **1975**, 18 (1975).
44. E. D. Weil, *J. Fire Flammability/Fire Retardant Chem.* **1**, 125 (1974).
45. B. Mikofalvy, U.S. Pat. 3,489,706 (1970) (to B. F. Goodrich).
46. P. Kraft and R. Brunner, U.S. Pat. 3,691,127 (1972) (to Stauffer).
47. P. Kraft and P. S. Yuen, Ger. Pat. Appl. 2,452,369 (1975) (to Stauffer).
48. Vistron, Fr. Pat. 1,584,020 (1969).
49. J. K. Craver, U.S. Pat. 3,163,627 (1964) (to Monsanto).

50. P. Robitschek and T. Bean, U.S. Pat. 2,931,746 (1960) (to Hooker).
51. P. Z. Li, Z. V. Mikhaelova, and L. V. Bykova, *Plast. Massy* **2**, 12 (1964).
52. Y. Shichijo, H. Sato, T. Iwasa, and Y. Uchida, U.S. Pat. 3,824,222 (1974) (to Asahi).
53. P. Kraft and L. Smalheiser, Br. Pat. 1,382,625 (1975) (to Stauffer).
54. A. J. Papa, *Ind. Eng. Chem. Prod. Res. Devel.* **9**, 478 (1970).
55. L. Naturman, *SPE J.* **1961**, 965 (1961).
56. Virginia-Carolina Chem. Co., Br. Pats. 954,792 (1964) and 999,588 (1965).
57. T. M. Beck and E. N. Walsh, U.S. Pat. 3,235,517 (1966) (to Stauffer).
58. J. Wortmann, F. Dany, and J. Kandler, U.S. Pat. 3,850,859 (1974) (to Hoechst).
59. M. Lewis, U.S. Pat. 3,534,073 (1970) (to Swift).
60. W. Reeves, *Text. Chem. Color,* **1**, 365 (1969).
61. Ciba-Geigy, Ger. Pat. Appl. 2,136,407 (1972).
62. R. Aenishänslin, C. Guth, P. Hofmann, A. Maeder, and H. Nachbur, *Text. Res. J.* **39** (4), 375 (1969).
63. R. Aenishänslin and N. Bigler, *Textilveredlung* **3**, 467 (1968).
64. C. R. Williams, in *Proceedings of the 1973 Symposium on Textile Flammability*, LeBlanc Research Corp., E. Greenwich, R.I. (1973), p. 67.
65. B. J. Eisenberg and E. D. Weil, *Text. Chem. Color.* **6**, 140 (1974).
66. J. P. Bruce, *Am. Dyest. Rep.* **62** (10), 68 (1973).
66a. K. Masuda and M. Tomita, *J. Fire Retardant Chem.* 3,53, 164 (1976).
67. B. J. Eisenberg and E. D. Weil, paper presented at AATCC National Meeting, Chicago, Oct. 1975.
68. S. L. Madorsky, *Thermal Degradation of Organic Polymers*, Wiley-Interscience, New York (1964), pp. 238.
69. J. E. Hendrix, G. L. Drake, Jr., and R. H. Barker, *J. Appl. Polym. Sci.* **16**, 41, 257 (1972).
69a. A. B. Pepperman, Jr., M. F. Margavio, and S. L. Vail. *J. Fire Retardant Chem.* **3**, 276 (1976).
70. E. D. Weil, in *Flame Retardancy of Polymeric Materials* (W. C. Kuryla and A. J. Papa, eds.), Vol. 3, pp. 185–243, Marcel Dekker, New York (1975).
71. W. A. Reeves, R. M. Perkins, B. Piccolo, and G. L. Drake, Jr., *Text. Res. J.* **40**, 224 (1970).
72. J. J. Willard and R. E. Wondra, *Text. Res. J.* **40**, 203 (1970).
73. I. J. Gruntfest and E. M. Young, *Am. Chem. Soc. Div. Org. Coat. Plast. Papers* **21**, (2), 113 (1962).
74. J. J. Anderson, *Ind. Eng. Chem. Prod. Res. Dev.* **2**, (4), 260 (1963).
75. I. Einhorn, Thermal Degradation and Flammability Characteristics of Polymeric Materials, Polymer Conference Series, University of Utah, Salt Lake City, June 15–20, 1970.
76. D. W. Mitchell and E. M. Murphy, *Proceedings of the Conference on Foamed Plastics*, Publ. PB 181576, NASA-NRC (April 1973).
77. M. A. Boult, R. K. Gamadia, and D. H. Napier, *Inst. Chem. Eng.* (London) *Symp. Ser.* **33**, 56 (1972).
78. A. W. Benbow and C. F. Cullis, *Combust. Flame* **24**, 217 (1975).
79. D. H. Napier and T. W. Wong, *Br. Plast.* **4**, 45 (1972).
80. H. Piechota, *J. Cell. Plast.* **1**, 186 (1965).
81. A. J. Papa and W. R. Proops, *J. Appl. Polym. Sci.* **16**, 2361 (1972).
82. J. E. Kresta and K. C. Frisch, *J. Cell. Plast.* **11**, 68 (March–April 1975).
83. A. E. Sherr, H. C. Gillham, and H. G. Klein, *Adv. Chem. Ser.* **85**, 307 (1968).
84. S. K. Brauman, N. Fishman, A. S. Bradley, and D. L. Chamberlain, *J. Fire Flammability* **6**, 41 (1975); S. K. Brauman, *J. Fire Retardant Chem.* **4**, 18 (1977); S. K. Brauman and N. Fishman, *J. Fire Retardant Chem.* **4**, 93 (1977).
85. R. Sanders, *Text. Chem. Color.* **5**, 48 (1973).

86. T. J. Swihart and P. E. Campbell, *Text. Chem. Color.* **6**, 32 (1974).
87. W. A. Rosser, Jr., S. H. Inami, and H. Wise, *Combust. Flame* **10**, 287 (1966).
88. E. T. McHale, *Fire Res. Abstr. Rev.* **11** (2), 90 (1969).
89. D. Steutz, B. Barnes, A. DiEdwardo, and F. Zitomer, paper presented at University of Utah Polymer Conference, Salt Lake City, June 8, 1970.
90. W. Hastie and C. L. McBee, Mechanistic Studies of Triphenylphosphine Oxide-Poly-(ethylene terephthalate), and Related Flame Retardant Systems, National Bureau of Standards Final Report NBSIR 75-741, National Bureau of Standards, Washington, D.C. (Aug. 1975).
91. E. D. Weil, unpublished investigation.
92. D. J. Daigle, W. A. Reeves, and J. V. Beninate, *J. Fire Flammability* **1**, 178 (1970).
93. P. Rohringer, P. Stensby, and A. Adler, paper presented at the 167th National Meeting of the ACS, Los Angeles, Calif.
94. C. V. Stevens and S. B. Sello, in *Proceedings of the 1975 Symposium on Textile Inflammability*, LeBlanc Research Corp., E. Greenwich, R. I. (1975), p. 186.
95. B. Miller, results preliminarily reported in *Textile Research Institute Newsletter* (1975).
96. E. Pearce, research preliminarily reported in *NBS-ETIP Contract 4-35963 Report for July–Sept. 1975;* A. B. Deshpande, E. M. Pearce, H. S. Yoon, and R. Liepins, *J. Appl. Poly. Sci.:* Applied Polymer Symposium **31**, 257 (1977).
97. Y. Okamato, N. Kimura, and H. Sakurai, *Bull. Chem. Soc. Jpn.* **47** (5), 1299 (1974); also see Ref. 78.
98. E. D. Weil and A. M. Aaronson, paper presented at the University of Detroit Polymer Conference on Recent Advances in Combustion and Smoke Retardance of Polymers, Detroit, May 25–27, 1976.
99. Y. Uehara and E. Yanai, *J. Fire Flammability* **4** (1), 23 (1973).
100. P. O. Sceglov, T. A. Ioffe, M. P. Penkova, and M. A. Tjuganova, *Faserforsch. Textiltechn.* **24** (4), 147 (1973).
101. K. Ashida, F. Yamauchi, M. Kotoh, and T. Harada, *J. Cell. Plast.* **10**, 181 (July–Aug. 1974).
102. C. Hilado, paper presented at the International Conference on Fire Safety at the University of San Francisco, Calif., Jan. 12–16, 1976.
103. J. H. Petajan, F. D. Hileman, and L. H. Wojcik, *J. Polym. Sci. Polym. Lett. Ed.* **13**, 293 (1975).
104. J. M. Funt and J. H. Magill, *J. Fire Flammability* **6** (1), 28 (1975).

5

Flammability of Cotton–Polyester Blend Fabrics

Christine W. Jarvis and Robert H. Barker

Cotton–polyester blends pose a special flammability problem because the thermal and mechanical properties of the fibers are so different. Cotton tends to char on heating but generally maintains some structural integrity; polyester normally melts and flows at temperatures of ca. 260°C. If a mixture of the two fibers is burned, the molten polyester frequently tends to wick on the cotton char, resulting in the phenomenon of scaffolding detailed by Kruse.[1] Because of these effects, it is impossible to predict *a priori* the flammability of the blend on the basis of the behavior of the individual component fibers; thus, Tesoro and Meiser[2] have shown that needle-punched webs prepared from 50:50 blends of cotton and polyester exhibited oxygen indices lower than those of either fiber alone in a similar structure. The hazard in the case of apparel may be further increased by the tendency of the molten polyester to cling to the body of the person wearing the garment, thus causing severe burns on its own.

Hendrix and co-workers have also studied cotton–polyester blends using oxygen index techniques[3,4] and have concluded that the blends are inherently more flammable than 100% polyester if the latter is tested without external support. By studying the differential thermal analysis of these fibers and their blends, these workers concluded that the cotton is the initial source of fuel from the blend fabrics and is the component that dictates the limits of

Christine W. Jarvis and Robert H. Barker • College of Industrial Management and Textile Science, Department of Textiles, Clemson University, Clemson, South Carolina.

flammability. This conclusion was based on observations that the degradation of the cotton begins at temperatures well below those required for the pyrolysis of the polyester. After comparing the vertical downward burning rates for cotton and cotton–polyester blend fabrics. Hendrix *et al.* proposed that the role of the polyester component was to furnish additional fuel to the combustion as the polymer temperature was raised by heat produced from the burning of the cotton decomposition products. This additional fuel increases the vigor of the gas phase oxidation and thus results in a fabric which appears to burn more rapidly.

These conclusions have recently been challenged by Drews.[5] After observing the behavior of both treated and untreated cotton–polyester blends under the vertical test conditions of FF 5-73, Drews has proposed that the burning of these blends is primarily controlled by the polyester portion. After examining charred specimens from vertical tests using scanning electron microscopic techniques, he noted that the polyester melts a head of the progressing flame front and coats the cellulosic material, preventing its pyrolysis. The polyester then burns off before significant decomposition of the cotton occurs.

The rate of heat release from a series of cotton–polyester blends has also been shown to be a direct function of the amount of cotton in the blend, with pure cotton exhibiting the highest rate.[6] In previous investigations Drews had shown a correlation between the rate of heat release and the rate of flame propagation.

1. Theory of Flame-Retardant Action

The theory of flame-retardant action on fibrous polymers has been developed using the basic assumption that the fuel-generating and fuel-consuming processes are separated in space. The fuel production is assumed to proceed by a purely pyrolytic mechanism in the condensed phase with no involvement of thermooxidative processes. The fuel-consuming process, and thus the heat-generating process, has been assumed to occur only in the gas phase at a finite distance from the surface of the decomposing polymer. Several recent studies have challenged this assumption. Stuetz *et al.*[7] have proposed that there is significant involvement of condensed phase oxidative degradation in the case of many hydrocarbon polymers. Drews and Heckman[8] have produced data which they believe indicate a lesser, but nevertheless measurable, involvement of thermooxidative degradation in the burning of polyester. At present, it is not known whether this thermooxidative degradation is a significant factor in determining the flammability of polyester fabrics; however, there does seem to be evidence to support the proposal that the importance of condensed phase oxidation decreases as the oxygen content of the polymer increases. Thus, there seems to be no signicant role for thermooxidative processes in the decomposition of cellulose.

Therefore, it is reasonable to use the classical model for flame-retardant action with certain reservation. This model makes it possible to separate those retardants which act predominantly in the condensed phase from those which act predominantly in the vapor phase above the decomposing polymer. Retardants which act in the gas phase are thought to function as either inert diluents or free-radical inhibitors which alter the oxidation processes and decrease the heat returned to the polymer surface. Those retardants which act in the condensed phase appear to operate by several different mechanisms. They may inhibit the polymer pyrolysis so that the substrate does not break down to produce the volatile molecules necessary for flame propagation. More commonly, however, they act to alter rather than inhibit the thermal degradation reaction. The alteration is such that the mode of decomposition is changed and smaller quantities of flammable gas are produced. Finally, they may also exert their effect in a physical rather than chemical manner. In this case, they may act either as a heat shield to prevent the transfer of heat from the flame back to the polymer surface or as heat sinks, i.e., to absorb and dissipate the heat and thus render it unavailable for continued polymer pyrolysis. Larsen[9] has recently proposed that many organobromine compounds act in the latter manner because of their large heats of vaporization.

Of the wide variety of phosphorus-containing flame retardants known to be effective on cellulosic substrates, the vast majority are thought to operate by a predominantly condensed phase mechanism. Simple phosphorus compounds such as phosphoric acid, diammonium phosphate (DAP), or a tetrakishydroxymethylolphosphonium hydroxide (THPOH)–ammonia polymer have all been found to have esssentially the same efficiency characteristics in a study using static oxygen bomb techniques.[10] The limiting factor in determining the efficiency of these flame retardants is not their chemical structures but rather their ability to break down to form phosphorus oxides. These phosphorus oxides are then the common intermediates in the reaction of the retardants with the cellulosic substrate. The ultimate efficiency of such nonthermally stable systems is therefore given by the effectiveness of the interaction between cellulose and the phosphorus oxides. The only possibility for their efficiency to be enhanced would be for them to be converted into degradation products other than phosphorus oxides by preceding reactions with an additional reagent applied to the cellulose. This alternate path constitutes the basis for the phosphorus–nitrogen synergistic effects observed by many workers.[11] In this case, the phosphorus and nitrogen compounds react on heating to form a P–N polymer. The phosphorus atoms in this polymer have enhanced reactivity toward the cellulose and are more efficient as flame retardants. This conclusion has been confirmed by model compound studies.[12]

The situation becomes considerably more complex when the flame retardants are thermally stable up to the point of polymer degradation. In such systems, the nature of the phosphorus-containing functional group is important in determining the flame-retardant efficiency of the system. To study this effect, compounds of the type $\phi_n P(O\phi)_{3-n}$ and $\phi_n P(O)(O\phi)_{3-n}$ having $n = 0, 1, 2,$ and 3 have been synthesized and applied to cotton fabric.[11] Calorimetric evaluation of these compounds shows that the efficiency of the flame-retardant system increases as the number of oxygen atoms bonded to the phosphorus increases, as shown in Figs. 1 and 2. An exception to this behavior is observed with triphenylphosphate, which probably represents a special case because of its volatility. Therefore, for thermally stable flame-retardant additives, the efficiency of the system can be limited by the chemical nature of the phosphorus compound.

Substantiation for this proposal has recently been found by a study of phosphoramides and related compounds.[11] Simple phosphoramides such as $(CH_3NH)_3P{=}O$ and $(\phi NH)_3P{=}O$ exhibited efficiencies that were significantly different from those P–O compounds studied previously. In all cases,

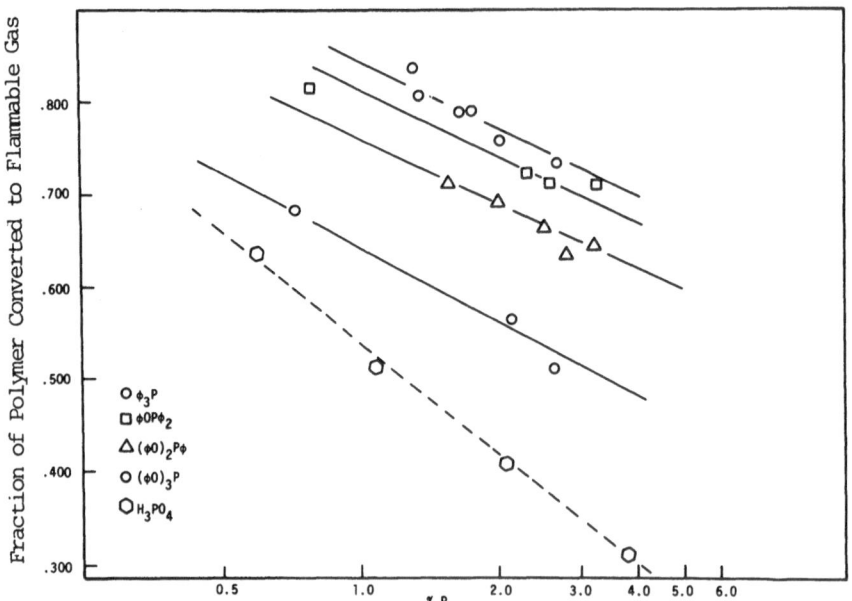

FIGURE 1. Flame-retardant efficiencies of trivalent phosphorus compounds.[11]

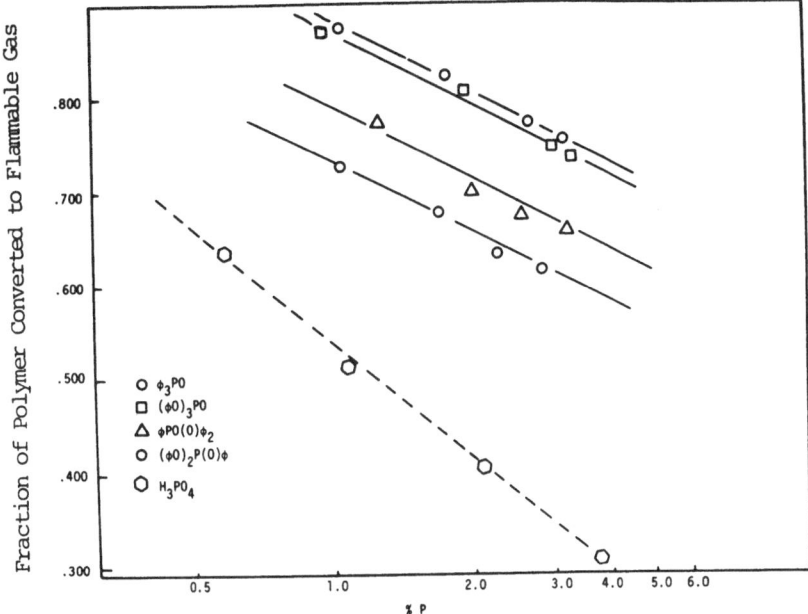

FIGURE 2. Flame-retardant efficiencies of penetrated phosphorus compounds.[11]

the P–N systems were more efficient flame retardants. The volatility of the methyl compound was a major factor in determining its effectiveness, but this could be altered by various fixation techniques. In these systems, there apparently is a competition between volatilization of the retardant and its reaction with the cellulose. Added evidence for this interpretation is found in recent results from a study[13] of the efficiencies of a series of model flame retardants of the type $\phi_n P(O)(NH\phi)_{3-n}$ and $(\phi O)_n P(O)(NH\phi)_{3-n}$.

All of these results, when viewed in the context of other experiments involving product analysis and model compound studies, allow the postulation of chemical mechanisms to explain the interaction of phosphorus flame retardants with cellulose. The efficiency of such reagents appears to be related to the reactivity of the phosphorus atom. Reactivity in phosphorylation reactions and efficiency in catalyzing the degradation of the cellulose are both important. These reactions inhibit the formation of flammable volatiles from the cellulose and promote formation of char and noncombustible volatiles.[14]

Although such correlations are fairly straightforward in most cases, the static oxygen bomb calorimetric technique is also capable of discerning unusual and unexpected effects. For example, it has recently been found that for structures of type (1) the flame-retardant efficiency of the phosphorus compound is in the indicated order and depends strongly on the chemical nature of the moiety attached to the other end of the carbon chain.[11]

$$(CH_2O)_2 - \overset{\overset{\displaystyle O}{\|}}{P} - CH_2 - CH_2R$$

(1)

$$R = C(O)NHCH_2OH > C(O)NH_2 \gg CH_3 > C(O)OCH_3$$

The reasons for these unusual structural effects are not obvious from the calorimetric data alone, but these results have been interpreted in terms of an interaction between the phosphorus- and nitrogen-containing groups within the same molecule. This interaction occurs prior to reaction with the cellulose so that the species which reacts with the cellulose is most probably a phosphoramide rather than a phosphonate.

The chemistry is much less well defined with polyester than with cellulosics. This is due partly to the thermoplasticity of the synthetic polymer and partly to the fact that the pyrolysis chemistry appears to be more complex and less thoroughly studied. Examination of several simple model esters has led to the determination of a number of features of polyethylene terephthalate (PET) degradation.

The PET decomposes by random chain scission at the ester links, and the principal weakness in the polyester chain seems to be the β-methylene groups.[15] The vinyl esters which result from this initial beta elimination reaction would appear to be amenable to a variety of grafting and cross-linking reactions which should lead to char formation. In practice, however, few flame retardants have been found which affect this reaction and significantly increase the char yield from the PET. For this reason, most emphasis has centered on flame retardants capable of vapor phase inhibition of the oxidation of the polyester pyrolysis products.

These factors considerably complicate the problem of flame-retarding cotton–polyester blends since one fiber would appear to be most efficiently treated by condensed phase action whereas the other would appear to be most amenable to vapor phase activity. None of the studies to date have demonstrated this unequivocally, but all the data are compatible with such an interpretation. For example, Loss et al.[16] have compared the action of a phosphonate and an oligomeric phosphonium compound on cotton and blend fabrics. They found that both compounds had lower efficiencies on 65:35 polyester–cotton blends than on 100% cotton but that the relative efficiencies were different. The phosphonate exhibited higher efficiency than the phosphonium compound on 100% cotton, whereas the reverse was found in the case of the blend. Although these workers did not attempt to explain their results, it seems likely that the phosphonate is more effective in reacting with the cellulose to provide condensed phase retardancy. Similarly, the larger phosphonium salt should penetrate the cellulose to a lesser extent and react more slowly with the cellulose, allowing a portion of the phosphorus to volatilize and provide some vapor phase activity. Of course, the latter mode of action would be particularly important when the fabric contains significant portions of polyester.

Similar conclusions can be reached on the basis of studies using phosphonates and phosphine oxides. Tesoro[17] has used oxygen index measurements to compare the activity of $(CH_3O)_2P(O)CH_2CH_2C(O)NHCH_2$ OH and $(CH_3)_3P(O)CH_2CH_2C(O)NHCH_2OH$ on 100% cotton and blends. In her studies, the retardants were applied to the fabric along with melamine resins and fixed by a pad-dry cure process. Under these conditions the phosphine oxide was significantly more effective on both 65:35 and 50:50 blends. The reasons for these differences are clear from her findings that the efficiency of the phosphine oxide was the same on both 100% cotton and the 50:50 blend, whereas that of the phosphonate was much lower on the blend than on 100% cotton. The fact that the phosphine oxide was essentially substrate independent is indicative of vapor phase activity, while the findings that the efficiency of the phosphonate was directly related to the chemical nature of the substrate is strong evidence for condensed phase activity.

Similar results were reported by Drews and Barker.[13] The efficiency of triphenylphosphine oxide and $(CH_3O)_2P(O)CH_2CH_2C(O)NHCH_2OH$ were evaluated. The phosphonate was found to have essentially the same effect regardless of whether or not it was chemically fixed on 100% cotton fabrics. However, with polyester–cotton blends, the situation was quite different. Apparently, the reactivity of the phosphonate toward cellulose is high enough to prevent its loss by volatilization when it is present in low concentrations, but when it is overloaded, as it would be in the cotton portion of a 50:50 or 65:35 blend, some volatilization may occur. This process could provide some vapor phase active species to retard the polyester portion and account for the observation of higher efficiency of the unfixed retardant on the blends. In support of this interpretation, a large difference was found in the effectiveness of the phosphonate and triphenylphosphine oxide on 100% cotton fabric, but the two retardants seemed to be almost equally effective on 65:35 blends. Due to the low reactivity of the triphenylphosphine oxide, there would seem to be good reason to believe that it exerts essentially all of its effect in the vapor phase. These results were therefore interpreted as demonstrating the importance of volatility of the flame retardants with polyesters included in the fabrics.

Because of these effects and the potential complications which may arise from the different modes of flame-retardant action on cotton and polyester in blend materials, the interaction of flame-retardant fibers has recently been studied extensively by Yeh.[18] In an attempt to determine the effect of adding phosphorus flame retardant to the cotton portion of a blend, Yeh investigated a series of blends with various blend levels in which the cotton had been impregnated with phosphoric acid. Examination of the fabrics by isoperibol calorimetry led to the results shown in Fig. 3. As can be seen, the

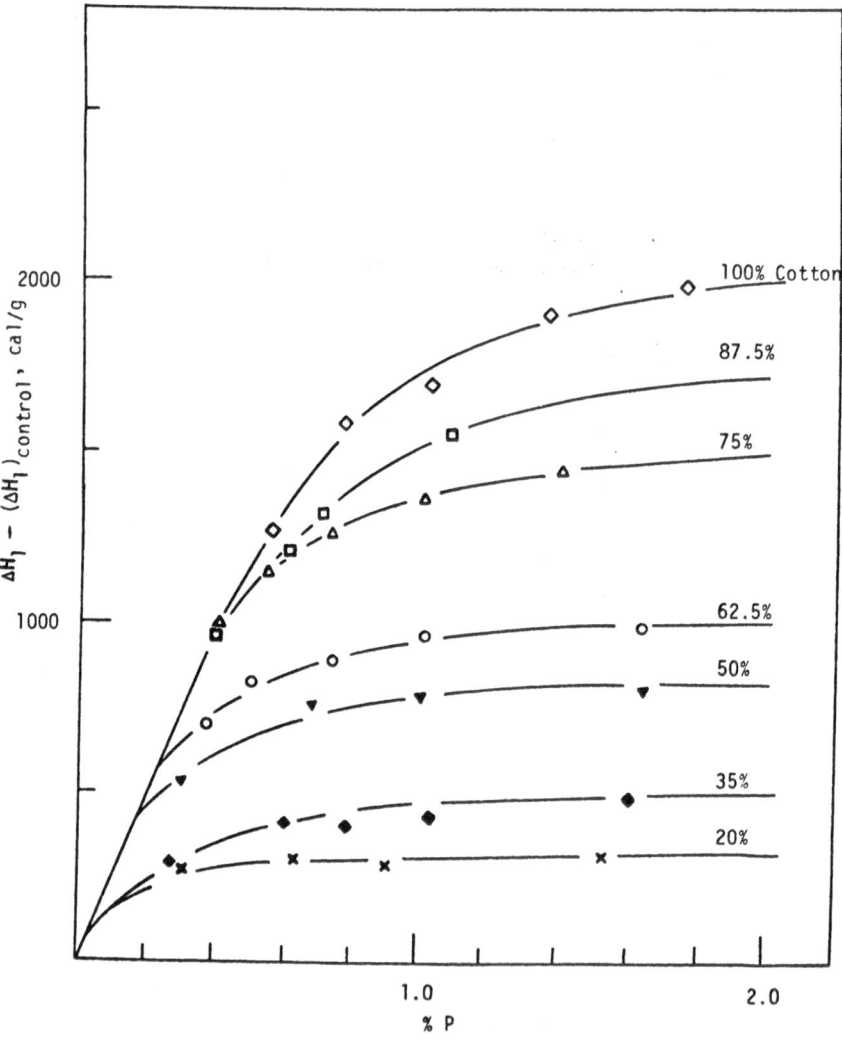

FIGURE 3. Net heat reduction of H_3PO_4-treated polyester/cotton blends.

response to the treatment of cotton-rich blends was vastly different from that of the polyester-rich blends. In the first case, increasing amounts of retardant significantly reduced the heat output from the burning fabrics, whereas in the latter case a point was quickly reached beyond which increasing amounts of retardant produced essentially no effect. The nature and origin of this effect are not well understood, but presumably a similar effect could exist with blends having only the polyester portion treated. For this reason, Yeh and Valente[19] have also studied a series of blends containing a brominated flame retardant in the polyester portion. The specific system used was the duPont Dacron® copolymer type 900F, which contains bromine in the form of ethoxylated tetrabromobisphenol A. In this series, the bromine seemed to act completely in the vapor phase, reducing the heat evolution and producing only slight alterations in the char yield. This effect was not highly substrate dependent, and there was no evidence of unusual interaction between the flame-retardant polyester and the cotton.

The interactions of the two flame-retardant fibers were then evaluated. A series of 65:35 and 50:50 polyester–cotton blend fabrics in which the polyester portion was Dacron® 900F were treated with phosphoric acid at add-ons ranging from 1.5 to 6%. The treated blends were then studied by isoperibol calorimetry, and the heat release and char yield data obtained were compared with those obtained from normal PET–cotton blends.

The char yield data are given in Table 1. The char yields which would be expected from the cotton portion of the blends if they were not affected by the presence of the polyester were calculated from previous data on phosphoric-acid-treated 100% cotton fabrics and are included in Table 1 for comparison. The 50:50 900F–cotton blend char yields were found to be

TABLE 1

Char Yields of H_3PO_4-Treated 900F®–Cotton and Polyester–Cotton Blends

Sample	% P cotton[a]	% Residue		
		900F®–cotton	PET–cotton	Calc.[b]
50:50 blend control	—	4.6	5.2	5.2
	1.16	15.7	14.6	16.0
	2.86	23.4	19.8	23.7
	3.80	55.1	21.6	26.0
65:35 blend control	—	10.8	8.7	8.7
	1.86	15.9	13.6	14.0
	3.43	23.5	16.6	17.6
	4.49	26.2	17.8	19.3
	5.22	40.1	18.5	20.1

[a] P calculated assuming all P is located on cotton portion.
[b] Calculated from previous data on H_3PO_4-treated 100% cotton using c_i P cotton.

essentially the same as those calculated on the basis of the cotton content up to about 1.5% phosphorus. However, the char yield from the fabric containing 1.9% phosphorus was more than double that expected on the basis of char formation from the cotton portion only. Similar behavior was noted for the 65:35 900F–cotton blends. These fabrics gave slightly higher char yields than those calculated on the basis of the cotton portion. However, the largest difference again occurred at about 1.8% phosphorus content where the 900F–cotton blend gave twice the amount of char calculated on the basis of cotton content. Both blends (50:50 and 65:35) became very difficult to ignite at their highest level of treatment. On the basis of subjective observations of the burning characteristics in the calorimeter, it was estimated that a phosphorus content of 2.0% should be sufficient to effectively inhibit burning and that these fabrics should pass FF 3-71.

The flame-retardant contribution of the bromine in Dacron® 900F is probably due to a vapor phase mechanism and not significantly related to char formation. At the lower levels of add-on (up to 1.5% phosphorus for 50:50 and 1.0% for 65:35) where there is still sufficient heat generation from the cotton portion of the blend to sustain the combustion process, there is essentially no residue from the polyester remaining in the char. However, at higher phosphorus contents (1.8–1.9%), the diminished heat generation from the cotton portion coupled with the assumed vapor phase phase retardance from the 900F results in a large reduction in total heat generation and thus in the heat fed back to the substrate. This reduction decreases in the heat flux at the fabric surface to a point where it is not sufficient to sustain complete degradation of either the cotton or the polyester. Therefore, a large residue is left.

The dependence of the heat release on phosphorus content for both 900F–cotton blends and normal PET–cotton blends is presented graphically in Fig. 4. The 50:50 blends exhibit essentially identical heat release values up to about 0.5% phosphorus. At higher phosphoric acid add-ons, the heat release from the PET–cotton blend remained unchanged. Similar behavior was observed with the 65:35 blend, although the heat release of the untreated 900F system was about 400 cal lower than that containing normal polyester, indicating the effect of the bromine in the 900F. Similarly, the heat release from the untreated 65:35 blend was also about 400 cal lower than that of the 50:50 900F–cotton blends. This effect was presumably due to the larger amount of bromine in the 65:35 blend.

The effect of the bromine can be seen more clearly if the heat release (ΔH_1) is shown as the sum of the heat release from the cotton and the polyester portions of the blend as follows:

$$\Delta H_1 = Q_c + Q_p \tag{1}$$

where Q_c and Q_p are, respectively, the heat release from the cotton and polyester portions of the blend. Since the char data showed that the combus-

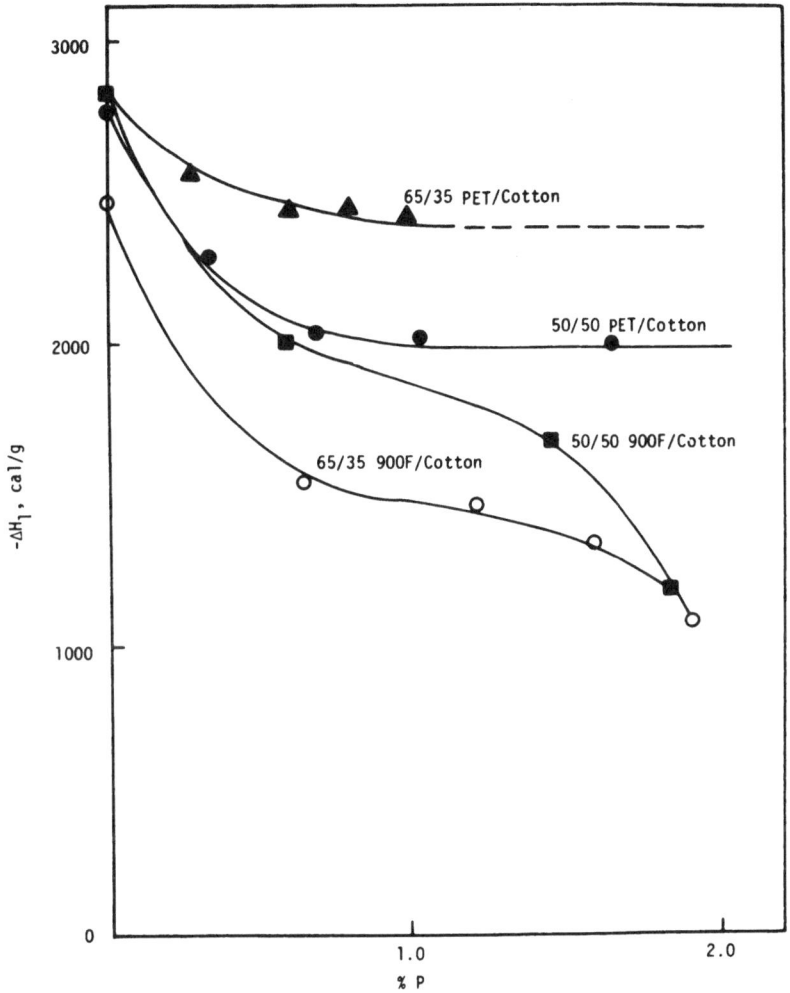

FIGURE 4. Heat release of H_3PO_4-treated 900F-cotton blends.[19]

tion of cotton in the blend was the same as that in the form of 100% cotton fabric, the Q_c can be calculated from data previously obtained on 100% cotton treated with phosphoric acid and Q_p can be calculated from Eq. (1). Plots of $\Delta H_1, Q_c$, and Q_p as a function of percent phosphorus are shown in Fig. 5 for the 50:50 blends and in Fig. 6 for the 65:35 blends. Both Figs. 5 and 6 show the drastic difference in heat release between 900F and regular PET. For either blend, Q_p increases and Q_c decreases with increasing phosphoric acid content, while the total heat (ΔH_1) remains essentially constant. However, Q_p in the 900F remains unchanged until the phosphoric acid content is increased to about 1.5% phosphorus. It then drops drastically to below 1000 cal at 1.8–1.9% phosphorus, while Q_c is less than 400 cal. These data indicate a distinct retardant effect by the bromine. The heat release values are in agreement with the char yield data and show that up to 1.5% phosphorus the heat release from 900F is essentially unaffected by the presence of the phosphoric acid. Such results were expected since phosphoric acid is a condensed phase retardant and has no significant retardant effect on polyester. The drastic retardant effect observed on 900F–cotton blends at a phosphorus content

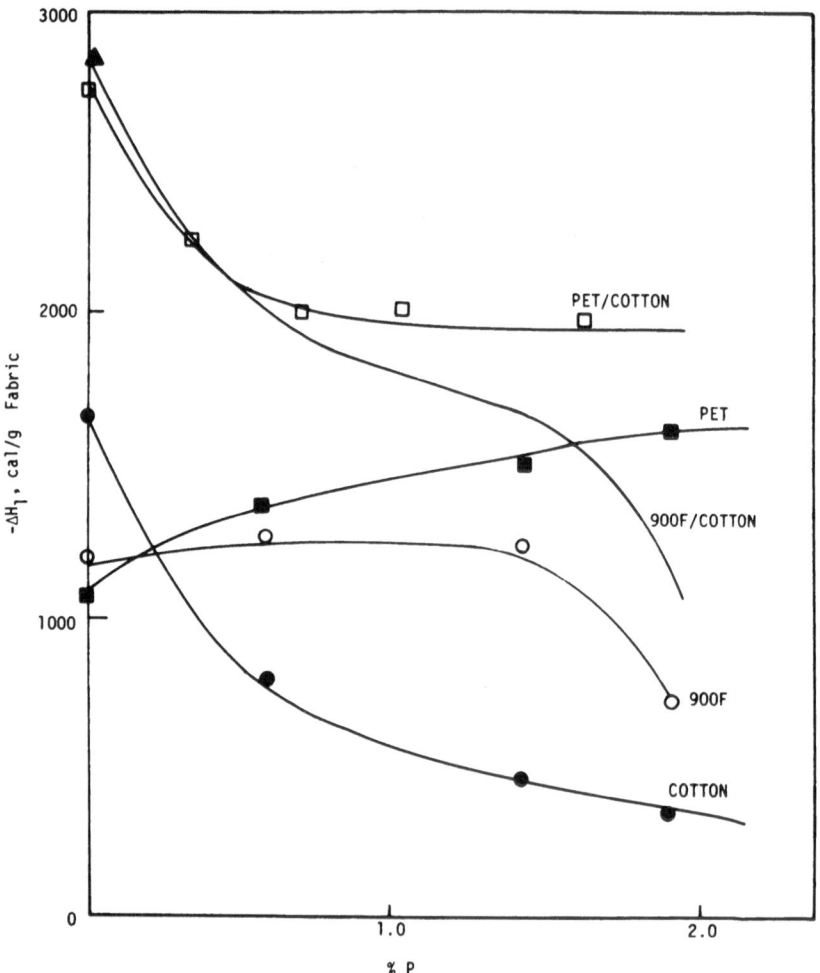

FIGURE 5. ΔH_1 of H_3PO_4-treated 900F–cotton and PET–cotton 50:50 blend fabrics.[19]

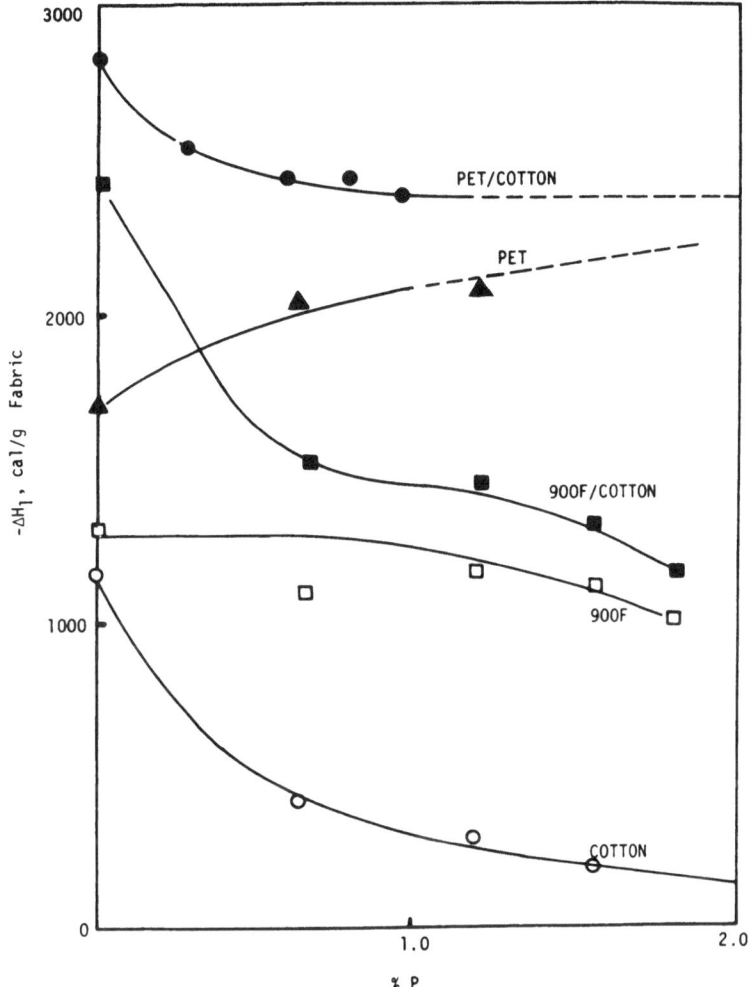

FIGURE 6. ΔH_1 of H_3PO_4-treated 65:35 900F–cotton and PET–cotton blend fabrics.[19]

above 1.5% is probably due to the synergistic effect which results from the reduction in heat release from the cotton portion and vapor phase retardant effect from the bromine. This should produce an insufficient heat feedback to the substrate, resulting in less than complete degradation of both polymers. This, in turn, of course, reduces the amount of fuel gas available to the flame, and the effect is considerably greater than additive.

In an effort to further evaluate the flame-retardant action of the bromine, experimental samples of 900F containing 7.5% bromine instead of the normal

6.0% were obtained from duPont in the form of 50:50 and 75:25 blends with cotton. These two fabrics were also treated with phosphoric acid at three different add-ons up to 1.8% phosphorus. The isoperibol calorimetric results on these blends in treated and untreated forms are shown in Table 2, together with similar samples based on the regular 900F fiber. These four series of blends contain different levels of bromine, ranging from 3.0 to 5.6 due to the different ratios of 900F and cotton. The data in Table 2 and Fig. 7 show that blends with higher bromine content tend to release less heat than ones with lower bromine content at the same phosphorus add-on.

The significance of bromine in these blends can be shown if the heat release data are further reduced. Figure 8 presents a plot of heat release versus bromine content at several levels of phosphorus. A linear decrease of ΔH_1 is observed with increasing bromine contents. Furthermore, since these data represent three different blends, the linear dependency of ΔH_1 on bromine content significantly demonstrates one of the important characteristics of vapor phase retardancy; i.e., the efficiency of vapor phase retardants should be independent of the nature of the substrate. Values for the 65:35 blend are consistently off the line. Yeh and Valente have proposed that this deviation may be due to an error in determining the phosphorus content in the treated samples.[19]

TABLE 2

Isoperibol Data of 900F®–Cotton Blends with Varying Bromine Content

Fabric	% Br	% P	% Residue	Rate (cal/cm-sec)	$-\Delta H_1$ (cal/g)
50:50 900F®–cotton	3.0	Control	4.6	43.7	2739
		0.58	15.7	38.8	2001
		1.43	23.4	24.4	1658
		1.90	55.1	21.3	1066
50:50 900F®–cotton	3.8	Control	7.2	52.7	2546
		0.47	17.7	35.1	1920
		1.28	21.9	23.2	1443
		1.67	35.5	20.1	1141
65:35 900F®–cotton	3.9	Control	10.8	44.2	2409
		0.65	15.9	32.7	1524
		1.20	23.5	15.2	1441
		1.57	26.2	17.3	1303
		1.83	40.1	11.2	1162
75:25 900F®–cotton	5.6	Control	11.1	39.3	2277
		0.72	21.0	15.5	1653
		1.28	37.6 ± 8.9[a]	—	1259 ± 208[a]
		1.68	Did not burn		
		1.80	Did not burn		

[a] Average of five runs; large error probably due to uneven treatment.

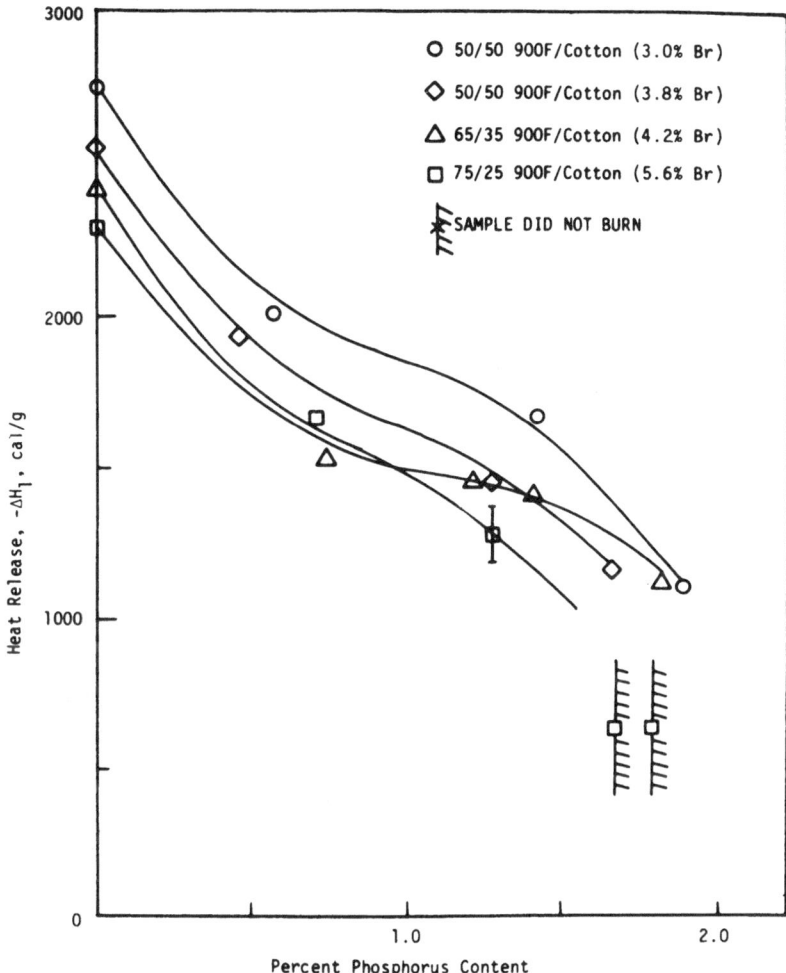

FIGURE 7. Heat release of various H_3PO_4-treated 900F–cotton blends.[19]

The fact that all of the lines in Fig. 8 seem to parallel each other was taken as an indication that the effectiveness of the bromine on these blends is identical irrespective of the amount of phosphorus present in the blend, suggesting that bromine and phosphorus are working independently of each other. There is apparently little or no bromine–phosphorus interaction or synergism such as has previously been suggested to exist in other systems.[20]

A series of experiments[6] was undertaken to determine quantitatively the effects and interactions of phosphorus with vapor phase active bromine additives whose levels of add-on could be easily varied. One system meeting

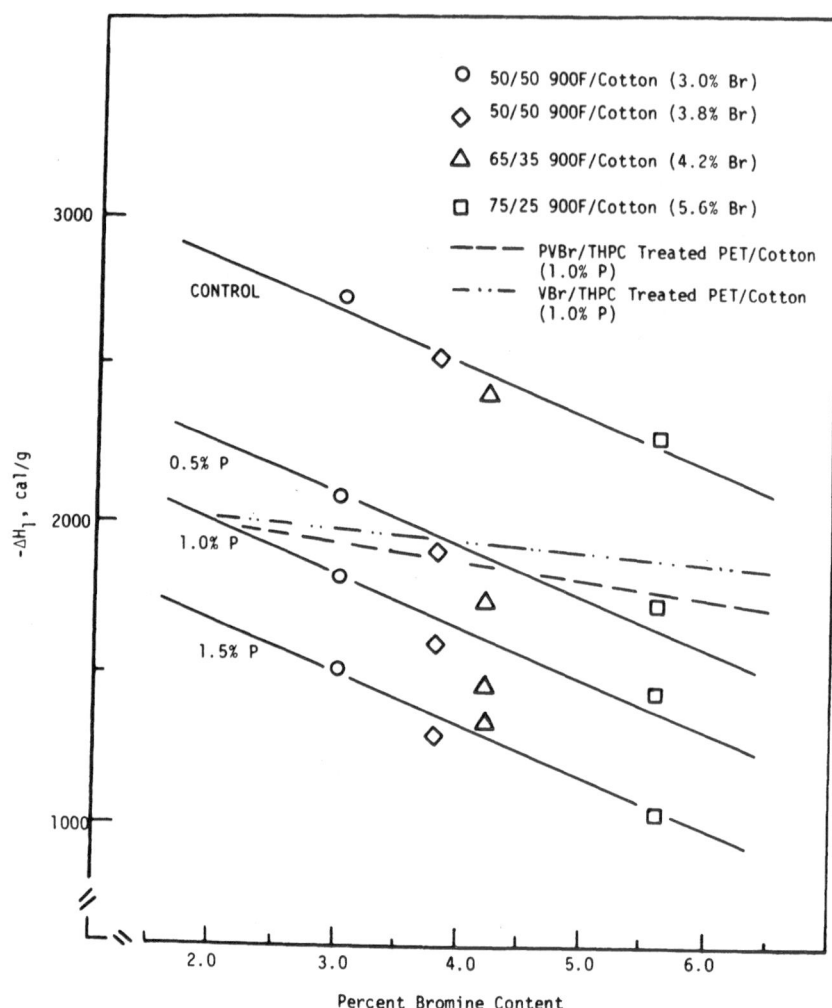

FIGURE 8. Heat release of various 900F–cotton and PET–cotton treated with H_3PO_4 or PVBr–THPC.[19]

this requirement is tris(2,3-dibromopropyl) phosphate (TBPP).* Because of durability considerations, this compound was not thought to be a practical candidate for commercial treatment of blends but was taken as an instructive model for the study. The thermal properties of a series of blend fabrics were impregnated with varying levels of TBPP. In an effort to determine the effect

---

Note added in proof: The flame retardant TBBP, commonly known as Tris, has recently been banned from use in children's sleepwear by the U.S. Consumer Product Safety Commission because of its reported carcinogenicity toward laboratory animals.

of the distribution of the flame retardant between the two fibers, a series of fabrics was padded with the retardant and heat-treated at 150°C. Half were then washed in acetone to remove as much of the TBPP as possible from the cotton portion of the blend. Oxygen index values were found to reflect only the amount of retardant present in the sample and not the method by which each level was achieved. Thus, the flammability appears only minimally dependent on flame-retardant distribution.

To determine the relative contributions of the two elements to the flame-retardant action of TBPP, triallylphosphate was prepared as a model retardant containing only phosphorus. The triallylphosphate was applied to samples of 65:35 and 50:50 polyester–cotton blends at levels ranging from 10 to 30% add-on, i.e., phosphorus contents of approximately 1–4%. The OI values of these fabrics were found to increase slowly with increasing phosphorus content up to about 2% phosphorus, beyond which the added phosphorus had essentially no effect on OI values.[6]

In an effort to characterize more completely the bromine–phosphorus system, Valente and Yeh[21] carried out experiments utilizing variable Br/P ratios. In this work, DAP was used as the model phosphorus compound and TBPP as the model bromine compound. The net heat reductions were calculated and plotted as a function of phosphorus content (Fig. 9). Values of heat reduction from 50:50 blend treated with DAP alone are included in Fig. 9.

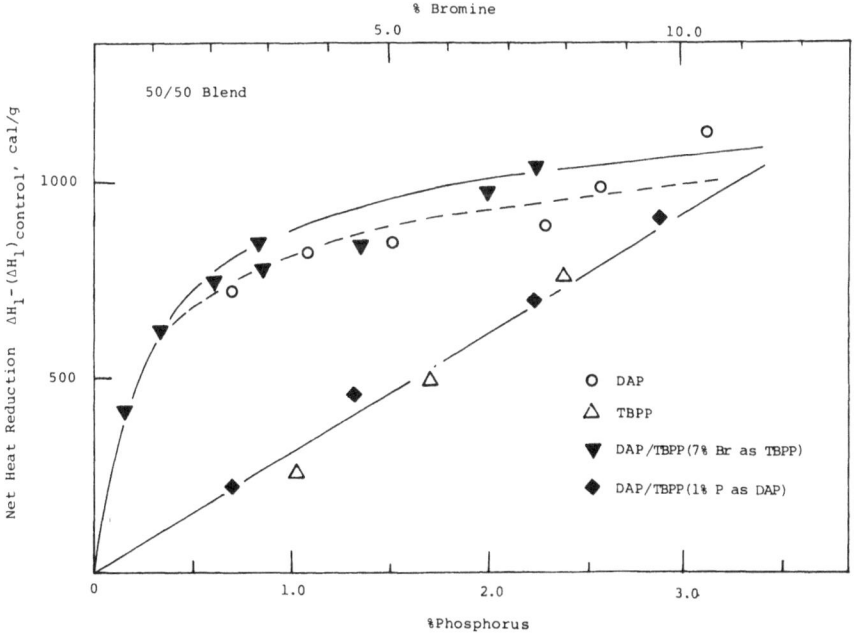

FIGURE 9. Net heat release of DAP-TBPP-treated 50:50 blend.

There is a consistent difference of about 70 cal between the curves for DAP
alone and DAP–TBPP combinations, with the single treatment of DAP
having the lower values. However, the standard experimental error estimated
for ΔH_1 is $+5\%$, or about 70 cal for these samples; the authors did not feel the
difference observed was enough to warrant an argument of synergistic effects
in DAP–TBPP systems. They also found that there was essentially no

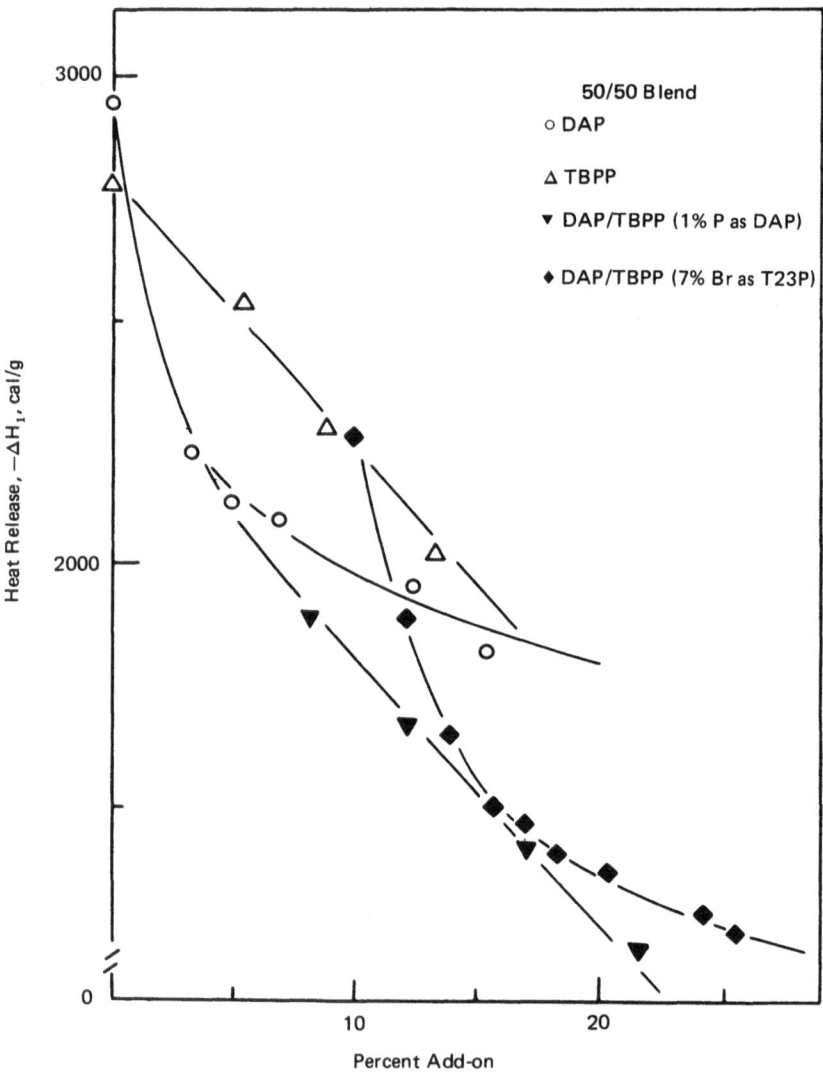

FIGURE 10. Calorimetric data for optimization of P–Br finishes.[19]

difference in heat reduction achieved regardless of which reagent was applied first.

Since the order of treatment is apparently not important in determining the efficiency of this system, these workers proposed that it should be possible to predict the effect of any particular formulation from the characteristics of each component alone. Experimental confirmation of this is shown in Fig. 10. The efficiency curves produced by adding DAP to the fabric have the same shape regardless of whether addition is made to a fabric containing 0% or 7% Br. Similarly, the curves produced by adding TBPP are parallel starting at 0% and 1% P. Thus, one can use these curves to arrive at those formulations of DAP and TBPP which will produce a given heat release. To do this, one need only select a point on the DAP curve corresponding to a specific P content, from this point draw a line parallel to the TBPP curve, and proceed down this curve to a point corresponding to the desired heat release. The amount of Br required can then be read directly from the abscissa.

Whether this is a general approach which would be valid for other P–Br retardant systems remains to be determined experimentally. However, if it should be found to be valid, it should constitute a powerful tool for the optimization of formulations for characteristics not related to flammability, such as cost, hand, ease of application, etc.

2. Flame-Retardant Treatments for Polyester–Cotton Blends

As would be expected on the basis of the theory of flame-retardant action, formulations incorporating both condensed phase active and vapor phase active flame retardants have been widely studied. In these treatments, one flame retardant is predominantly durable and effective on cotton, while the second is durable and effective on polyester. For such a system to be commercially feasible, the retardants must be compatible in a single formulation, and their cure or fixation processes must also be similar.

In one of the earliest attempts to utilize this approach, Hamalainen and Guthrie developed a one-bath treatment incorporating a bromoform adduct of triallylphosphazene with a THPC–amide finish. This formulation was applied as an emulsion in a pad-dry cure process.[22] The THPC–amide retardant penetrated the cotton fiber, forming an insoluble polymer during the heat cure and reacting to a modest degree with the cellulose of the cotton.

The bromoform adduct of hexaallylphosphazene was prepared by introducing bromoform during the polymerization of hexaallylphosphazene using a peroxide catalyst and polyvinyl alcohol emulsifier. As the reaction progressed, bromoform added to some of the allyl groups of the hexaallyl-phosphazene. Other allyl groups participated in vinyl polymerization, while the remaining unreacted allyl groups were left available for reaction on the fabric during the heat cure process. The emulsion contained about 30% by weight of the brominated phosphazene. This product was an aqueous emulsion which could be added directly to aqueous THPC–amide flame-retardant solutions. The amount of bromoform adduct needed was proportional to the amount of polyester in the blend fabric. Higher concentrations of the bromoform adduct were required for 50:50 polyester–cotton blends than for fabrics containing 35% or less polyester. It was found that fabrics weighing about 7 oz/yd and containing not more than 50% polyester required about 20% add-on of the combined retardant in order for the fabric to pass a vertical flame test. Some treated fabrics were reported to exhibit slight stiffness, which was attributed to the bromoform adduct emulsions. The THPC–amide portion of this retardant was exceedingly durable to both home and commercial-type laundering as well as to dry cleaning. The adduct polymer which coated both cotton and polyester fibers had moderate durability. Some of the adduct was lost gradually during each successive laundering, but a substantial part remained through 50 laundry cycles.

In a somewhat similar approach, Barber and co-workers[23] developed a one-bath process for fabrics containing up to approximately 50% polyester using the THPC–amine finish with TBPP. The emulsified combined retardant was applied by a pad-dry cure procedure. The fabrics were padded to a 70–80% wet pickup, then dried at about 95°C, and finally cured 3 min at 160°C. Fabrics impregnated with this retardant frequently yellowed when treated with sodium hypochlorite bleach solution due to the presence of melamine. Heavy fabrics had good esthetic properties; however, $3\frac{1}{2}$-oz or lighter materials generally exhibited some stiffness. Both the stiffness and discoloration could be essentially eliminated by omitting the trimethylol-melamine and using instead dimethyloldihydroxyethyleneurea (DMDHEU), dimethylolpropyleneurea, or dimethyloluron. Approximately twice as much of these materials was required compared to trimethylolmelamine due to the greater polymerization efficiency of the trimethylolmelamine and its greater degree of fire retardance per unit weight. These formulations were suitable for treatment of blends in which polyester content did not appreciably exceed 50% of the fabric weight.

The THPC part of the retardant penetrated the cotton, forming an insoluble polymer which was exceedingly durable. Only a small amount deposited on the surface of the polyester fiber was lost during laundering. The TBPP penetrated both the cotton and the polyester and was equally good as a fire retardant on both fibers. However, the TBPP was readily removed from the cotton fiber during laundering, although it was durable in the polyester. Loss of the phosphate from the cotton during laundering meant that only half of the bromopropylphosphate remained in the textile structure to contribute fire retardancy after about 15 laundry cycles.

The bromopropylphosphate was found to be as effective on 100% cotton as on 100% polyester fabric. On both, it raised the oxygen index about 0.05

units with approximately 11.5% add-on and without the THPC–amide retardant. With 15% THPC–amide and without the bromopropylphosphate, the oxygen index of cotton was about 0.27 and gradually dropped as the percentage of polyester was increased. Addition of 11.5% of the bromopropylphosphate to 13% THPC–amide raised the oxygen index of 100% cotton to about 0.35. Again the oxygen index gradually decreased as the polyester content was increased, paralleling the activity of the THPC–amide finish alone.

Although this approach has not received a great deal of commercial interest because of the lack of durability of TBPP on the cellulosic portion, a commercial modification of this finish which circumvents many of the durability problems is under development at present.[24]

Another phosphorus–bromine system utilizing THPC chemistry which is reportedly in the final stages of development at present is that based on a phosphonium salt with urea and a bromine-containing latex. As first reported by Donaldson and co-workers,[25] this system utilized polyvinylbromide (PVBr) with a THPC-urea finish. This retardant was also applied to fabrics by a pad-dry cure process. The THPC–urea component was developed primarily for cotton fabrics of various weights and represented one of the least expensive fire retardants for cotton based on THPC. The polyvinylbromide component was developed specifically for use with phosphorus-based fire retardants to extend their use to polyester–cotton blend fabrics. The PVBr was produced in emulsion so that it was compatible with aqueous fire-retardant formulations.

The mole ratio of THPC to crystalline urea in the THPC–urea retardant was found to have a significant influence upon durability to laundering. Since the polymerization functionality of THPC is 3 and that of urea is 4, one would expect the greatest insolubility of the polymer and greatest durability of retardant to occur when the mole ratio of THPC to urea is about 1.3:1. This was determined to be approximately correct. During the course of the reaction of THPC with urea, 1 mol of formaldehyde was released which could combine with the urea. Thus, a preferred mole ratio of about 1:1 was expected. This would be in the middle of the experimentally observed durable ratios. Some of PCH_2OH groups from THPC and NCH_2OH from methylolurea (formed *in situ*) react with cellulose. These reactions were deemed important in aiding durability to laundering and in cross-linking cellulose through polymer structure and thus imparting wrinkle resistance to treated cellulosic fabrics. With a ratio of 1:2, the add-on was very good, but the retardant was not very durable to laundering. At a mole ratio of 1:1, the add-on was just as good as with the 1:2 ratio, and durability was excellent. The durability seemed to drop when a ratio of 2:1 was used, but this was partly due to the lower add-on obtained with excess THPC.

Disodium hydrogen phosphate was employed as the catalyst in this treating system for several purposes. The main value appeared to be maintenance of proper pH for reaction of PCH_2OH groups with NH groups of urea. The buffering also helped to prevent acid degradation of the cellulose. Some of the phosphate may also have esterified the cellulose.

The polyvinylbromide concentration was found to be another important factor. Four percent PVBr appeared to be inadequate when used with 26% THPC–urea, assuming that the fabric must withstand 50 laundry cycles. However, 6.4% of the polymer appeared to be adequate since it gave a char length of 4.5 in. after 50 laundry cycles. With 30% THPC–urea in the formulation, the amount of polyvinylbromide could be reduced to as little as 3%. All of the formulations were reported to produce fabrics with essentially no loss in breaking strength, but tearing strengths reduced by about 50%. Other factors such as drying time and temperature, curing time and temperature, and use of softening agents were also important.

As it was originally formulated, this system would seem to have a great deal of commercial potential, but it suffered from two significant problems. The treated fabrics, particularly those of lighter weight, were frequently found to have a harsh hand. The finish was also found to discolor significantly if even slightly overheated during processing. In an attempt to circumvent these problems, Donaldson and co-workers[26] suggested the substitution of a vinylbromide–vinylchloride copolymer, P(VBr–VCl), for the polyvinylbromide latex. This produced a flame-retardant blend fabric which was soft and white when processed under pilot-plant conditions but which still produced some stiffening and discoloration under mill conditions.

An isoperibol calorimetry study of both the PVBr and P(VBr–VCl) systems showed that both bromine latices produced significant reductions in the heat evolved from the treated fabrics during burning[6]. These experiments also showed that there was little difference in the efficiency of the two retardant systems; however, it was shown that the efficiencies of these bromine-containing latices were much lower than those observed for bromine in many other forms. Comparison of the heat release data obtained on the vinylbromide systems with those from Dacron® 900F–cotton blends treated to a similar level of phosphorus showed a significant difference (Fig. 8). The heat release value for both the 900F and PVBr systems showed linear dependency on the bromine content. However, the slope obtained with the 900F system was considerably steeper, indicating that the bromine from the 900F is more effective. These data indicate that the phosphorus–bromine retardant system could be made considerably more effective by utilization of bromine in some form other than that in which it is incorporated into the backbone of the polymer chain and thus activated for dehydrobromination reactions.

In an attempt to develop a more thermally stable bromine-containing latex which would be compatible with the phosphonium compounds,

Barker and co-workers[27] have recently reported the use of a latex prepared by the emulsion polymerization of dibromopropylacrylate. When used in conjunction with a phosphonium salt precondensate, urea and trimethylolmelamine, this poly(2,3-dibromopropylacrylate) [P(DBPA)] imparted a good level of flame retardancy to a variety of 50:50 and 65:35 polyester-cotton blend fabrics. The lighter-weight fabrics were noticeably stiffened, and their tear strength was reduced; however, good esthetic properties were obtainable on the 10-oz denim and the 7½-oz twill. Because of the latex's greater thermal stability, no discoloration was observed on any of the fabrics treated.

A somewhat different approach to incorporating phosphorus and bromine in one system has recently been reported by Stahel and co-workers[28], who have developed a flame retardant based on a bromine-containing phosphazene. This material, when applied to polyester–cotton blends in conjunction with a resin such as hexamethylolmelamine, has been found to produce an effective and durable treatment with good color but a stiff hand.

In a completely different approach to the treatment of polyester–cotton blends, Mischutin[29] has reported the development of an all-bromine system utilizing decabromodiphenylene oxide in conjunction with antimony oxide (FR P-44® from White Chemical Co.). This system is bound to the fabric with an acrylic or urethane latex. The flame retardancy achieved in this manner is generally good, but many of the fabrics feel like coated materials, an undesirable property for many applications. Some problems were also reported due to roller buildup during processing, shade change, and incompatibility with permanent-press finishes. Valente and Yeh[21] have recently evaluated the efficiency of P-44 relative to a number of other retardants utilizing isoperibol calorimetry and calculating the efficiency of the treatments in terms of their ability to reduce the heat released from treated fabric as shown in Table 3. Of all of the systems evaluated, the P-44

TABLE 3

Efficiencies of Bromine-Containing Retardants

$$\Delta H_1 - (\Delta H_1)_{control} = A + B \, (\% \, Br)^a$$

Retardant	Fabrics	A	B (cal/g-% Br)	r
P-44	100% cotton, 65/35, 50/50 P/C blends	11	228.7 ± 14.7	0.9666
900F	75/25, 65/35, 50/50, 900F® Ct blends	− 8	126.5 ± 14.0	0.9880
P-53	100% cotton, 50/50 P/C blend	12	108.6 ± 7.4	0.9796
T23P	50/50 P/C blend	− 7	93.5 ± 0.9	0.9999
PVBr	50/50 P/C blend	−11	61.2 ± 3.4	0.9953
VBr-VCl	50/50 P/C blend	0	60.3 ± 5.5	0.9877

$^a \Delta H_1 - (\Delta H_1)_{control}$ = net heat reduction; A, B = regression constant; r = correlation coefficient.

appeared to be the most efficient, with a heat reduction of approximately 230 cal/g % bromine. Unfortunately, this system is not compatible with many phosphorus-based finishes because of the antagonistic interactions which occur between phosphorus and antimony.[20]

In an attempt to use an alternative approach to the optimization to the P-44 finish for apparel applications, Jarvis and co-workers[30] have used the P(DBPA) latex as a binder for decabromodiphenylene oxide with antimony oxide. The incorporation of the P(DBPA) into the P-44 formulation led to a marked increase in flame-retardant efficiency (approximately 20%) but decreased its durability to laundering. It was found later that this deficiency could be overcome by increasing the degree of cross-linking in the latex or by the addition of a small amount of trimethylolmelamine to the formulation.

This formulation could also be made compatible with a glyoxal resin to impart both flame resistance and durable press characteristics at the same time. When applied in a normal pad-dry cure process with a reactive silicone softener (Dow Corning 1111 emulsion), this system produced a flame-retardant finish with good permanent-press properties in hand, strength losses were minimized, and the level of finish required to pass FF3-71 was reported to be less than that previously recommended for P-44, even though the finish now contained the durable press resin. An additional processing advantage was also noted since there was no buildup of flame retardant on the pad rolls under either laboratory or mill conditions.

This technique of combining a bromine–antimony oxide flame retardant with a bromine-containing latex binder appears to be quite general. Preliminary studies reported by Jarvis and co-workers[30] showed that the P-44–P(DBPA) formulation was effective on a variety of fabrics including cotton, polyester, and nylon in addition to the blends. They have also reported results indicating that a number of other pigmentary bromine compounds and bromine-containing monomers can be incorporated into effective formulations.

On the basis of the above discussions, it can readily be seen that while there are no commercially feasible flame-retardant treatments available for a broad spectrum of cotton–polyester blend fabrics at present, there are several systems at various stages of development which presumably could be utilized on a commercial basis for specific flame retardant uses and specific polyester–cotton blend fabrics. In addition, it should also be noted that blends containing a predominance of cotton can frequently be treated by flame-retardant finishing methods devised for 100% cotton fabrics. Thus, finishes such as THPOH–ammonia are frequently useful on fabrics containing 65% or more cotton with 35% or less polyester. Such fabrics are currently being produced commercially.

3. References

1. W. Kruse, in *Proceedings of the Study Conference on Textile Flammability and Consumer Safety* (Gottlieb and Dutweiler, eds.), p. 137, Institute for Economic Social Studies, Ruschlikon-Zurich, Switzerland (1959).
2. G. C. Tesoro and C. Meiser, Jr., *Text. Res. J.* **40**, 430 (1970).
3. J. E. Hendrix, G. L. Drake, Jr., and W. A. Reenes, *Text. Res. J.* **41**, 360 (1971).
4. J. E. Hendrix, G. L. Drake, Jr., and W. A. Reenes, *J. Fire Flammability* **3**, 2 (1972).
5. M. J. Drews, in *Abstracts of Papers Presented, 172nd National Meeting, American Chemical Society, San Francisco, Calif., August 1976,** Port City Press, Baltimore (1976).
6. R. H. Barker and M. J. Drews, Development of Flame Retardants for Polyester/Cotton Blends, NBS-GCR-ETIP 76-22, National Bureau of Standards, Washington, D.C. (Sept. 1976).
7. D. E. Stuetz, A. H. Diedwordo, F. Zitomer, and B. P. Baines, *J. Polym. Sci. Polym. Chem. Ed.* **13**, 585 (1975).
8. G. E. Heckman, masters' thesis, Clemson University, Clemson, S.C. (1976).*
9. E. Larsen, in *Abstracts of Papers Presented, 172nd National Meeting, American Chemical Society, San Francisco, Calif., August 1976*, Port City Press, Baltimore (1976).
10. K. Yeh and R. H. Barker, *Text. Res. J.* **41**, 932 (1971).
11. M. J. Drews, K. Yeh, and R. H. Barker, *Textilveredlung* **8**, 180 (1973).
12. J. T. Langley, Ph.D. dissertation, Clemson University, Clemson, S.C. (1976).*
13. M. J. Drews and R. H. Barker, in *Abstracts of Papers Presented, 167th National Meeting, American Chemical Society, Los Angeles, Calif., April 1974*, Port City Press, Baltimore (1974).
14. M. J. Drews and R. H. Barker, in *Abstracts of Papers Presented, 165th National Meeting, American Chemical Society, Dallas, Tex., April 1973*, Creative Printing, Hyattsville, Md. (1973).
15. H. V. R. Iengar and P. O. Ritchie, *J. Chem. Soc. London* (**1957**) 2556.
16. R. Loss, P. Hofmann, and H. Machbur, *Textilveredlung* **8**, 194 (1973).
17. G. C. Tesoro, *Text. Chem. Color.* **5**, 235 (1973).
18. K. Yeh, in *Abstracts of Papers Presented, 163rd National Meeting, American Chemical Society, Boston, Mass., April 1972*, Creative Printing, Hyattsville, Md. (1972).
19. K. Yeh and J. A. Valente, paper presented at ETIP meeting, Clemson University, Clemson, S.C. (Sept. 1976).*
20. E. D. Weil, in *Flame Retardancy of Polymeric Materials* (W. C. Kuryla and A. J. Papa, eds.), vol. 3, pp. 185–244, Dekker, New York (1975).
21. J. Valente and K. Yeh, paper presented at ETIP meeting, Clemson University, Clemson, S.C. (Sept. 1976).*
22. C. Hamalainen and J. D. Guthrie, *Text. Res. J.* **26** (2), 141 (1956).
23. R. P. Barber, F. Moussalli, F. Marascia, J. Bridgemen, J. Flipo, J. Jermoe, G. Moser, and R. Richardson, *Am. Dyest. Rep.* **56**, 373 (1967).
24. R. E. Smith, paper presented at ETIP meeting, Clemson University, Clemson, S.C. (Sept. 1976).
25. D. J. Donaldson, F. L. Normand, G. L. Drake, Jr., and W. A. Reeves, *J. Fire Flammability Flame Retardant Chem. Suppl.* **2**, 102 (1975).
26. D. J. Donaldson, F. L. Normand, and G. L. Drake, Jr., paper presented at ETIP meeting, Clemson University, Clemson, S. C. (Sept. 1976).*

*Material may also be found in Ref. 6.

27. Palmetto Section, AATCC, *Text. Chem. Color.* **9**, 28 (1977).
28. F. Stahel, C. Schouten, and M. J. Drews, paper presented at ETIP meeting, Clemson University, Clemson, S.C. (Sept. 1976).*
29. V. Mischutin, paper presented at ETIP meeting, Clemson University, Clemson, S.C. (Sept. 1976).
30. C. W. Jarvis, R. H. Barker, and V. Mischutin, *J. Coated Fab.* **6**, 182 (1977).

*Material may also be found in Ref. 6.

6

Factors Affecting the Combustion of Polystyrene and Styrene

R. V. Petrella

1. Introduction

The widespread use of polystyrene and its copolymers in modern society predicates an increasing awareness of the combustibility associated with the styrenic structure. The 1975 consumption of styrenic polymers has been estimated at 1.8 million metric tons (3.98 billion lb) of which 67% is the homopolymer.[1] The problems pertinent to the pyrolysis and combustion of styrene polymers are only slightly modified by the addition of comonomers such as acrylonitrile, butadiene, and methyl methacrylate. A basic knowledge of the modes of pyrolytic decomposition and the high-temperature reactions of the products with the oxygen in the atmosphere is necessary to a rational approach to the problem of flame inhibition.

Many new flame-retardant chemicals have been developed for use in plastics. In the thermoplastics area, the subject of this chapter, the great majority of these flame retardants are of the additive type, usually containing halogen and sometimes a synergist. Our purpose in this chapter is not to list the many flame retardants on the market or to categorize them as

R. V. Petrella • Olefin Plastics Department, The Dow Chemical Company, Midland, Michigan.

being especially efficient for a particular styrenic polymer system. A number of texts exist in which this information is available, especially the publications of Lyons,[2] Hilado,[3] and Kuryla and Papa.[4] Our purpose herein is to present a discourse on the pyrolysis and combustion of polystyrene and styrene monomer and to show the effects of selected flame inhibitors upon the course and extent of the pyrolysis and combustion under different conditions of temperature and stoichiometry.

2. Physical and Thermal Properties of Polystyrene and Styrene

The general physical properties of unmodified polystyrene as tabulated by Boundy and Boyer are given in Table 1.[5]

The glass-transition temperature has been reported as being between 80 and 100°C depending on the purity of the polymer and the mode of polymerization.[3,5a,6] The heat capacity data of Stull[7] indicate a T_g of 90°C as shown by the inflection point in Fig. 1. The figure also shows that the heat capacity increases as the temperature is raised until a maximum is reached at ~200°C. Thereafter, the heat capacity decreases as the temperature is further increased. The self-ignition and flash ignition temperatures of polystyrene have been reported to be 490 and 350°C, respectively.[8]

The thermal properties of a typical unmodified polystyrene are shown in Table 2.[7]

The thermal degradation of polystyrene, whether carried out in conventional oven-type apparatus[9-12] with heating rates of 0–100°C/min or in the extremely rapid flash pyrolysis (heating rate of 20°C/μsec) appa-

TABLE 1
General Physical Properties of Polystyrene

Property	ASTM test method	Values
Fabrication		
Bulk factor	D392-38	1.9–2.3
Flow or mobility range (°F)	D569-44T	240–280
Injection molding temp. (°F)		325–550
Injection molding pressure (psi)		10,000–30,000
Injection molding mold shrink. (in./in.)		0.003–0.006
Compres. molding temp. (°F)		265–350
Compres. molding pressure (psi)		1000–10,000
Compres. molding mold shrink. (in./in.)		0.002–0.006
Extruding temp. (°F)		375–550

TABLE 1 (*Continued*)

Property	ASTM test method	Values
Miscellaneous test values		
Specific gravity	D792-44T	1.052–1.065
Haze (crystal) (%)	D672-44T	10
Luminous transmittance (dependent on color) (%)	D672-44T	0–93
Refractive index (η_D^{25})	D542-42	1.59–1.60
Thermal coefficient of exper. linear (max./°C)	D696-42T	$6–8 \times 10^{-5}$
Thermal conductivity (cal/cm²/sec/°C/cm)	C177-45	$2.4–3.3 \times 10^{-4}$
Specific heat (cal/g)		0.32
Heat distortion, 264 psi (°F)	D648-45T	180–190
Flammability (in. min)	D635-44	0.5–1.0
Water absorption, 24 hr (%)	D570-42	0.03–0.04
Durability		
Effect of light		Slight
Heat resistance (max. recommended for continuous service) (°F)		150–185
Electrical values		
Dielectric str., short time ($\frac{1}{8}$ V/mil)	D149-44T	500–700
Dielectric str., step by step ($\frac{1}{8}$ V/mil)	D149-44T	400–600
Volume resistivity (Ω-cm)	D257-46	$10^{17}–10^{19}$
Dielectric constant, 60 cps	D150-46T	2.45–2.65
Dielectric constant, 1000 cps	D150-46T	2.45–2.65
Dielectric constant, 10^6 cps	D150-46T	2.45–2.65
Dissipation factor, 60 cps	D150-46T	$10–30 \times 10^{-5}$
Dissipation factor, 1000 cps	D150-46T	$10–30 \times 10^{-5}$
Dissipation factor, 10^6 cps	D150-46T	$10–40 \times 10^{-5}$
Loss factor, 60 cps	D150-46T	$25–70 \times 10^{-5}$
Loss factor, 1000 cps	D150-46T	$25–70 \times 10^{-5}$
Loss factor, 10^6 cps	D150-46T	$25–70 \times 10^{-5}$
Arc resistance (sec)	D495-42	70–135
Mechanical test values		
Impact str., Izod notched 77°F (ft-lb/in.)	D256-45T	0.25–0.40
Compres. str., stress at yield (upper), (psi)	D695-44T	11,500–16,000
Compres. str., strain at yield (upper), (%)	D695-44T	4.5–5.5
Deformation under load at 50°C, 4000 psi (rigid materials) (%)	D621-45T	1.0
Flexure, stress at fracture (psi)	D650-42T	8000–19,000
Flexure, strain at fracture (in.) (%)	D650-42T	0.10–0.20
Tension, stress at fracture (psi)	D638-46T	5,000–8,500
Tension, strain at fracture (%)	D638-46T	1.0–3.5
Modulus of elasticity (apparent)		
Flexure (psi)	D650-42T	$4.0–5.0 \times 10^5$
Compression (psi)	D695-44T	$3.0–5.6 \times 10^5$
Tension (psi)	D638-46T	$4.0–5.0 \times 10^5$
Shear strength (psi)		6000–8000

TABLE 2
Thermal Properties of Polystyrene

T, temperature (°K)	C_p°, heat capacity (cal/deg/mol)	$H_T^\circ - H_{298.18}^\circ$, heat content (cal/mol)	S_T°, entropy (cal/deg/mol)	$-\left(\dfrac{F^\circ - H_{298.15}^\circ}{T}\right)$, free-energy function (cal/deg/mol)	Formation from elements		
					Heat, ΔH° (cal/mol)	Free energy, ΔF° (cal/mol)	$\log_{10} K_p$
298	30.43		32.23	32.23	8040	38,927	−28.535
300	30.63	56	32.42	32.24	8016	39,114	−28.496
400	48.59	3900	43.36	33.61	7108	49,668	−27.139
500	60.40	9346	55.46	36.77	7210	59,320	−26.367

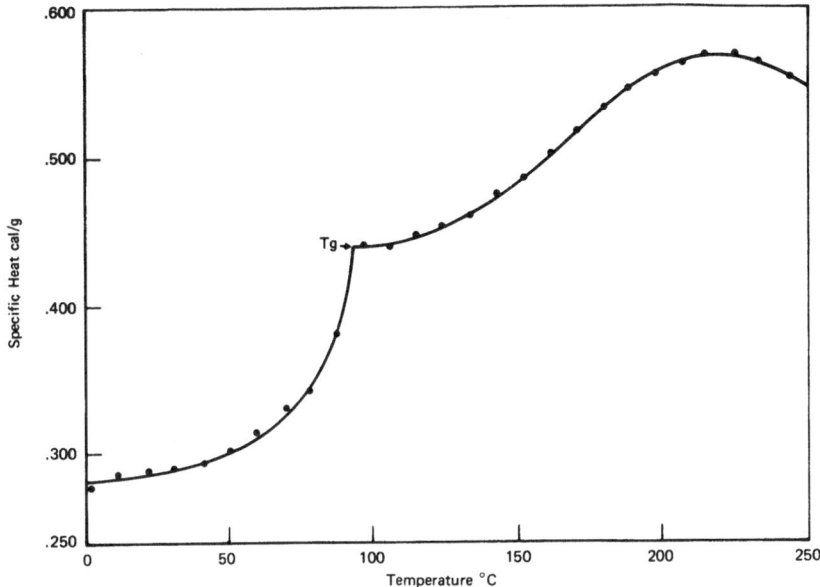

FIGURE 1. Specific heat of polystyrene at various temperatures.

TABLE 3

Physical Properties of Styrene Monomer

Formula	C_8H_8
Molecular wt.	104.14
Density at 25°C (g/ml)	0.9019
Pounds per gal at 25°C	7.5
Refractive index at 25°C	1.5439
Viscosity at 25°C (cP)	0.730
Surface tension at 25°C (dynes/cm)	31.7
Coefficient of expansion at 25°C (cc/°C)	0.0009719
Boiling point (°C)	145.2
Heat of vaporization at 25°C (cal/g)	102.65
Freezing point in air at 1 atm (°C)	−30.628
Heat of fusion (cal/g).	25.4
Specific heat at 25°C (cal/g/°C)	0.416
Heat of combustion at 25°C (kcal/g)	10.086
Critical temperature (°C)	373
Critical pressure (atm)	40.0
Critical density (g/ml)	0.30
Flash point (°C)	31
Fire point (°C)	34
Explosive limits (vol. % in air)	1.1–6.1
Heat of polymerization (cal/g)	160.2
Volume shrinkage on polymerization (%)	17

ratus,[13-15] yields 30–80% monomer as the primary product. It is therefore prudent that the physical and thermal properties of styrene monomer
be listed as it is the primary gas phase fuel involved in the combustion of
styrenic polymers. The physical properties (as shown in Table 3) are for the
monomer in the liquid state.[5] The thermal properties are listed for both
the liquid and gaseous state as reported by Stull[7] (Tables 4 and 5). The heat
of polymerization can readily be calculated from the heats of formation of
the polymer and liquid state. The value of 16.7 kcal/mol or 160.2 cal/g is
consistent with experimentally derived values of the heat of polymerization.

3. The Pyrolysis and Combustion of Polystyrene and Styrene

3.1. The Pyrolysis and Combustion of Polymers

Combustion or burning, as it is more generally referred to, is a chemical
process whose exothermicity is such that once the process is initiated it
proceeds to its thermodynamically defined and kinetically controlled end
independently of the extent and duration of the source of initiation. This
chemical process is not a single-step process but is made up of several sequential steps. Some of the steps are exothermic in nature, and some are
endothermic. The basic conditions which must be met are that the sum total
of the reaction steps be exothermic and that the steps proceed rapidly enough
that the exothermic steps supply sufficient energy to sustain the reaction
chain. The energy and mass transport subprocesses occurring in the combustion process are beyond the scope of this chapter and will only be referred
to in terms of their interactions, with the chemistry of the particular steps
being discussed.

The initial steps leading toward the combustion of polystyrene involve
the response of the solid surface to the heat flux of the ignition source. The
source can be a match or a heated filament, a radiant panel, or the flame from
a previously established fire. The temperature of the surface rises as the
heat flux continues to be directed at the polystyrene. At a surface temperature of about 90°C the polymer starts to lose its rigidity. The thermal conductivity of the polymer carries the heat into the body of the polymer, and
its internal temperature rises more slowly than does the surface. As the temperature of the surface increases, the polymer starts to degrade, and the
viscous surface bubbles as gaseous pyrolysis products are expelled from it.
At around 490°C the pyrolysis products react with the oxygen in the air
to produce the observed combustion products. The oxidation of the pyrolysis products is the *critical* exothermic step in the combustion process. Pro-

TABLE 4
Thermal Properties of Liquid Styrene

Styrene (liquid)

G_{fw} = 104.152 g	M.P. = 242.52 °K ΔH_m = 2645 cal/mol
T_G = °K	B.P. = 418.4 °K ΔH_v = 8870 cal/mol
P_G = atm	S.P. = °K ΔH_8 = cal/mol
$(H_{298}^\circ - H_0^\circ)$ = 4985 cal/mol	T.P. = °K ΔH_1 = cal/mol
	T.P. = °K ΔH_1 = cal/mol

T, temperature (°K)	C_p°, heat capacity (cal/deg/mol)	$H_T^\circ - H_{298.15}^\circ$, heat content (cal/mol)	S_T°, entropy (cal/deg/mol)	$-\left(\dfrac{F^\circ - H_{298.15}^\circ}{T}\right)$, free-energy function (cal/deg/mol)	Formation from elements		
					Heat, ΔH° (cal/mol)	Free energy, ΔF° (cal/mol)	$\log_{10} K_p$
298	43.37		57.16	57.16	24,716	48,170	−35.311
300	43.50	80	57.44	57.18	24,716	48,308	−35.195
400	52.48	4,847	71.09	58.98	24,731	56,199	−30.708
500	65.97	10,720	84.14	62.70	25,260	64,030	−27.989

TABLE 5
Thermal Properties of Gaseous Styrene

Styrene (ideal gas state)			
G_{fw} = 104.152 g	M.P. = 242.52°K	ΔH_m = 2645 cal/mol	
T_G = °K	B.P. = 418.4 °K	ΔH_v = 8870 cal/mol	
P_G = atm	S.P. = °K	ΔH_5 = cal/mol	
$(H^\circ_{298.15} - H^\circ_0)$ = 4985 cal/mol	T.P. = °K	ΔH_1 = cal/mol	
	T.P. = °K	ΔH_1 = cal/mol	

T, temperature (°K)	C°_P, heat capacity (cal/deg/mol)	$H^\circ_T - H^\circ_{298.15}$, heat content (cal/mol)	S°_T, entropy (cal/deg/mol)	$-\left(\frac{F^\circ - H^\circ_{298.15}}{T}\right)$, free-energy function (cal/deg/mol)	Formation from elements: Heat, ΔH° (cal/mol)	Formation from elements: Free energy, ΔF° (cal/mol)	Formation from elements: $\log_{10} K_p$
298	29.18		82.48	82.48	35,220	51,124	−37.476
300	29.35	52	82.65	82.48	35,192	51,221	−37.317
400	38.32	3,443	92.35	83.75	33,831	56,795	−31.033
500	45.94	7,675	101.76	86.41	32,719	62,679	−27.398
600	52.14	12,583	110.70	89.73	31,803	68,727	−25.035
700	57.21	18,059	119.13	93.34	31,087	74,984	−23.413
800	61.40	23,991	127.04	97.06	30,515	81,267	−22.202
900	64.93	30,313	134.48	100.80	30,093	87,693	−21.296
1000	67.92	36,965	141.49	104.53	29,833	94,023	−20.550
1100	70.48	43,895	148.10	108.20	29.683	100,413	−19.952
1200	72.66	51,055	154.31	111.77	29,651	106,943	−19.478
1300	74.54	58,415	160.21	115.28	29,595	113,354	−19.478
1400	76.16	65,955	165.80	118.69	29,607	119,711	−18.689
1500	77.57	73,645	171.10	122.01	29,657	126,287	−18.401

vided a sufficient quantity of pyrolysis gases have been transported from the surface, the ensuing exothermic reactions are more than sufficient to maintain the combustion process. If sufficient heat is fed back to the polymer, the process will sustain itself without the external heat source. The combustion process is a closed-loop situation. The flame supplies heat to perpetuate the pyrolysis processes, which in turn supply the reactants to sustain the flame. The role of a chemical flame retardant is to interrupt this closed-loop process by means of alternate chemical reactions which will minimize the exothermic interaction of the pyrolysis products with the oxygen in the atmosphere by substituting less reactive species in place of the pyrolysis products or by reacting with the pyrolysis products themselves or by preventing the degradation to volatile pyrolysis products.

3.2. The Pyrolysis of Polystyrene

The pyrolysis of polystyrene has been treated in detail by Madorsky[16] and Jellinek.[17] The major pyrolysis products have been shown to be styrene monomer, dimer, and trimer, with the monomer being the largest fragment. The generally accepted mechanism for the thermal degradation of polystyrene involves the scission of a backbone C—C bond resulting in two free radicals or in two smaller chain units as a result of hydrogen transfer.

1. Scission into free radicals:

$$\text{(1)}$$

2. Scission accompanied by transfer of a tertiary hydrogen:

$$\text{(2)}$$

The free radical (2) can remove a tertiary hydrogen intermolecularly or intramolecularly, causing a new chain scission and the formation of a new radical and a new olefin similar to (4).

The high yield of monomer shows the tendency of the free radicals to unzip to monomer:

$$
\underset{(2)}{\text{—C—C—C—C—C}\!\mid\!\text{C—C—}} \longrightarrow \underset{(5)}{\text{—C—C—C—C—C·}} + \underset{(6)}{\text{C=C}} \tag{3}
$$

The dimer and trimer can be formed by an intramolecular transfer of a tertiary hydrogen and the unzipping of a free radical:

$$
\underset{(2a)}{\text{—C—C—C—C—C—C—C·}} \longrightarrow \underset{(7)}{\text{—C—C—C}} + \underset{\text{dimer}}{\text{C=C—C—C—H}} \tag{4}
$$

The importance of the tertiary hydrogen in the unzipping of the chain to form the monomer can best be illustrated by comparing the monomer "yield" of polystyrene to poly-α-methylstyrene, which has no tertiary hydrogen:

$$
\left(\begin{array}{c} \text{H} \quad \text{H} \\ \text{C—C} \\ \phi \quad \text{H} \end{array}\right)_n \qquad \left(\begin{array}{c} \text{CH}_3 \ \text{H} \\ \text{C—C} \\ \phi \quad \text{H} \end{array}\right)_n \tag{5}
$$
$$
\text{polystyrene} \qquad \text{poly-}\alpha\text{-methylstyrene}
$$

The monomer yield for polystyrene pyrolyzed in a vacuum in the temperature range 200–500° is 40%, while the monomer yield from poly-α-methylstyrene is 95–100%. The role played by hydrogen atoms in the degradation of polymers is treated by Madorsky.[16]

3.3. Inhibiting the Pyrolysis of Polystyrene

Pyrolysis inhibition of polystyrene has been attempted, with limited success, either by slowing down the rate of pyrolytic degradation, by insulating the receding polymer surface from the heat flux of the ignition source, or by modifying the pyrolysis process itself.

The use of phosphorus flame retardants is predicated on the formation of polyphosphoric acids at the polymer surface via

$$
\text{HO—P—} \longrightarrow \left(\begin{array}{c} \text{O} \\ \| \\ \text{P—O} \\ \text{OH} \end{array}\right)_n \tag{6}
$$

These viscous liquids are thought to act as barriers to both the feedback from the flame and to the transpiration of pyrolysis products to the flame front. The most commonly used phosphorus flame retardant seems to be tris(2,3-dibromopropyl) phosphate. In recent years the claim has been made that the polyphosphoric acids can act as Lewis acids and promote Diels–Alder reactions at the polymer surface, leading to enhanced cross-linking and, hence, reduced volatilization.

The phosphorus flame retardant must be thermally stable at polystyrene processing temperatures (100–225°C) but must undergo rapid degradation between these temperatures and the flash ignition temperature of 350°C. Kinetically this implies that the flame retardant should have a high activation energy and a large preexponential factor, as stated by Schwarz in Ref. 4 (Vol. 2, p. 112).

The alternative to the *in situ* cross-linking mechanisms attributed above to phosphorus flame retardants is to copolymerize the styrene with multifunctional but physically similar monomers such as divinyl- and trivinylbenzene. The cross-linked polystyrene is more thermally stable than uncrosslinked polystyrene. A fundamental change in the nature of the thermal degradation of highly cross-linked polystyrene has been cited by Madorsky[16] as leading to enhanced char formation. Whatever the mechanisms involved, the copolymerization of styrene with divinyl- or trivinylbenzene leads to products with enhanced char formation and increased thermal stability. It is interesting to note that although the degree of volatilization from polymer degradation decreases as the DVB and TVB levels are increased, the activation energies of the thermal degradation processes are between 53 and 61 cal/mol. These values are not too far from the 55 cal/mol found for polystyrene. The modes of thermal degradation are different, but their temperature dependencies are similar. The brittleness associated with highly cross-linked polystyrene severely limits the use of this polymer in load-bearing applications.

3.4. The Pyrolysis and Combustion of Polystyrene

Significant amounts of styrene monomer are produced during the pyrolysis of polystyrene. The thermal degradation and ultimate combustion of this material should play a large role in facilitating our understanding of the combustion of polystyrene. A knowledge of the fate of styrene monomer in combustion situations can help point the way toward the development of more efficient flame-retardant chemicals.

Mass-polymerized polystyrene was flash-pyrolyzed in the apparatus shown in Fig. 2. This apparatus is similar to that employed by Norrish *et al.*

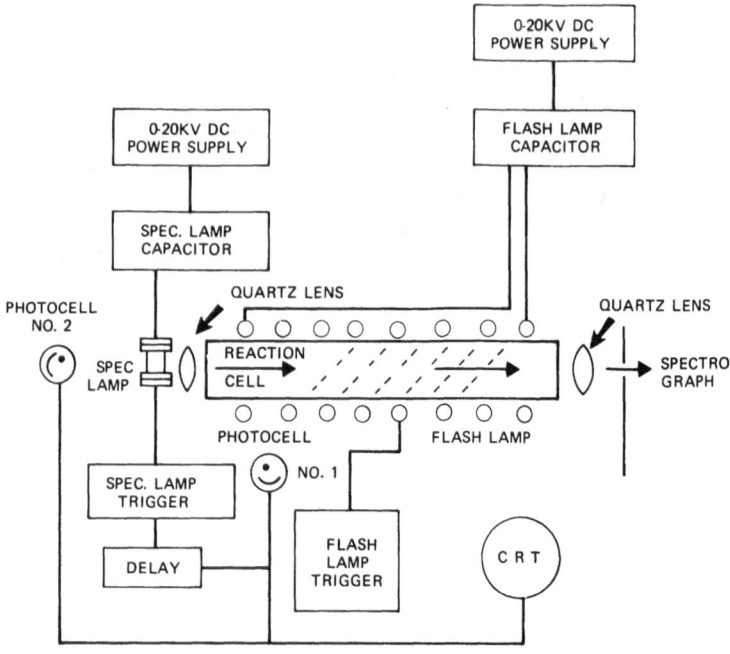

FIGURE 2. Schematic diagram of the flash photolysis–pyrolysis apparatus.

in their studies of the combustion of several hydrocarbon fuels.[18-21] The pyrolysis variation of this apparatus follows Nelson and Kuebler[22,23] and has been previously used to study the combustion of solid propellants.[24,25]

Samples of polystyrene ranging in weight from 1 to 8 mg were flash-pyrolyzed in vacuum at flash energies of 700–1500 J. The monomer yields ranged from 20 to 58%, the yield and flash energy both increasing. The temperature of the graphite surface from which the polymer was pyrolyzed was calculated to be between 350 and 1500°C. The calculations were carried out in the manner suggested by Nelson and Kuebler. The temperature was also determined by observing the absorption intensity of FeI lines resulting from the thermal pyrolysis of ferrocene and iron carbonyl, separately pyrolyzed under the same conditions used in the polymer studies. The spectroscopic data used in these calculations were from tables by Pearce.[26]

The flash lamps used in these studies had a discharge time of 50 μsec as observed by a heavily filtered photocell. The discharge time at half-peak height was 20 μsec.

When flash pyrolysis studies were carried out in an oxygen environment, the OH (hydroxyl) radical was seen to form quickly, as shown in Fig. 3. The C_3O_2 (carbon suboxide) moiety was also seen during these studies.

The appearance of the OH radical was taken to indicate the onset of high-temperature chain-propagating combustion. The time interval between the initiation of the flash lamp discharge and the appearance of the OH radical is defined as the induction time to combustion and is a measure of the cool-flame reaction prior to true ignition. The C_3O_2 can be taken to indicate the fragmentation of the polymer chain and is indicative of the cool flame nonchain branching reactions.

The addition of 0.5% pentabromomonochlorocyclohexane (FR-651 A®, Dow Chemical Company) drastically alters the flash-induced combustion of polystyrene, even in the overoxidized situation. Figure 4 shows the time–intensity curve for the OH radical and the suboxide formed during the combustion of the polystyrene containing the flame retardant. The first thing that one notices is the increased induction time to combustion as evidenced by the appearance of the OH radical at 300 μsec compared to its appearance at 50 μsec in the uninhibited system. Also the C_3O_2 appears to form before the OH, indicating a large amount of cool-flame reaction before true ignition is achieved.

It can be concluded that the halogen delays the onset of ignition by interfering with the H_2–O_2 combustion scheme.

While the study reported above seeks to ascertain the combustion of

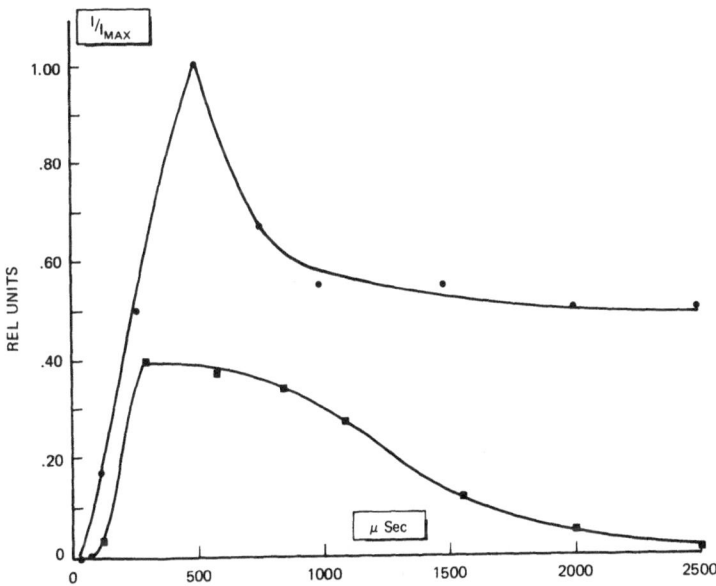

FIGURE 3. Time–intensity curves for selected species observed during the flash pyrolytic combustion of 4 mg of polystyrene in 80 mm oxygen; flash energy = 1500 J. ● OH; ■ 3048 C_3O_2.

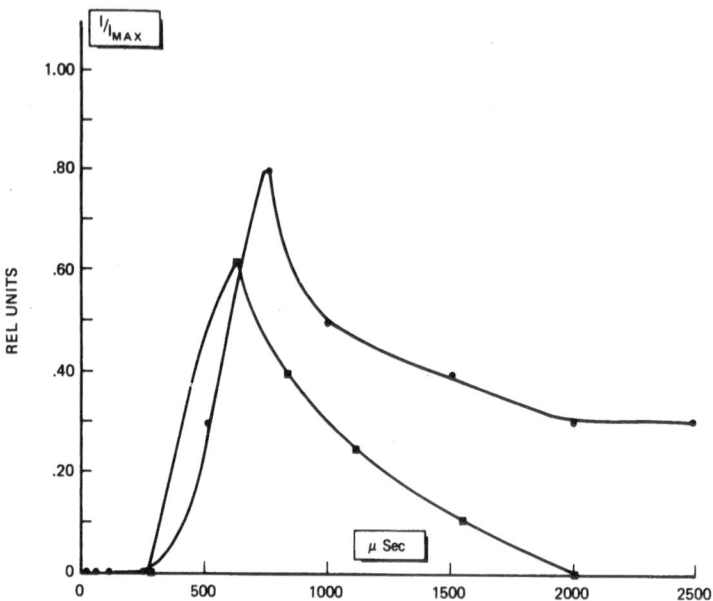

FIGURE 4. Time–intensity curves for the OH radical and the suboxide (C_3O_2) formed in the combustion of 4 mg of polystyrene containing 0.5% by weight of SE-651 in 80 mm oxygen; flash energy = 1500 J. ● OH; ■ 3048 Å C_3O_2.

polystyrene in terms of the concentration of reactive intermediates, the work of Tewarson and Pion[27,28] deals with the energy balance relationships in polymeric combustion. The apparatus used in the Factory Mutual tests is shown in Fig. 5. The apparatus is similar in design to the limiting oxygen index (LOI) apparatus of Fenimore except that the Factory Mutual equipment has a load cell to measure the mass burning rate of the sample. The radiant heaters supply external heat to the polymer surface. The mass burning rate of the sample is measured as a function of the oxygen content of the atmosphere and/or of the amount of external heat supplied to the sample surface. The mass burning rate or "burning intensity" is related to the energy balances as follows:

$$\dot{Q}_T'' + \dot{Q}_E'' = \dot{Q}_G'' + \dot{Q}_L'' \tag{7}$$

where

$\dot{Q}_T''$ (cal/cm^2-sec) = total heat flux from the flame to the surface of the polymer

$\dot{Q}_E''$ (cal/cm^2-sec) = externally applied heat flux

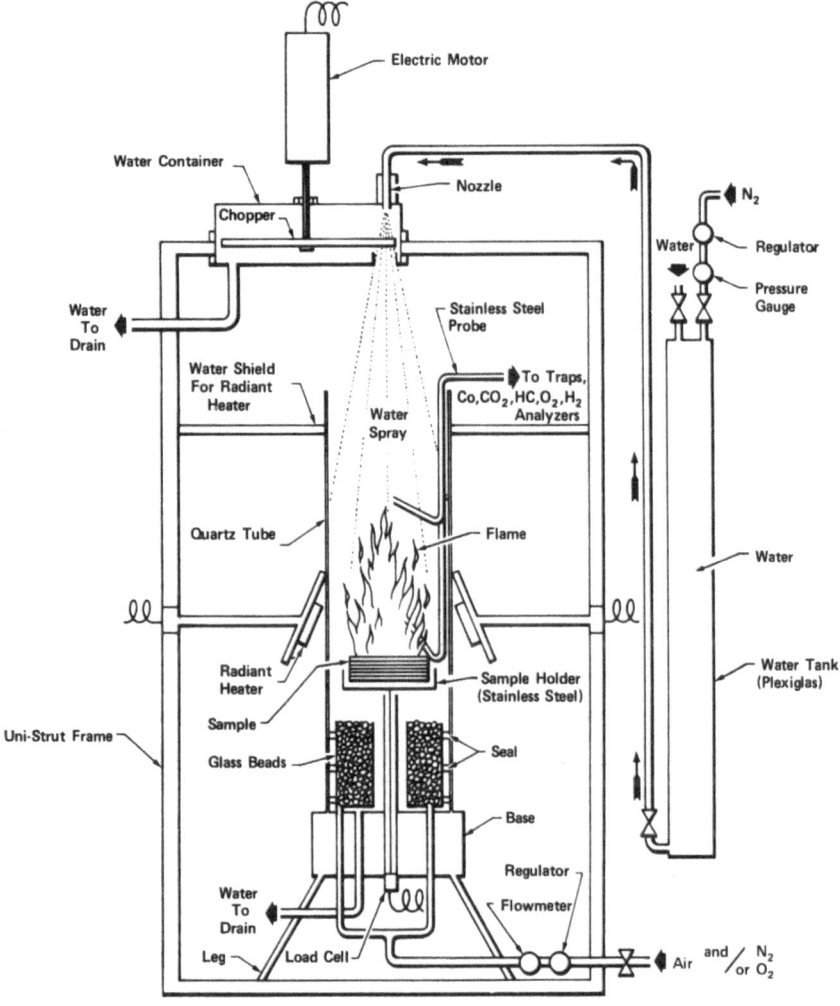

FIGURE 5. Factory Mutual flammability apparatus.

$\dot{Q}_G''$ (cal/cm^2-sec) = heat flux required to
                                        gasify–pyrolyze the sample
$\dot{Q}_L''$ (cal/cm^2-sec) = total heat flux lost by the sample

The term $\dot{Q}_G''$ in Eq. (7) can be expressed as

$$\dot{Q}_G'' = \dot{m}'' \cdot L_G$$

where

$\dot{m}''$ (g/cm^2-sec) = mass burning rate

L_G (cal/g) = heat of gasification–pyrolysis–depolymerization of the polymer initially at ambient temperature

If we further define the term $\dot{Q}_T''$ in Eq. (7) as

$$\dot{Q}_T'' = \epsilon \cdot N_{O_2}$$

where

ϵ = a constant

N_{O_2} = mole fraction of oxygen

then Eq. (7) can be rewritten as

$$\dot{m}'' = \left(\frac{\epsilon}{L_G}\right)N_{O_2} + \frac{\dot{Q}_E''}{L_G} - \frac{\dot{Q}_L''}{L_G} \tag{8}$$

At a constant oxygen mole fraction, the burning rate of a polymer is directly related to the amount of external heat impinging on the surface. Small laboratory-scale tests have $\dot{Q}_E''$ values much smaller than $\dot{Q}_L''$ values; therefore, the mass burning rate of the polymer is considerably lower than would be the case in a full-scale fire, where nearby burning objects or hot walls would contribute considerably to the heat incident upon the sample.

Some interesting relationships can be drawn from this work. Table 6 shows the heat balance for two polystyrene samples, for styrene monomer burning in normal air, and for a polycarbonate which will not burn in air. These data show that materials which burn readily in air have $\dot{Q}_T'' > \dot{Q}_L''$; that is, more heat is received from the flame than is lost by the sample. The solid polystyrene flame transmits 16% less heat to the surface than does the monomer, while the foam flame radiates about half as intensely as does the monomer flame. Heat losses from the solid polystyrene are nearly 2.5 times as large as through the foam, as would be expected from the K factors for each. The data suggest that considerable monomer is formed during

TABLE 6

Heat Flux Transmitted from the Flame ($\dot{Q}_T''$), Heat Flux Lost by the Sample ($\dot{Q}_L''$), and Heat of Gasification–Pyrolysis (L_G)

	$\dot{Q}_T''$ (cal/cm^2-sec)	$\dot{Q}_L''$ (cal/cm^2-sec)	L_G (cal/g)
Styrene monomer	1.74	1.04	153
Polystyrene solid	1.47	1.20	420
Polystyrene foam	0.82	0.56	324
Polycarbonate	1.24	1.77	495

the pyrolysis of the polystyrene samples. The heats of gasification for both polystyrene samples are larger than the heat of polymerization (160 cal/g), suggesting that some degradation of the monomer occurs during pyrolysis.

3.5. The Pyrolysis and Combustion of Styrene Monomer

Pyrolysis studies carried out on different styrenic polymers revealed the monomer yield to lie between 30 and 100%. Flash pyrolysis studies of the combustion of polystyrene showed it to be similar in nature to that of the monomer. The main difference between the two combustion systems (monomer and polymer) was the lengthened time to induction for the combustion of the polymer as well as the appearance of monomer during the induction time.

The appearance of significant amounts of monomer during the pyrolysis of polystyrene suggests that the behavior of the monomer in high heat fluxes, as found in the flash pyrolysis studies or in large-scale fires, will significantly influence the combustion process.

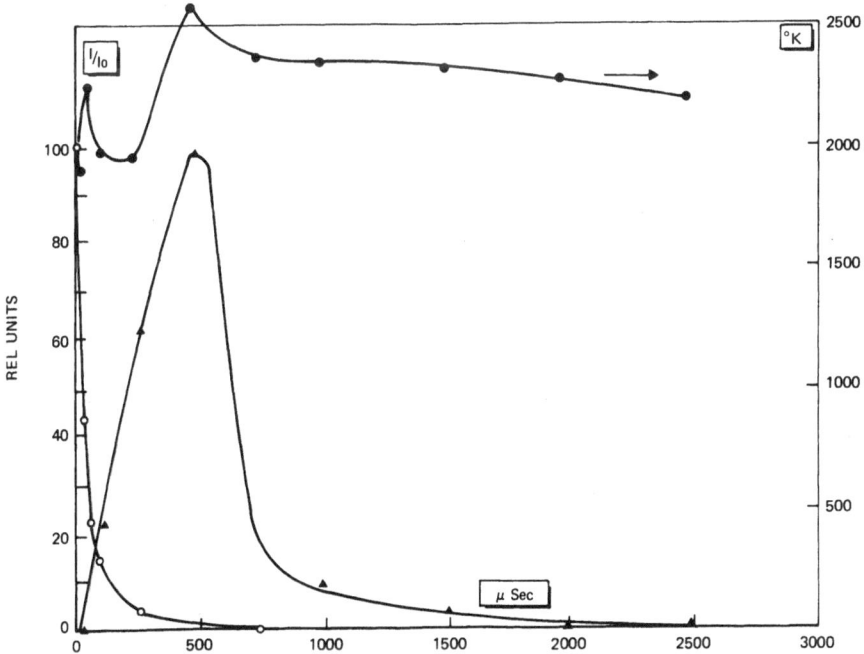

FIGURE 6. Disappearance of styrene monomer and the intensity of the OH radical formed in the flash photolytic combustion of 2 mm styrene and 20 mm oxygen (left ordinate); flame temperature for same combustion (right ordinate); flash energy = 1500 J. ▲ OH; ○ styrene; ● flame temperature.

Flash photolytic studies of the combustion of styrene monomer in oxygen at several initiation temperatures show the combustion of styrene to be a complex function of the combustion stoichiometry and the flame temperature.[14]

Figures 6 through 9 show the effect that the initiation temperature has upon the rate of pyrolysis of styrene monomer, as indicated by its decrease in absorption intensity, and upon the time history of the OH radical formed during the combustion of the monomer. The flame "temperatures" were calculated from the absorption intensities of separate rotational lines of the (0,0) vibrational system of the $A^2\Sigma^+-X^2\Pi$ electronic transition of the OH radical by the method of Dieke and Crosswhite.[29] The combustion mixtures are stoichiometrically balanced to CO_2 and H_2O as the desired combustion products.

The figures show that at the highest flash energy the styrene disappears slower than at lower initiation temperatures. The initial rate of pyrolysis of styrene is greatest at 1000 J, but the overall disappearance of monomer

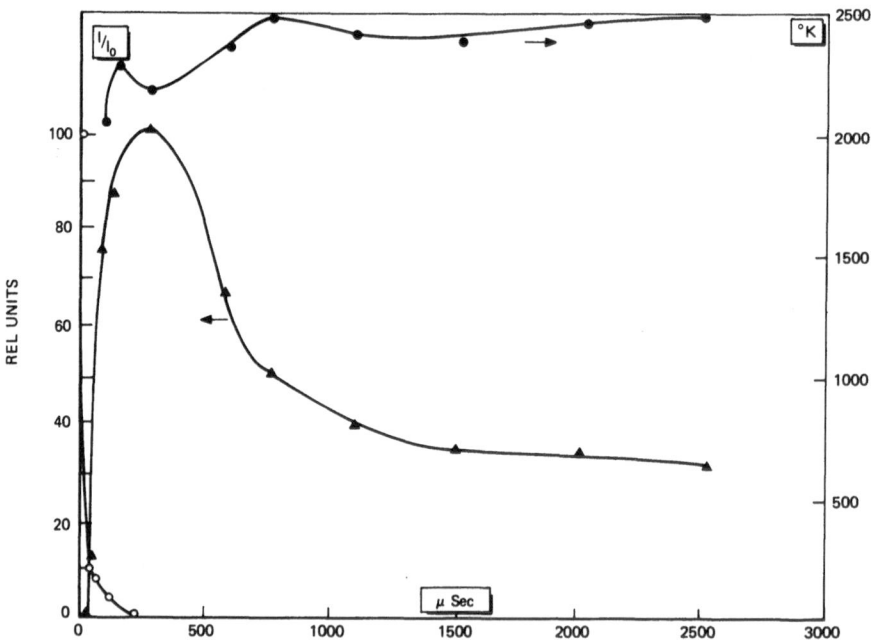

FIGURE 7. Disappearance of styrene monomer and the intensity of the OH radical formed in the flash photolytic combustion of 2 mm styrene and 20 mm oxygen (left ordinate); flame temperature for same combustion (right ordinate); flash energy = 1000 J. ▲ OH; O styrene; ● flame temperature.

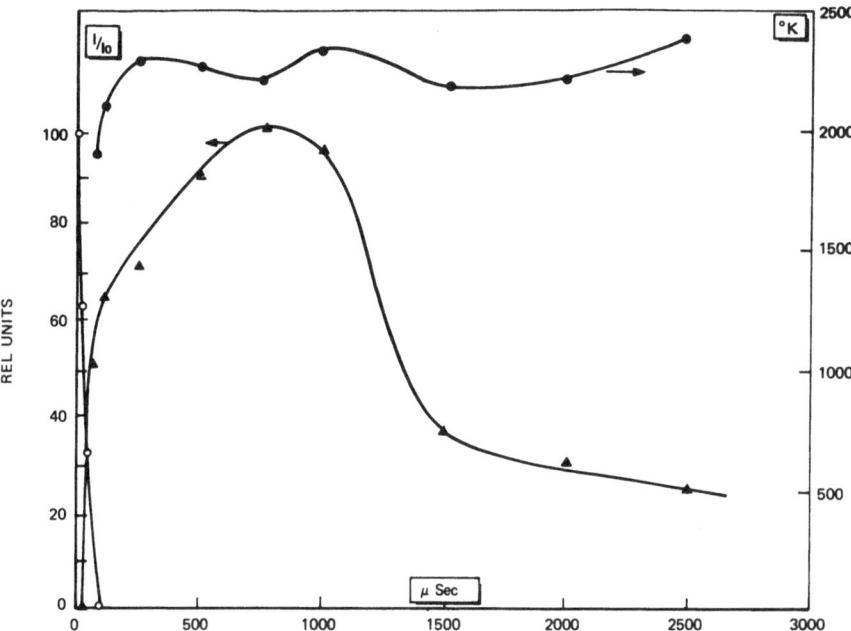

FIGURE 8. Disappearance of styrene monomer and the intensity of the OH radical formed in the flash photolytic combustion of 2 mm styrene and 20 mm oxygen (left ordinate); flame temperature for same combustion (right ordinate); flash energy = 750 J. ▲ OH; ○ styrene; ● flame temperature.

is fastest at 750 J. At 500 J the pyrolysis, even though at a low temperature, is faster than at the highest temperature. The pyrolysis of the monomer is a result of the absorption of the flash lamp energy by the styrene molecule. The molecule cannot dissipate the massive absorption of energy in less than a few molecular vibrations, and as a result scission of bonds takes place and radicals are formed. This process is similar to, but much faster than, the thermal scission resulting from conventional oven-type pyrolysis studies. It was just this similarity in bond-breaking activity that led Norrish and Porter and Ward to use flash photolysis as a means of studying fast combustion-type reactions.[21,30]

Another factor in favor of the photolytic simulation of monomer degradation is the fact that large amounts of radiation are emitted in polystyrene fires.[31,32] A considerable amount of radiation is available for absorption by the monomer as it leaves the polymer surface and diffuses toward the flame front.

The flame temperatures shown in Figs. 6–9 are all well above 2000° K. This is higher than the 1450° K observed for the combustion of unmodified

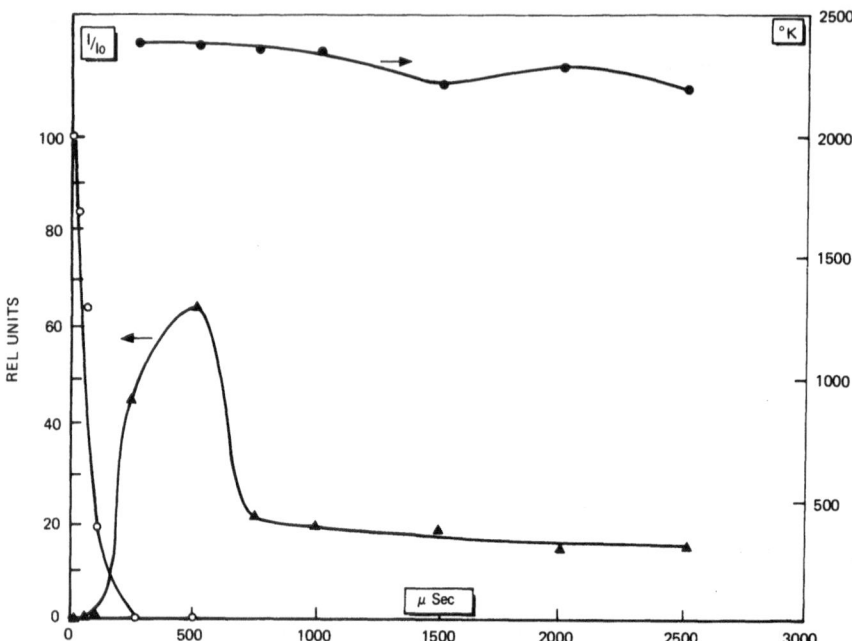

FIGURE 9. Disappearance of styrene monomer and the intensity of the OH radical formed in
the flash photolytic combustion of 2 mm styrene and 20 mm oxygen (left ordinate); flame temperature for same combustion (right ordinate); flash energy = 500 J. ▲ OH; ○ styrene; ● flame
temperature.

TABLE 7
Summary of Rate Constants for the Photolytic Decomposition of
Styrene Monomer in Oxygen

Fuel/oxygen	Flash energy (J)	k	T (°K)
10:1	1000	4.38×10^4	2100
7.5:1	1000	3.20×10^4	1900
5:1	1000	3.72×10^4	710
4:1	1000	3.66×10^4	470
10:1	750	2.45×10^4	1900
7.5:1	750	2.56×10^4	1200
5:1	750	2.83×10^4	650
10:1	500	1.61×10^4	800
7.5:1	500	1.41×10^4	650
5:1	500	1.20×10^4	450
4:1	500	9.54×10^3	380
10:1	406	6.49×10^3	500
7.5:1	406	7.22×10^3	400
5:1	406	5.25×10^3	330
10:1	1500	3.25×10^4	2250

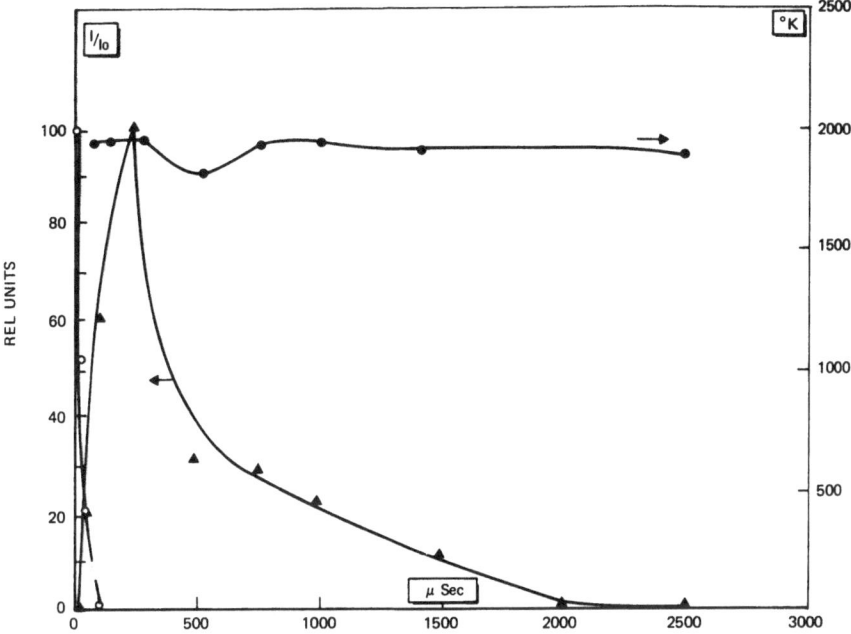

FIGURE 10. Disappearance of styrene monomer and the intensity of the OH radical formed in the flash photolytic combustion of 2 mm styrene and 15 mm oxygen (left ordinate); flame temperature for same combustion (right ordinate); flash energy = 1000 J. ▲ OH; ○ styrene; ● flame temperature.

polystyrene burning in a free ambient atmosphere. However, the same mechanism should prevail in either case.

Figures 10 through 12 show the time history of the OH radical and the disappearance of styrene monomer as a function of the reaction stoichiometry at a constant flash energy of 1000 J. These figures along with Fig. 7 show that as the oxygen/fuel ratio decreases, the flame temperature is lowered and becomes unstable at an oxygen/fuel ratio of 5:1. At the richest mixture the system does not burn completely. Table 7 is a summary of the data from the flash photolytic combustion of styrene in oxygen. The very rapid initial disappearance of the monomer, coincident with the lifetime of the flash (50 μsec), coupled with the observation that most of the monomer has disappeared before the OH radical has attained sufficient concentrations favors a unimolecular mechanism for the photolytic pyrolysis of the monomer. The pseudo-first-order rate constant for the disappearance of the monomer as shown in the third column of Table 7 appears to be a more sensitive function of the flash energy than of the reaction stoichiometry. The temperatures shown in the fourth column are the "flame" temperatures derived from the rotational distribution of the OH radicals observed at the end of the flash initiation.

Figure 13 shows the "flame" temperature as a function of the flash energy for successively richer mixtures of styrene burning in oxygen. The rapid increase in temperature above 500 J for the two leanest mixtures indicates that the ignition is achieved at about 600–700° K. A plot of ln k against the reciprocal of the flame temperature for the different oxygen/fuel ratios is shown in Fig. 14. A "pseudo"-activation energy can be calculated from this plot. The appearance of three separate slopes (labeled I, II, and III) for the leanest system suggests different mechanisms to be operable at flash energies above 750 J. This is not a too unreasonable assumption as the combustion duration is quite short at 1500 J. The absorption spectrum at this energy also reveals the presence of the species C_2 and C_3. Significant quantities of a fine gritty soot were removed from the reaction cell after each combustion.

The combustion studies carried out at 1000 J showed only trace amounts of C_2 and C_3. A fairly intense absorption spectrum attributed to the phenyl

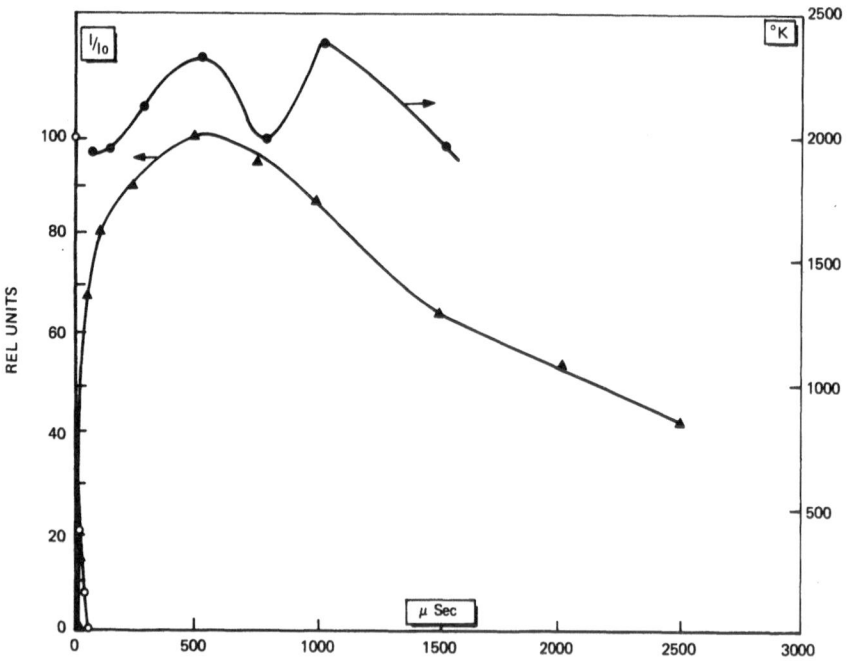

FIGURE 11. Disappearance of styrene monomer and the intensity of the OH radical formed in the flash photolytic combustion of 2 mm styrene and 10 mm oxygen (left ordinate); flame temperature for same combustion (right ordinate); flash energy = 1000 J. ▲ OH; ○ styrene; ● flame temperature.

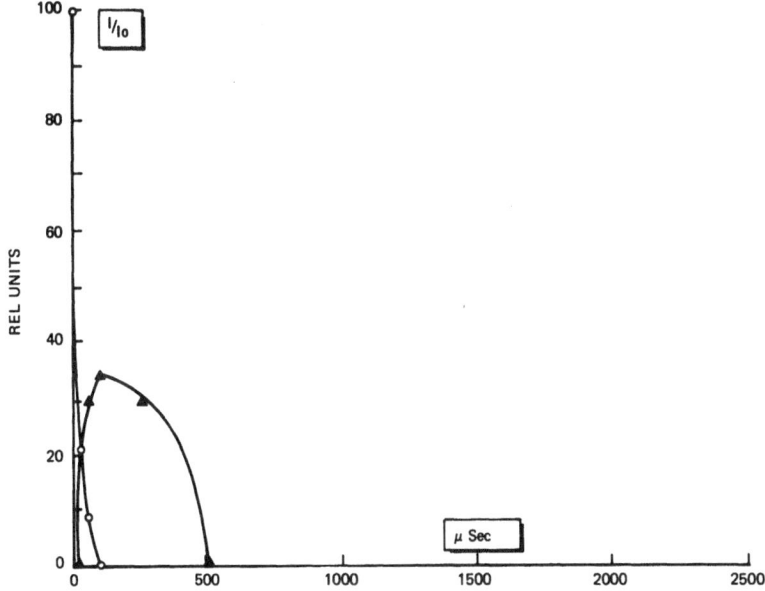

FIGURE 12. Disappearance of styrene monomer and the intensity of the OH radical formed in the flash photolytic combustion of 2 mm styrene and 8 mm oxygen; flame temperature for same combustion (right ordinate); flash energy = 1000 J. ○ sytrene; ▲ OH.

radical was observed early in the combustion.[30] An intense absorption spectrum of CH was also observed during the combustion. Except for the richest mixture, little if any soot was recovered from the reaction cell.

As the flash energy was lowered, the amount of soot formed during the combustion increased as the mixture became successively richer. The soot was soft and appeared to be stringy in nature. A strong odor of styene was noted upon examination of the soot. At the lowest flash energies, little decomposition of the monomer was observed.

The activation energies calculated for the three regions of the leanest combustion are:

Region I, $E_a = -14.9$ kcal
Region II, $E_a = 23.1$ kcal
Region III, $E_a = 1.8$ kcal

The activation energies for the other systems are:

Oxygen/fuel = 7.5:1, $E_a = 1.5$ kcal
Oxygen/fuel = 5:1, $E_a = 2.4$ kcal
Oxygen/fuel = 4:1, $E_a = 5.4$ kcal

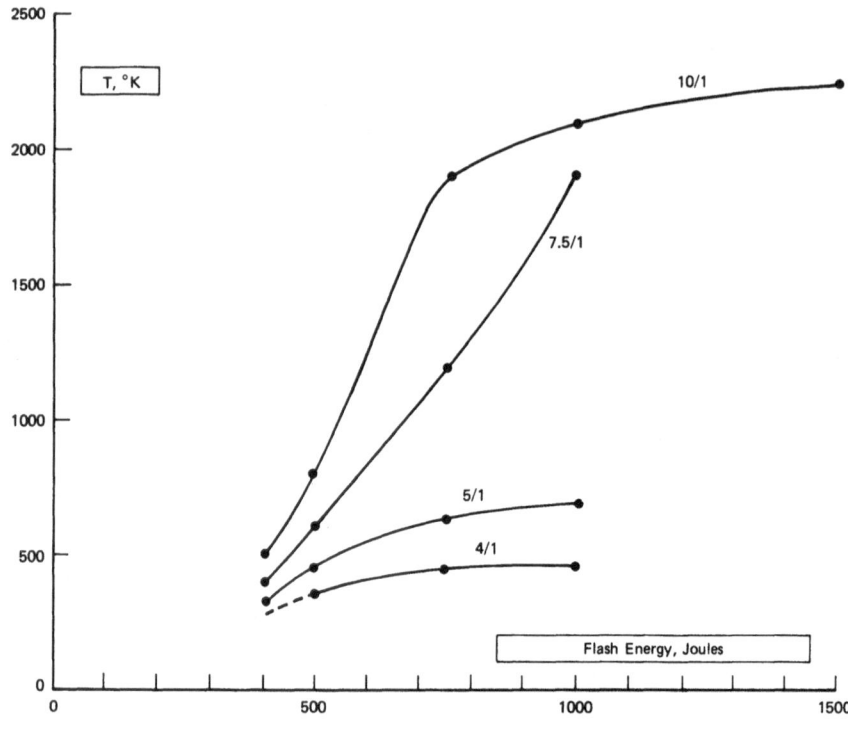

FIGURE 13. "Flame" temperature vs. flash energy.

Except for the uncertain activation energy in region I for the leanest mixture, the activation energies are considerably less than the 55 kcal for the pyrolysis of polystrene as reported by Madorsky.[16]

It can be concluded from the curves that a similarity of reaction order exists for the combustion of monomer below 750 J and that any monomer found during the pyrolysis of polystyrene will rapidly dissociate. The decomposition of the monomer is highly temperature dependent both as to its mode and rate of decomposition as shown by the spectroscopic observation noted above.[14]

4. The Mechanisms of Flame Retardation

4.1. Theories of Flame Retardancy

The five generally accepted mechanisms by which flame retardants function under combustion conditions are:

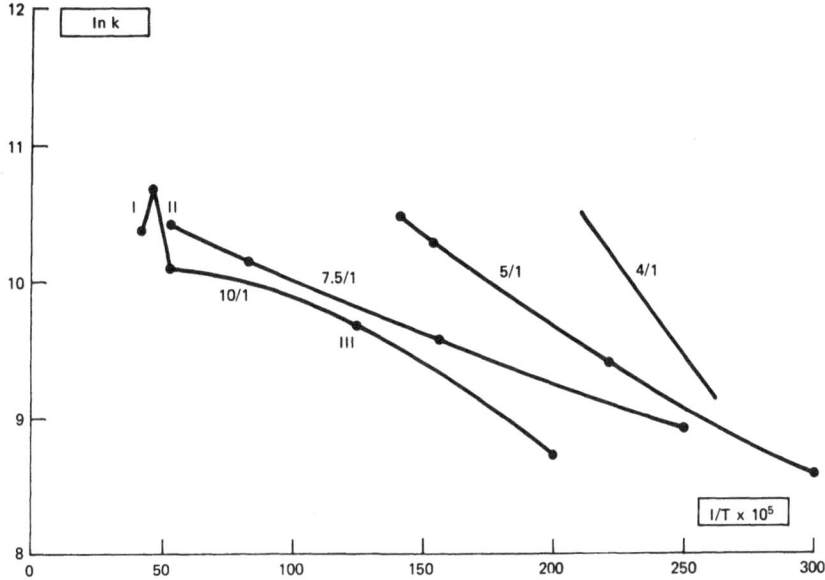

FIGURE 14. ln k vs. $1/T \times 10^5$ for the combustion of styrene in oxygen.

1. *The Gas Theory.* Large volumes of noncombustible gases are produced which dilute the oxygen supply to the flame and/or dilute the fuel concentration needed to sustain the flame.
2. *The Thermal Theory.* The endothermic degradation of the flame retardant lowers the polymer surface temperature and retards pyrolysis of the polymer. Also the degradation products of the agent can react endothermically with flame species and lower the temperature of the flame.
3. *The Chemical Theory.* The flame retardant dissociates into free-radical acceptors which compete with the chain-propagating steps of the combustion process.
4. *The Coating Theory.* The pyrolyzing flame retardant forms a protective liquid or char barrier which minimizes transpiration of polymer degradation products to the flame front and/or acts as an insulating layer to reduce the heat transfer from the flame to the polymer surface.
5. *The Physical Theory.* The flame retardant acts as a thermal sink to increase the heat capacity of the combustion system or to reduce the fuel content to a level below the lower limit of flammability. Either or both mechanisms will inhibit the combustion and extinguish the flame.

The flame-inhibiting effects of $CaCO_3$ and $Al_2O_3 \cdot 3H_2O$ are examples of the gas theory in action as the agents dissociate to noncombustible gases which dilute the flame reactions:

$$CaCO_3(s) \longrightarrow CaO(g) + CO_2(g) \tag{9}$$

and

$$Al_2O_3 \cdot 3H_2O(s) \longrightarrow Al_2O_3(s) + 3H_2O(g) \tag{10}$$

The endothermic decomposition of these inorganic salts are also examples of the thermal and physical theories of flame retardation.

The coating theory is exemplified by the actions of organophosphorous compounds which degrade to polyphosphoric acids. Intumescent coatings are also examples of the coating theory.

The physical theory of flame retardation is generally held in disfavor but has recently received renewed interest primarily due to the work of Larsen.[33,34]

The most widely held theory of flame retardation is the chemical theory. The theory presupposes that the flame retardant dissociates into species which can remove the free-radical intermediates from the flame reaction or which can replace the reaction species with those which do not propagate the flame.

The halogen-containing flame retardants operate by the chemical theory mechanism. Their specific applications to polystyrene systems are dictated by the following:

1. The chemical composition of the agent should be such that it decomposes at a slightly lower temperature than the flash ignition temperature of the polymer pyrolysis products.
2. It should have a minimal effect upon the physical properties of the polymer at the recommended dose level during fabrication and end use.
3. It should be compatible with the other components of the polymer system.
4. It should not present a health or environmental problem during fabrication, use, and disposal.

Specific flame retardants for a given end use for polystyrene polymers and copolymers have been tabulated recently[2-4] and will not be reported here. We shall concern ourselves with the mechanisms by which selected flame-retardant agents inhibit hydrocarbon flames in general and polystyrene flames in particular.

4.2. The Mechanisms of Halogen Flame Retardants

The combustion of any hydrocarbon fuel can be visualized as being comprised of two concurrent and interacting combustion systems. One system is the well-known hydrogen–oxygen combustion system, and the other is the hydrocarbon pyrolysis system. The former system is characterized and is propagated by the species $\cdot O$, O_2, $\cdot H$, H_2, and $\cdot OH$ and at low temperatures also by $HO_2 \cdot$. The latter system is complex, possibly comprising all combinations of carbon with hydrogen and/or oxygen with the constraint that carbon be tetravalent, oxygen divalent, and hydrogen monovalent. Fortunately, most hydrocarbon combustion systems are investigated by following the concentrations of the parent molecule(s), selected easily observed intermediates (including $\cdot OH$ and CO among others), and the final combustion products.

Most hydrogen-containing combustion systems, including the hydrocarbons, are generally described and quantified in terms of the H_2–O_2 reaction scheme and the perturbations imposed on it by the structure and stability of the hydrogen-containing fuel.

Hydrocarbons, including styrene, pyrolyze to species capable of reacting with the oxygen in the air and initiating the hydrogen–oxygen reaction scheme which propagates the hydrocarbon combustion.[35,36]

The hydrogen–oxygen combustion contains the following reactions:

$$\cdot H + O_2 \rightleftharpoons \cdot OH + O\cdot \tag{11}$$

$$\cdot O + H_2 \rightleftharpoons \cdot OH + H\cdot \tag{12}$$

which dominate the combustion process because of their chain-branching nature. The reaction

$$\cdot OH + CO \rightleftharpoons CO_2 + H \tag{13}$$

is the primary reaction by which carbon monoxide is converted to carbon dioxide. The important thing to note is that if the hydrogen–oxygen reaction scheme can be slowed down or stopped the combustion of the hydrocarbon fuel can likewise be kept at a minimum.

Halogenated flame retardants are postulated to function primarily by the following mechanisms, which produce HX, the actual flame-inhibiting specie,

$$M - X \rightleftharpoons M' + X\cdot \tag{14}$$

where M is the parent fire-retardant molecule and X is the halogen, usually Cl or Br, followed by

$$RH + \cdot X \rightleftharpoons R\cdot + HX \tag{15}$$

or by

$$\cdot X + H_2 \rightleftharpoons HX + H\cdot \qquad (16)$$

It would be advantageous if the flame retardant decomposed to HX rather than to X, as step (15), which actually promotes the combustion, would be negated. The "ideal" FR (flame-retardant) agent should decompose by

$$MX \rightleftharpoons HX + N \qquad (17)$$

where N is the residue of the agent which lost the HX. A multi-halogen-containing agent should repeat step (17) as many times as there are halogens in the molecule. Unfortunately, this is not the case, and 100% utilization of the halogen is not realized. The hydrogen halide has been shown to be the actual flame inhibitor in gas phase experiments.[37-39]

The flame-inhibiting effects of HX are manifested through reactions which inhibit the chain-branching step of the hydrogen–oxygen combustion systems, such as

$$\cdot H + HX \rightleftharpoons H_2 + X\cdot \qquad (18)$$

and

$$\cdot OH + HX \rightleftharpoons H_2O + X\cdot \qquad (19)$$

The effectiveness of the hydrogen halides have been shown to be HI $>$ HBr $>$ HCl $>$ HF[38,39] on a volumetric basis.

Practical considerations of stability at processing temperatures and end-use applications preclude the use of iodine compounds as flame retardants. The great reactivity of the fluorine atom produced by reactions (18) and (19) preclude the use of fluorine FR agents as the net result would be no diminution of active chain-branching centers in the flame-inhibiting process.

The remaining choice of whether to use a bromine- or a chlorine-containing flame retardant is left to the formulator to resolve in terms of thermal stability, compatibility, and economics. Bromine-containing flame retardants are being used increasingly in polystyrene systems.

Recent studies have shown that for bromine-containing systems reaction (18), not (19), is the predominant inhibiting reaction in the flame because of the prevailing (H_2/OH) ratio in the flame front and also because reaction (18) is twice as fast as reaction (19).[40,41]

If we assume that the primary step in the inhibition of hydrocarbon flames is the removal of "active" hydrogen atoms and their replacement by "inactive" halogen atoms, then the effectiveness of a halogen flame retardant can be measured by the rate at which HX competes with O_2 for hydrogen atoms, that is, between reactions (18) and (11):

$$\cdot H + HX \rightleftharpoons H_2 + X\cdot$$

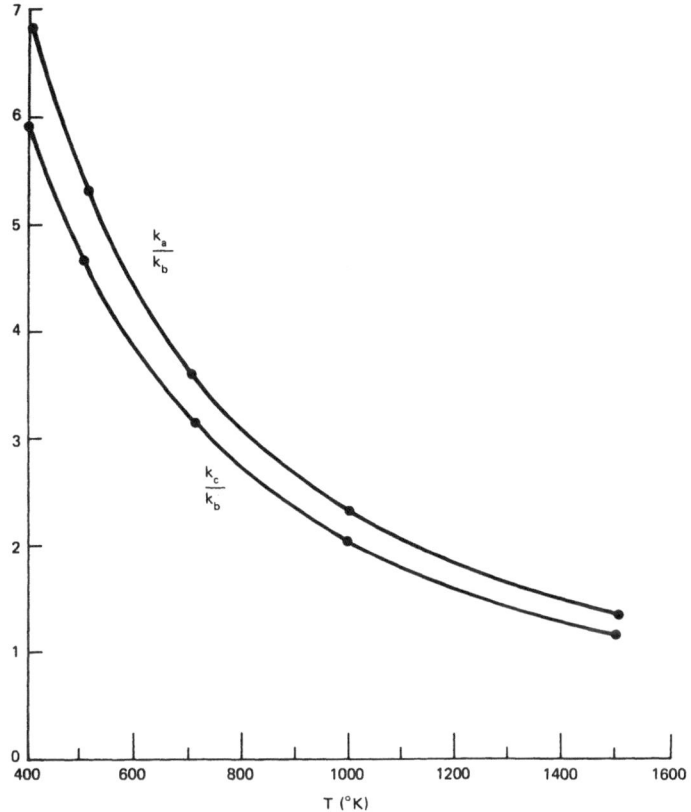

FIGURE 15. Reaction of H atom with O_2, HBr, and HCl. (a) H + HBr = H_2 + Br; (b) H + O_2 = OH + O; (c) H + HCl = H_2 + Cl.

and

$$\cdot H + O_2 \rightleftharpoons \cdot OH + O\cdot$$

Figure 15 shows that both HCl and HBr can compete effectively with O_2 for hydrogen atoms in the temperature range of interest in polymer combustion. Reaction (11) propagates the combustion by producing two free radicals for each H atom consumed. Reaction (18) produces an "inactive" free radical (inactive in that it is not a part of the hydrogen–oxygen reaction scheme) for each hydrogen atom consumed. The data are from Fristrom and Westenberg.[42]

Another way of looking at the relative efficiencies of HBr and HCl as flame inhibitors is to compare the forward reaction shown in (18) to its reverse, which would regenerate H atoms consumed in (18) via

$$H_2 + X\cdot \rightleftharpoons HX + H\cdot \tag{18a}$$

For HBr this is

$$\frac{k_{(18)}}{k_{(18a)}} = K_{HBr} = 0.374 \exp\left(\frac{16{,}760}{RT}\right)$$

while for HCl the ratio is

$$\frac{k_{(18)}}{k_{(18a)}} = K_{HCl} = 0.583 \exp\left(\frac{1097}{RT}\right)$$

where K_{HBr} and K_{HCl} are the equilibrium constants for the reactions. Numerical calculations show that in the temperature range 500–1500° K, K_{HBr} falls from 8×10^6 to 1×10^2, while K_{HCl} falls from 1.75 to 0.8, showing the greater effectiveness of HBr in removing H atoms from the reaction zone. The curves shown in Fig. 15 also show the greater effectiveness of HBr as a flame inhibitor. Both HBr and HCl are very effective at low temperatures coincident with the ignition temperatures for polymers. This is also the reason for the effectiveness of HBr and HCl in small-scale (low-heat) tests.

Laboratory studies carried out on the flash photolytic combustion of hydrogen and oxygen inhibited with halogen compounds showed the following sequence of effectiveness on a volumetric basis:

$$Br_2 > HI \geqslant HBr > HCl$$

The seemingly great effectiveness of Br_2 can be attributed to the observation that for this system, and probably for any system containing large amounts of hydrogen, 2 moles of HBr are formed for each mole of Br_2 present:

$$Br_2 + H_2 \rightleftharpoons 2HBr \tag{20}$$

The unpublished flash photolysis work is in essential agreement with the studies carried out by Nicholas and Norrish,[43] who showed that the hydrogen halide was formed during the induction time to combustion when Cl_2 or Br_2 was incorporated into the H_2–O_2 combustion system. They also showed that the appearance of the hydroxyl radical was delayed until almost all of the bromine was consumed to form HBr and that the HBr had a greater inhibiting effect than did HCl.

The effectiveness of halogen flame retardants for hydrogenic fuels can be summarized as follows:

1. The effectiveness follows the sequence $I > Br > Cl > F$ on a molar basis.
2. The actual flame retardant seems to be the hydrogen acid (HX).
3. The effectiveness of the acid is related to how well it removes H atoms from the flame zone by the reaction $HX + \cdot H \rightleftharpoons H_2 + X\cdot$, which competes with $\cdot H + O_2 \rightleftharpoons \cdot OH + O\cdot$.

5. The Effects of Halogen Flame Retardants on the Combustion of Polystyrene and Styrene

The addition of halogenated flame retardants to polystyrene polymers modifies the flammability in that the burning rate of the treated polymers in ambient air decreases in some direct proportion to the amount of retardant added.

The limiting oxygen index (LOI*) test of Fenimore and Martin[44] relates the agent content of a polymeric system to the volume percent oxygen, in an oxygen–nitrogen atmosphere, necessary to sustain combustion of the polymer. The greater the oxygen content required (the LOI), the more resistant the polymer is to sustained combustion under the defined conditions. An example of the effect that different amounts of a halogen flame retardant have upon the flammability of high-impact polystyrene and ABS, as determined by the LOI technique, is shown in Figs. 16 and 17, respectively, along with the UL-94* ratings for the polymers. The effect of antimony oxide can also be seen in the figures.

The figures show that the oxygen index increases as the halogen content of the polymer increases and that the effect is enhanced by the addition of the antimony oxide. The synergistic effect of Sb_2O_3 and other materials will be discussed in the next section.

The multitude of different flame retardants used in polystyrene polymers, to various degrees of success and satisfaction, have been adequately treated elsewhere and will not be repeated here.[2-4] Instead we shall direct our attention to the interactions of the pyrolyzing polymer and nascent monomer with the halogen components of the agents incorporated into the polymers.

Flash photolytic studies of the combustion of polystyrene and styrene monomer in oxygen in the presence of gaseous halogen compounds showed that the different halogen materials inhibit the pyrolysis–combustion in various ways. The halogen compounds included in the study were HI, HCl, HBr, and Br_2. The agents had no effect upon the depolymerization–gasification of the polymer to monomer but did have an effect upon the reactivity of the monomer. Thus, we can conclude that the combustion-inhibiting effect of halogen flame retardants is a gas phase phenomenon.

To quantitatively measure this effect, the combustion of styrene monomer in a stoichiometric amount of oxygen was studied at various levels of halogen concentration. The presence of HBr slowed down the rate of pyrolysis of the monomer. It also lengthened the induction time to combustion in proportion to the amount of HBr present. The HBr seems to affect the initiation steps to combustion more than it does the propagation steps, as once the sample is ignited it burns in a manner similar to that for uninhibited

*Small-scale test.

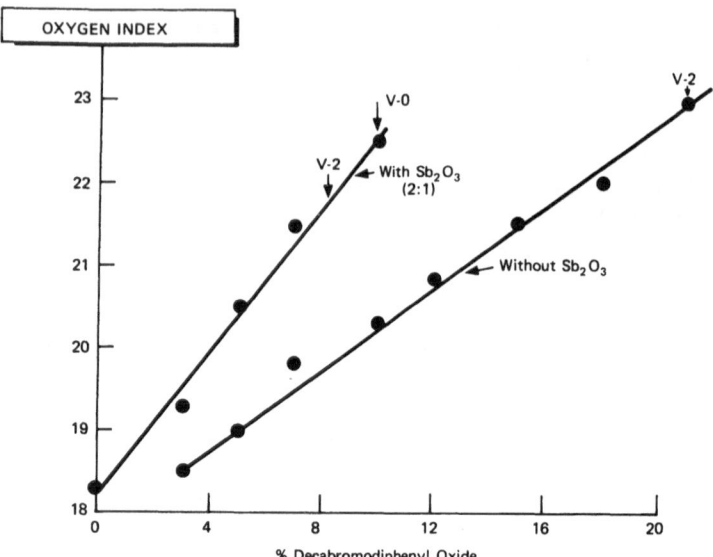

FIGURE 16. Oxygen index and UL-94 ratings of an impact polystyrene containing decabromo-diphenyl oxide as the flame-retardant agent and Sb_2O_3 as a synergist. (This numerical assessment of combustibility is not intended to reflect hazards presented by the materials under actual fire conditions.)

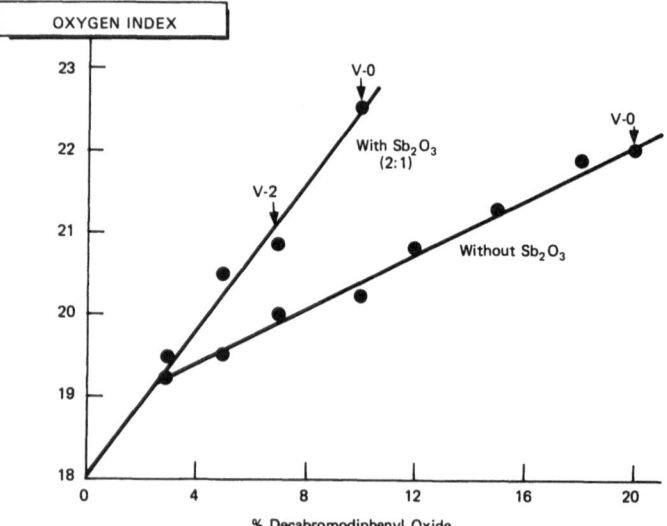

FIGURE 17. Oxygen index and UL-94 ratings of ABS containing decabromodiphenyl oxide as the flame-retardant agent and Sb_2O_3 as a synergist. (This numerical assessment of combustibility is not intended to reflect hazards presented by the materials under actual fire conditions.)

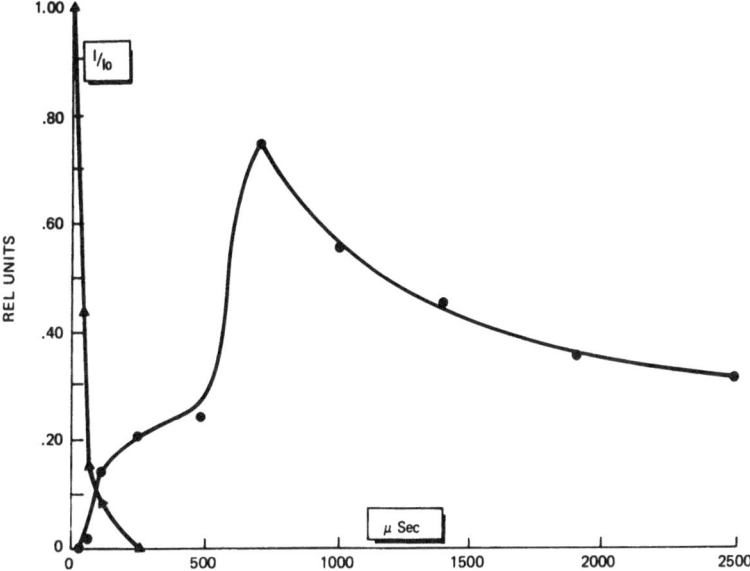

FIGURE 18. Disappearance of styrene and the time history of the OH radical formed during the combustion of 2 mm styrene + 20 mm O_2 + 4 mm HBr; flash energy = 1000 J. ● OH; ▲ styrene.

styrene. An example of this can be seen in Fig. 18. This system differs from that shown in Fig. 7 only by the presence of 4 Torr HBr. The inhibiting effect of HBr can be attributed to its reaction with H atoms via (18) to inhibit the H_2–O_2 reaction:

$$\cdot H + HBr \rightleftharpoons H_2 + Br \cdot \qquad (18b)$$

The styrene monomer pyrolyzed at an accelerated rate in the presence of HI. This can be due to the attack of free iodine upon the monomer supplementing the thermal pyrolysis of the monomer. The low H—I bond energy (71 kcal) would produce considerable atomic iodine easily in the pyrolysis. The accelerated rate of pyrolysis would result from

$$C_8H_8 + \cdot I \rightleftharpoons C_8H_7 \cdot + HI \qquad (15a)$$

The combustion-inhibiting effect of HI seems to be based on its ability to destroy the chain-propagating steps rather than the initiation steps as the induction time to peak is little affected by the HI, as seen in Fig. 19. On a volumetric basis HI is nearly twice as effective a flame inhibitor as HBr.

Hydrogen chloride seems to function more as an inert diluent to the styrene–oxygen combustion than as an active flame inhibitor. A similar behavior was noted in the hydrogen–oxygen combustion system.

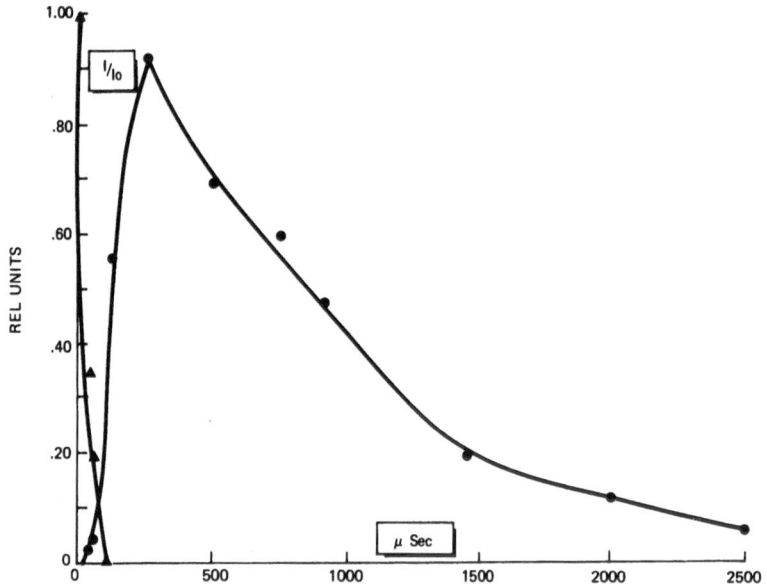

FIGURE 19.   Disappearance of styrene and the time history of the OH radical formed during the combustion of 2 mm styrene + 20 mm O_2 + 1 mm HI; flash energy = 1000 J. ● OH; ▲ styrene.

The addition of bromine to the styrene–oxygen combustion system shows the bromine to be both a pyrolysis promoter and a flame inhibitor. The monomer disappears at an accelerated rate in the presence of bromine. The rate of disappearance of the monomer increases as the bromine level increases at low bromine levels. This can be seen in Figs. 20 and 21, which show the effect of 1 and 2 Torr Br_2 on the styrene–oxygen, respectively.

The bromine has no effect upon the high-temperature ignition of the styrene, as the time to peak OH concentration is little affected by the bromine (compare Fig. 7 to Figs. 20 and 21). However, the bromine has a marked effect upon the propagation of the flame, as indicated by the rapid decline in OH concentration from the peak value. Figure 21 shows the inhibiting effect of the bromine on the flame and the unsteady nature of the flame as indicated by the OH· radical. It was impossible to observe combustion at bromine levels in excess of 2 Torr.

The pyrolysis-promoting nature of low levels of halogen has been documented (Ref. 36, p. 682). The bromine most likely reacts early in the pyrolysis with both the styrene via

$$C_8H_8 + \cdot Br \ \rightleftharpoons\ C_8H_7\cdot + HBr \qquad (15b)$$

and with the hydrogen from the pyrolysis of the monomer via

$$H_2 + Br\cdot \ \rightleftharpoons\ HBr + H\cdot \qquad (16a)$$

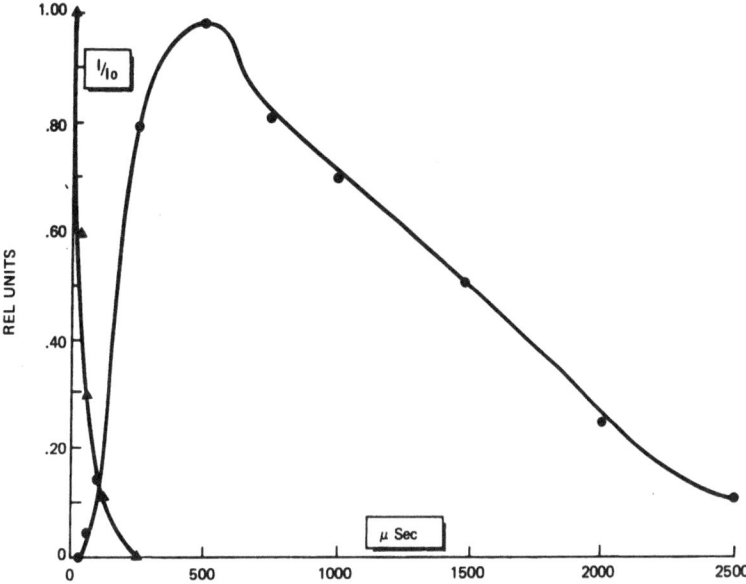

FIGURE 20. Disappearance of styrene and the time history of the OH radical formed during the combustion of 2 mm styrene + 20 mm O_2 + 1 mm Br_2; flash energy = 1000 J. ● OH; ▲ styrene.

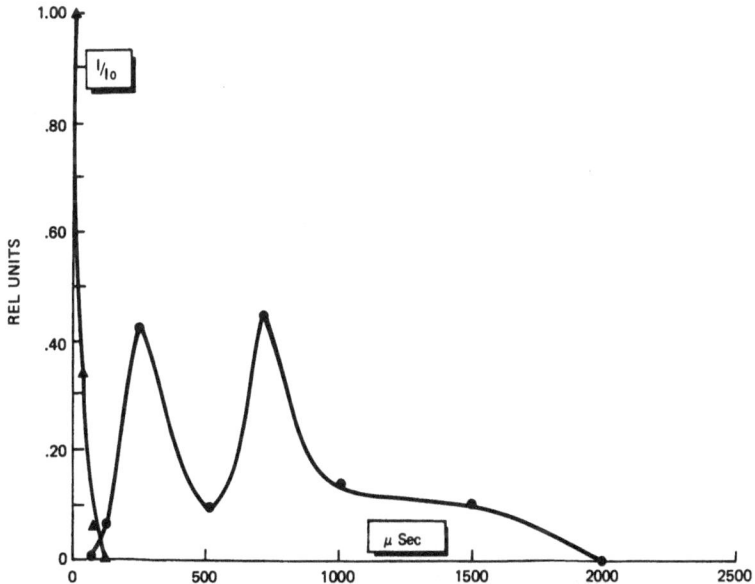

FIGURE 21. Disappearance of styrene and the time history of the OH radical formed during the combustion of 2 mm styrene + 20 mm O_2 + 2 mm Br_2; flash energy = 1000 J. ● OH; ▲ styrene.

and

$$\cdot H + Br_2 \rightleftharpoons HBr + Br\cdot \qquad (21)$$

to produce HBr which after ~ 200 μsec acts as the flame inhibitor. The great effectiveness of bromine as a flame inhibitor in the styrene combustion system can be attributed to reactions (16a) and (21), which produce 2 moles of HBr for each mole of Br_2 reacting.

Aliphatic bromine compounds tend to dissociate at a lower temperature than do aromatic bromine compounds.[45] They also tend to produce HBr at lower temperatures than do aromatic compounds. The lower thermal stability and earlier formation of HBr are the probable reasons for the greater effectiveness of aliphatic bromine compounds over aromatic bromine compounds to inhibit polystyrene combustion, as shown by Green and Versnel.[46]

The pro-combustion tendencies of small amounts of halogen compounds has been observed in other combustion situations and is not unique to polymeric combustion.[36,47,48] The flammability limits of propane are widened by the addition of less than 5% CH_3Br.[48]

Two general rules, exclusive of health and environmental considerations, should be observed in choosing a halogen flame retardant for use in styrenic polymers.

1. The agent should dissociate at a temperature considerably above the processing temperature of the polymer *but* slightly below the decomposition temperature or flash ignition temperature of the polymer.
2. The agent should be so chosen that HBr is the favored pyrolysis product of the polymer–agent system. If the polymer yields considerable hydrogen upon pyrolysis, then the agent need only supply bromine. HBr will be formed prior to ignition, as the reaction of hydrogen with bromine is considerably faster than with oxygen.[43] If the polymer yields little or no hydrogen upon pyrolysis, the agent should decompose to HBr rather than bromine, as the latter will act as a combustion promoter by abstracting hydrogen from the pyrolysis products by

$$RH + Br \rightleftharpoons R\cdot + HBr \qquad (15c)$$

   thereby producing an additional free radical for each bromine atom produced.

6. The Role of Synergists in the Combustion of Styrenic Materials Inhibited by Halogen Compounds

The flame-inhibiting behavior of many halogen agents was found to be greatly enhanced by a number of compounds. These compounds are given the general name "synergists." The most widely used synergist is antimony oxide. An example of the synergistic effect of Sb_2O_3 is seen in Figs. 16 and 17 where the oxygen index for impact styrene and ABS are significantly increased by the addition of one part Sb_2O_3 to two parts agent. The addition of the synergist to impact polystyrene gave a UL-94* rating of V-0 at an agent level of 10%, while only a V-2 rating could be achieved without the synergist at an agent level of 20%. The synergistic effect of Sb_2O_3 in ABS is equally dramatic.

The work of Eichhorn[49] is still considered the definitive study on the effects of free-radical synergists in polystyrene foams with halogen-containing additives. Little information on the mechanism of synergists exists. The work cited[49] can be considered the primer for those interested in the relationships among types of synergists, halogen compounds, and the practical temperature limits of the halogen–synergist combination in polystyrene foams.

It has recently been suggested that antimony halides function as gas phase inhibitors of combustion through two separate but related steps.[50] The antimony halides result from the reactions between the oxide and the halogenated agent in or near the polymer surface. The antimony halides react with the hydrogen atoms in the cooler region of the flame to produce HX and a lower halide. This reaction continues until the metal is devoid of all halogen. The metal can react with the oxygen or OH to produce the oxide, SbO, which acts as a catalyst for the recombination of H atoms. The HX meanwhile undergoes reactions with H atoms to produce X and H_2, as previously stated. The combined effect of the antimony and halogen is to remove H atoms from the reaction zone and effectively inhibit the combustion process. Flash photolytic studies of polystyrene containing pentabromomonochlorocyclohexane and antimony oxide have shown SbO, SbOCl, Sb_2, and Sb to be present in the gas phase early in the combustion. No SbH_3 was detected in the final products, nor was SbH observed during the combustion.

The sequence of reactions involving antimony can be visualized as

$$Sb_2O_3 + 6HX \rightleftharpoons 2SbX_3 + 3H_2O \qquad (22)$$

$$SbX_3 + H \rightleftharpoons SbX_2 + HX \qquad (23)$$

$$SbX_2 + H \rightleftharpoons SbX + HX \qquad (24)$$

$$SbX + H \rightleftharpoons Sb + HX \qquad (25)$$

followed by

$$Sb + O + M \rightleftharpoons SbO + M \qquad (26)$$

$$Sb + OH + M \rightleftharpoons SbOH + M \qquad (27)$$

*Small-scale test.

$$SbO + H \rightleftharpoons SbOH \qquad (28)$$

$$SbOH + H \rightleftharpoons SbO + H_2 \qquad (29)$$

$$SbO + X \rightleftharpoons SbOX \qquad (30)$$

where reactions (23)–(25) produce the HX and reactions (26)–(29) show the recombination of H atoms on the oxide surface.

The demand for higher thermal stability has severely limited the use of free-radical synergists such as the peroxides, hydroperoxides, and azo compounds. These materials are still found in polystyrene foams such as foaming-in-place beads or expandable packaging foams. The steps by which these organic free-radical-producing synergists function is still an area to be explored. A few experimental studies have been carried out in this area to investigate the mechanisms. Many more studies have been carried out to find new synergists or improved synergism by combining known synergists with new flame-retardant agents.

The four leading hypotheses to explain the synergistic effects of free-radical initiation are the following:

1. The free-radical initiator increases the rate of decomposition of the halogen flame retardant or initiates it at a lower temperature. There is some evidence that free-radical initiators do accelerate the decomposition of organic bromides and chlorides.[49]
2. The free-radical initiator working alone or in conjunction with the halogen compound lowers the temperature at which oxidation of the polymer occurs. In this way the decomposition of the polymer might be more closely matched to that of the flame retardant.[49]
3. The free-radical initiator accelerates the decomposition of the polymer and promotes reaction of the degradation products with the halogen compound in the condensed phase. The resulting gaseous products will be halogenated fragments instead of containing just carbon and hydrogen.[49]
4. The free-radical initiator decreases the melt viscosity of the molten polymer. The resulting low-viscosity fluid flows rapidly away from the flame front, carrying heat with it and effectively removing some of the polymer from the heat flux of the flame.[51]

Whatever the mechanisms by which selected synergists interact with the flame-retardant agent and the polymer substrate, the ability of the synergist to increase the flame inhibition behavior of the agent is recognized and used extensively in practice. A systematic study of the mechanisms by which various synergists interact with the great multitude of flame-retardant agents will be a costly and arduous job. A study of the effects rather than the chemistry, such as carried out by Deets[52] for the antimony–halogen interaction, is pertinent and pragmatic.

7. The Use of Phosphorus Compounds as Flame Retardants for Polystyrene

Phosphorus compounds other than a few containing halogen such as tris(2,3-dibromopropyl) phosphate are not extensively used as flame retardants in polystyrene. The phosphorus compounds are principally used in polymeric systems which form a solid lattice upon pyrolysis. The lattice acts as a support for the viscous polyphosphoric acid syrup, which provides the primary benefit of the agent. The phosphorus agents show greater efficiency in highly cross-linked polystyrenes.

A further discussion of the structures and flame-inhibiting mechanism of phosphorus-based flame retardants will not be attempted here as this information can be found in Chapter 2 in Volume 1 written by Lewin and Sello.

Phosphorus atoms could be the carrier of strong Lewis acid moities which could function as *in situ* cross-linking agents at pyrolysis conditions. The acid portion of the molecule, when activated by the heat, would initiate cross-linking in the polystyrene, resulting in enhanced char formation and decreased volatilization. The char would serve as a physical support for the formation of the viscous polyphosphoric acid. Such a flame retardant, if constructed, would be its own synergist. This may be the greatest opportunity for phosphorus chemistry to penetrate the polystyrene flame-retardant area.

8. Test Methods and Their Usefulness to Combustion Studies

The most commonly used laboratory test methods to rate the degree of flame inhibition imparted to a styrenic polymer by either structural modification or incorporation of flame-inhibiting chemicals depend on whether the polymer is in the solid or foamed configuration.[53]

The flammability characteristics of cellular plastics used in building construction and of low-density cellular plastics used in furniture are tested under numerous test methods and standards. Included among these are ASTM D-568, 635, 757, 1433, 1692, E-84, 162, and 286; UL 94 and 723; and NFPA 255. [The Federal Trade Commission considers that these standards are not accurate indicators of the performance of the tested materials under actual fire conditions and that they are valid only as a measurement of the performance of such materials under specific, controlled test conditions. The terminology associated with the above tests or standards, such as "nonburning," "self-extinguishing," "noncombustible," or "25 (or any other) flame spread," is not intended to reflect hazards presented by such products under actual fire conditions. Moreover, some hazards associated with numerical flame-spread ratings for such products derived from test methods and standards may be significantly greater then those which would be expected of other products with the same numerical rating.]

The ASTM D-635* or its equivalent UL-94* is the test most commonly used for solid plastics. The test employs a horizontal sample ignited by a Bunsen flame. The flammability ratings are based on many factors, including the rate of flame spread, whether the sample drips, and whether the drip causes a flame capable of igniting a sample of cotton under the test specimen.

The limiting oxygen index test—or ASTM D-2863, as it is now known—has been used to test the relative flammability of many materials. It has been used for self-supporting solids, foams, fibers, liquids, and even gases. The general acceptability and ease of the test method have led many people to abuse the test and to give its rating significance far beyond their range of application. Materials which melt during combustion or materials whose melt carries the flame down the specimen must be given special treatment in order to make their data relevant. This problem was described by Fenimore and Martin in 1966, but seemingly to no avail.[44] The uses and abuses of LOI test data have been discussed many times and will not be included here except to point out two facts. First, the LOI test is a test designed to measure flame propagation, not ignition. Second, as pointed out by Friedman[54] and deRis,[55] among others, the flammability ratings observed in small-scale laboratory tests are not indicative of how the materials will perform in full-scale tests where large heat fluxes and enhanced radiation transfer have a dominating influence.

Studies similar to those carried out by Tewarson and Pion[27,28] should be undertaken to determine the scaling laws and factors between small-scale laboratory tests and large-scale controlled burning tests, such as the FM corner and crib tests.

Further insights into the chemistry of the initiation of the combustion phenomenon can be obtained if very fast heating rates, such as are available through flash pyrolytic[13] and laser-induced ignitions, are coupled to fast analytical apparatus such as the time-of-flight mass spectrometer, kinetic spectroscopy, or the electron spin resonance techniques.

The Aminco-NBS smoke density chamber is receiving considerable interest as a versatile testing apparatus. In addition to measuring smoke generated from burning polymers, it is being used as an animal exposure chamber and as a closed combustion chamber for the analysis of pyrolysis and combustion products. It is being fitted with vertically mounted heaters to study the combustion of low-viscosity and liquid polymers. It is being fitted

*Small-scale test.

with extremely high-flux heaters. It is being vented at a controlled rate. The NBS chamber is a versatile piece of equipment that does lend itself to these many diverse modifications and tests. However, the temptation to make it the universal test, as was tried with the LOI test, should be avoided. A little prudence is called for, lest the mistakes of the past be made again.

9. Conclusions

The combustion of styrenic polymers seems to be a two-step process. The first step is the thermal degradation of the polymer to volatile pyrolysis products of which the monomer seems to be the largest fraction. The second step is the gas phase combustion of the pyrolysis products. The heat released in this second step is conveyed to the polymer surface, primarily by radiative transfer, where it causes additional pyrolysis and volatilization to sustain the flame. Any interruption or diminution of the steps in this closed-loop process will decrease the flammability of the polymer.

Halogenated flame retardants operating in the gas phase on the flame-propagating species are the most efficient flame retardants for styrenic polymers. Bromine compounds are the most effective of the halogen compounds based on chemical activity in the flame front and thermal stability to withstand processing conditions. Hydrogen bromide (HBr) seems to be the actual flame-inhibiting specie. The activation energy for dissociation of the monomer expelled from the surface is small compared to the activation energy for pyrolysis of the solid polymer. Little additional energy is needed to cause the pyrolysis products to ignite.

An important factor in the choice of flame retardants for styrenic polymers is whether the polymer yields hydrogen upon pyrolysis. If it does, the agent need only produce bromine to generate the HBr needed for flame inhibition. If the polymer yields little or no H_2 upon pyrolysis, the agent should furnish the HBr for maximum efficiency.

In situ cross-linking agents are needed to improve the flame retardance of styrenic polymers. This is an area where phosphorus compounds may be of great value.

Antimony compounds are and will continue to be the synergist of choice for halogenated flame retardants in polystyrene fabricated by injection molding or extrusion. Organic free-radical initiators cannot stand these processing temperatures and will be of limited value.

There exists a need to correlate small-scale laboratory tests to large-scale fire situations. The large thermal and radiative fluxes associated with styrene fires cannot be ignored in laboratory tests.

10. References

1. Anon., *Mod. Plast.* **53** (1), 44 (Jan. 1976).
2. J. W. Lyons, *The Chemistry and Uses of Fire Retardants*, Wiley, New York (1970).
3. C. J. Hilado, *Flammability Handbook for Plastics*, 2nd ed., Technomics, Westport, Conn. (1974).
4. W. C. Kuryla and A. J. Papa, *Flame Retardance of Polymeric Materials*, Dekker, New York (1973).
5. R. H. Boundy and R. F. Kryer, *Styrene, Its Polymer, Copolymer and Derivatives*, ACS Monograph No. 115, Holt, Reinhart and Winston, New York (1952).
5a. R. F. Boyer, *Encyclopedia of Polymer Science and Technology*, Vol. 13, Wiley, New York (1970).
6. D. W. Van Krevelen, *Properties of Polymer*, American Elsevier, New York (1972).
7. D. R. Stull, The Dow Chemical Company, unpublished data (1958).
8. M. Lewin, S. M. Atlas, and E. M. Pearce, eds., *Flame-Retardant Polymeric Materials*, Vol. 1, Plenum Press, New York (1975).
9. H. Staudinger, M. Brunner, K. Frey, P. Garbsch, R. Singer, and S. Wherli, *Ber.* **62B**, 241 (1929).
10. S. L. Madorsky, *J. Polym. Sci.* **9**, 133 (1952).
11. S. L. Madorsky, *J. Polym. Sci.* **11**, 491 (1953).
12. H. H. G. Jellinek, *J. Polym. Sci.* **9**, 13 (1949).
13. R. V. Petrella, T. L. Spink, and L. T. Finlayson, *Rev. Sci. Instrum.* **37**, 1500 (1966).
14. R. V. Petrella and G. D. Sellers, *Combust. Flame* **16**, 83 (1971).
15. R. V. Petrella, The Dow Chemical Company, unpublished data (1968).
16. S. L. Madorsky, *Thermal Degradation of Organic Polymers*, Wiley-Interscience, New York (1964).
17. Polymer Degradation Mechanism, National Bureau of Standards Circular 525, National Bureau of Standards, Washington, D.C. (1953).
18. R. G. W. Norrish, G. Porter, and B. A. Thrush, *Proc. Roy. Soc. London Ser. A* **216**, 165 (1953).
19. K. H. L. Erhard and R. G. W. Norrish, *Proc. Roy. Soc. London Ser. A* **234**, 178 (1956).
20. A. B. Callear and R. G. W. Norrish, *Proc. Roy. Soc. London Ser. A* **259**, 309 (1960).
21. R. G. W. Norrish, *Chem. Br.* **1**, 289 (1965).
22. L. S. Nelson and N. A. Kuebler, *J. Chem. Phys.* **37**, 47 (1962).
23. L. S. Nelson and N. A. Kuebler, *Appl. Opt.* **1**, 77S (1962).
24. R. V. Petrella and T. L. Spink, *J. Chem. Phys.* **47**, 488 (1967).
25. R. V. Petrella and T. L. Spink, *J. Chem. Phys.* **48**, 1445 (1968).
26. W. J. Pearce, in *Optical Spectrometer Measurements of High Temperature* (P. J. Dickerman, ed.) University of Chicago Press, Chicago (1961), pp. 142–151.
27. A. Tewarson and R. F. Pion, *Fire Technol.* **11**, 274 (1975).
28. A. Tewarson and R. F. Pion, *Combust. Flame* **26**, 85 (1976).
29. C. H. Dieke and H. M. Crosswhite, The Johns Hopkins University, Department of Physics, Bumblebee Report No. 87, Baltimore (1948).
30. G. Porter and B. Ward, *Proc. Chem. Soc. London* **1964**, 288.
31. J. deRis and L. Orloff, in *Fifteenth Symposium (International) on Combustion*, The Combustion Institute, Pittsburgh (1975), pp. 175–182.
32. A. Tewarson, Heat Release Rates from Burning Plastics, paper presented at the Flammability and Combustion of Non-Metallic Materials Symposium of the 172nd National Meeting of the American Chemical Society, San Francisco, Sept. 1976.
33. E. R. Larsen, *J. Fire Flammability Fire Retardant Chem. Suppl.* **1**, 4 (1974).

34. E. R. Larsen, *J. Fire Flammability Fire Retardant Chem. Suppl.* **2**, 5 (1975).
35. G. J. Minkoff and C. F. H. Tipper, *Chemistry of Combustion Reactions*, Butterworth's, London (1962).
36. V. Ya. Shtern, *The Gas-Phase Oxidation of Hydrocarbons*, Macmillan, New York (1964).
37. W. A. Rosser, H. Wise, and J. Miller, in *Seventh Symposium (International) on Combustion*, The Combustion Institute, Pittsburgh (1959), p. 175.
38. R. N. Butler and R. F. Simmons, *Combust. Flame* **12**, 447 (1968).
39. W. E. Wilson, J. T. O'Donovan, and R. M. Fristrom, in *Twelfth Symposium (International) on Combustion*, The Combustion Institute, Pittsburgh (1968), p. 929.
40. G. A. Takacs and G. P. Glass, *J. Phys. Chem.* **77**, 1060 (1973).
41. N. Cohen, R. R. Giedt, and T. A. Jacobs, *J. Chem. Kinet.* **5**, 425 (1973).
42. R. M. Fristrom and A. A. Westenberg, *Flame Structure*, McGraw-Hill, New York (1965), pp. 341–379.
43. J. E. Nicholas and R. G. W. Norrish, *Proc. Roy. Soc. London Ser. A* **309**, 171 (1969).
44. C. P. Fenimore and F. J. Martin, *Combust. Flame* **10**, 135 (1966).
45. R. B. Ludwig and S. Bergman, Flame Retardant Additives for Styrenic Resins, presented at the Fire Retardant Chemicals Association Conference at New York, 1975, Article 20-22
46. J. Green and J. Versnel, *J. Fire Flammability Fire Retardant Chem. Suppl.* **1**, 185 (1974).
47. A. E. Finnerty, The Effects of Halons on the Auto Ignition Temperature of Propane, Interim Memorandum Report No. 415, Aberdeen Proving Grounds, Maryland (July 1975).
48. R. V. Petrella and H. R. Frick, Mechanism of Extinguishment of Jet Fuel Fires, presented at the Energy Conversion Conference, University of Denver, Nov. 20–21, 1975.
49. J. Eichhorn, *J. Appl. Polym. Sci.* **8**, 2497 (1964).
50. J. W. Hastie, *Combust. Flame* **21**, 49 (1973).
51. C. P. Fenimore, *Combust. Flame* **12**, 155 (1968).
52. G. L. Deets, *J. Fire Flammability Flame Retardant Chem. Suppl.* **1**, 26 (1974).
53. C. J. Hilado, *Flammability Test Methods Handbook*, Technomics, Westport, Conn. (1973).
54. R. Friedman, *J. Fire Flammability* **2**, 240 (1971).
55. J. deRis, in *Conference—Polymers Materials for Unusual Service Conditions*, NASA Ames Research Center, Moffet Field, Calif. (Nov. 1972).

7

Phenolic Fibers

J. Economy

1. Introduction

Development of new and improved flame-resistant materials has been a major goal of fiber scientists during the past decade. However, in spite of legislated standards for flame-resistant textile materials, meaningful progress has been achieved only in selected areas. One of the major obstacles has been the incompatibility between the required structural properties for flame resistance and those necessary for preparation of filaments. Ideally, a flame-resistant fiber would consist of a highly aromatic cross-linked structure which on exposure to flame would char in high yield and produce relatively nontoxic gases. However, routes to preparing such fiber did not exist until recently. Consequently, most of the progress in this area has been achieved by either incorporating halogen atoms into the polymer chain or through use of processible aromatic polyamides. Unfortunately, when exposed to a flame these systems may produce large amounts of smoke and significant concentrations of toxic gases such as HCl and HCN.

Phenolic resins have traditionally been recognized as outstanding flame-resistant materials. When exposed to a flame, phenolic resins tend to char, giving off water and CO_2. In fact, phenolic resins are used in the heat shields of reentry vehicles because of their ability to resist temperatures of several thousand degrees centigrade for brief periods of time. The ability of phenolic resins to char and ablate, i.e., provide cooling through evaporation of small molecules, is exactly what is desired in flame-resistant materials. Phenolic

---

J. Economy • IBM Research Laboratory, San Jose, California.

resins have also found use commercially as flame-resistant foams; however, the brittleness of these foams sharply limits their use to applications such as insulation for roofs.[1] In the heat shields the problem of brittleness is overcome by incorporating reinforcing agents such as glass or carbon fibers.

In light of the above, the report in 1969[2] that cross-linked phenolics could be formed into filaments which displayed mechanical properties more similar to textile fibers rather than brittle resins was greeted with considerable surprise. The program to develop a phenolic fiber was prompted in the early 1960's by the observation that melts of novolac resins could easily be fiberized by the turning action of a stirrer. The possibility of drawing continuous filaments of uncured resin was tested, and it was found that the resin did indeed have sufficient melt extensibility to permit drawing of very fine continuous filaments. A technique for curing the filaments was devised which depended on an acid-catalyzed diffusion reaction with formaldehyde. The initial motivation to develop these fibers was based on their potential use as a precursor for carbon fibers and possibly as a filter medium for highly corrosive environments. However, the unexpected high elongations of the phenolic fibers along with their unusual flame resistance suggested a far broader potential as a textile fiber for protective clothing. The fact that the fibers gave off only small amounts of smoke or carbon monoxide appeared to represent a major advance in safety for textiles used in enclosed areas (the volatile components consist essentially of carbon dioxide and water). The fibers also displayed a relatively high moisture regain of 6–7%, an important requirement for good comfort and wearing characteristics. The discovery of this product was first announced in the spring of 1969, and an active program to scale up and market the fibers under the trade name Kynol was begun in 1973–1974 through a joint-venture company involving the Carborundum Co., Mitsubishi Chemical Industries, and Kanebo. The importance of this announcement is evidenced by the fact that from 1972 to 1975 over 100 patents were issued in Japan alone on modifications of this fiber.

In this chapter, we shall make an attempt to present the first comprehensive description of this development. Toward that end discussions on the preparation, properties, and uses of the phenolic fiber are included. In addition, data on the effect of smoke and toxic gases evolved during combustion of phenolic fibers are described and compared to other flame-resistant fibers.

2. Preparation of Phenolic Fibers

The successful preparation of phenolic fibers was first reported in 1969; however, at the time details of the synthesis were not disclosed. Between 1972 and 1973 the two most critical patents were issued to Economy et al.,[3] in which the preparation of the fibers was described. The abstract of the patent on "Fibers from Novolac" reads as follows: "A novolac melt is fiberized to produce a thermoplastic uncured novolac fiber, and the novolac is cured by heating the fibers in a formaldehyde environment in the presence of an acid catalyst to obtain an infusible, cured novolac fiber." In the second patent on "Etherified or Esterified Phenolic Resin Fibers and Production Thereof" the abstract disclosed the following: "Infusible, cured phenolic resin fibers are reacted with a suitable reagent to block at least 50% of the phenolic hydroxyl groups of the cured resin by esterification or etherification whereby infusible cured phenolic resin fibers are obtained which have improved oxidation resistance and colorfastness and which are white."

One of the unique features common to both of the patents is the fact that these fibers are prepared by chemical reaction of a precursor fiber with a reactive species. In the first case, uncured novolac fiber is treated with formaldehyde and an acid catalyst to introduce methylene cross-links, while in the latter, the phenolic hydroxyls in the cured novolac fiber are esterified by reaction with an organic anhydride. These reactions proceed by diffusion of a reactive species into the fiber so that the rate of reaction is controlled by the rate of diffusion. Reaction conditions must therefore be used which preserve the integrity of the fiber and yet permit the diffusion of the reactive species at an acceptable rate. In addition, substantial changes in fiber weight of up to 35% may occur as in the case of the esterification using acetic anhydride. Such weight changes greatly increase the possibility for formation of internal and surface defects. Techniques for minimizing such defects depend on (1) the use of suitable swelling agents, (2) optimizing the cross-link density of the precursor fibers, and (3) limiting the fiber's diameter to under 5 (denier).*

The concept of chemically converting a precursor fiber to a completely new fiber composition appears to represent a general method for preparation of organic and inorganic fibers.[4] Details for the synthesis of boron nitride,[5] boron carbide,[6] and niobium carbonitride[7] fibers have recently been described along with techniques for optimizing their strength properties.

2.1. Selection of Precursor Resin and Spinning Characteristics

In the selection of a starting material for preparing cured phenolic fibers one could possibly use either a novolac or resole resin. The resole has the advantage that it is available as a viscous solution and can be drawn directly into a filament. However, such filaments will fuse unless they are chilled in an ice bath. A technique has been reported by Yamada and Yamamoto[8] for producing carbon fibers from a mixture of resole, furan resin, and novolac where the solution-spun fibers are collected in a chilled hydrochloric acid bath.

---

*Denier is defined as the diameter of a filament weighing 1 g per 9000 m.

However, this approach is not suitable for producing cured phenolic fibers because of the tendency for such filaments to fuse during curing and the relatively weak and brittle condition of the cured fiber. Also, drawing of such filaments from a spinnerette is greatly complicated by the tendency of the resole to advance in molecular weight at the temperatures necessary for spinning.

Use of a novolac resin as precursor offers the advantages of a thermally stable melt and the potential to draw and collect filaments directly from the melt. The ease with which filaments could be drawn from a low-molecular-weight novolac melt was rather unexpected since it was commonly accepted in the field that synthetic fibers could not be melt-drawn at an M_n of less than several thousand. In retrospect it is not clear whether this lower limit was established as a result of difficulties in melt drawing or from problems in cold drawing such filaments to achieve optimum properties. Regardless, it was immediately apparent that filaments could indeed be drawn from novolacs with M_n values in the range 500–900. Furthermore, the novolac melt displayed excellent attenuation so that filaments could be drawn from a 1.9-mm orifice down to a fiber diameter of 10–15 μm at a rate of about 1000 m/min. This corresponds to an attenuation of over 1000:1.[3] The dimensional stability of the fiber during attenuation is demonstrated by the fact ribbons, dogbones, and "T"-shaped filaments could readily be drawn from the appropriate orifice.[9] In fact, ribbons with aspect ratios of 10:1 could be drawn and cured to give excellent mechanical properties. See Table 1 for mechanical properties of fibers with different cross sections. Very fine filaments of 5–10 μm could also be drawn and displayed significantly improved mechanical properties over larger-diameter fibers (see Sec. 3.2). Such fibers were rather difficult to handle as staple and could not be processed on a typical cotton card.

The drawn fibers in the uncured state were relatively brittle and displayed tenacities of only 0.3–0.6 g/d (gram per denier). However, this was sufficient to permit further handling of the fiber either as a staple or a

TABLE 1
Effect of Fiber Cross Section on Mechanical Properties

Shape	Cross-sectional area (μm^2)	Tenacity[a] (g/d)	Elongation[a] (%)
Ribbon			
15 μm $\times$ 8.5 μm	100	1.9	37
29 μm $\times$ 4.9 μm	142	1.9	47
60 μm $\times$ 6.9 μm	390	1.5	39
T-shaped	530	1.4	17

[a]Each value is an average of ten tests.

continuous tow. Some orientation of the low-molecular-weight novolac chains during drawing could be detected from birefringence studies; however, the birefringence disappears after the fiber is cured.

2.2. Curing Process

As noted earlier, curing of the novolac fiber occurs by a diffusion-controlled process in which formaldehyde and an acid catalyst such as hydrochloric acid diffuse into the fiber and react to form methylene cross-links. The rate of reaction, as shown in Table 2, is very strongly dependent on temperature; however, it is apparent that a final temperature of at least 90°C is required to achieve adequate curing. The curing step can be carried out in either a gaseous environment using vapors of formaldehyde and the acid catalyst or in an aqueous formalin solution containing an acid catalyst.[3] The formaldehyde and hydrochloric acid are relatively small molecules and diffuse into the fiber without the need of a swelling agent. Possibly some swelling of the low-molecular-weight novolac fiber may occur as a result of the water and or formaldehyde; however, the degree of swelling or softening of the fiber surface is not sufficient to lead to problems of fusion between tightly packed fibers.

In a typical curing process carried out in aqueous media, uncured fiber which was collected on a spool was cured directly in an aqueous mixture of 36% HCl and 37% formaldehyde. The temperature of the solution was brought to 40°C and then increased to 60°C over a period of 3 hr and finally brought up to 100°C for several hours. Even though the fibers were tightly packed on the spool, there was no evidence of fusion between fibers.

As shown in Table 2, the time of curing at 100°C is approximately 100 min for a vapor phase process. Most of the curing reaction probably occurs at about 100°C. This is consistent with the approximate temperature range

TABLE 2
Effect of Temperature on Curing Time

Temperature (°C)	Approximate curing time (min)
25	> 18,000
50	4,200
75	600
100	100
125	30
150	15
175	6
200	3

for the well-known acid-catalyzed reaction of phenol and formaldehyde to produce the novolac resin. The steps leading to increase in molecular weight and eventual cross-linking appear to be initiated by the reactions of formaldehyde with the novolac resin at the end groups. The relative reactivity of the terminal phenolics in a novolac resin is ten times that of the other units in the chain. Hence, the initial reaction between formaldehyde and the novolac is to form extended chains.

Further reaction with formaldehyde results in actual cross-links between these extended chains. It is not clear to what degree the structure is cross-linked since the curing reaction is carried out in the solid state and the mobility of the active intermediates is greatly reduced. A similar argument is very likely applicable to the manufacture of cross-linked novolac resins with hexamethylene tetraamine; i.e., once the resin has gelled the mobility of the reactive sites is greatly reduced and the actual number of cross-links may be significantly lower than the estimated percentage. The relative intractibility of cured phenolic resins is probably due in part to the presence of rigid aromatic rings and clusters arising from hydrogen bond interaction between the phenolic hydroxyls. The reason for pursuing this kind of rationale arises from the need to explain the unusually high elongations of the phenolic fibers which can range up to 70–80% and appear to be inconsistent with a highly cross-linked network.

As might be expected for a diffusion-controlled process, the diameter of the precursor filament is a critical factor. Thus, in filaments larger than 5–6 d (25 μm) significantly longer curing cycles are required in order for the diffusing species to react with uncured resin in the center of the fiber. Furthermore, the cross-linked surface structure may act to reduce the rate of diffusion of the reactants. An even more compelling reason for limiting the fiber diameter to under 5 μm can be found in the sharp reduction in mechanical properties associated with larger-diameter filaments.

2.3. Acetylation of Phenolic Fibers

Perhaps the most serious drawback to a wider commercial usage of phenolic fibers is the inherent gold color of the fiber, which limits the shades and colors for dyeing. Consequently, a considerable effort was made to identify the cause of the gold color and then devise techniques to eliminate it. Intuitively, it appeared that for the phenolic structure the most likely source for chromophoric units would derive from quinoid formation. Hence, by blocking the phenolic hydroxyl with an appropriate ester or ether quinoid formation might be impeded. In fact, it was found that by simply acetylating the hydroxyl groups it was possible to obtain a white colorfast fiber.[3,10] It should be noted that use of this technique to form a white phenolic molding compound is not possible because of the need to diffuse the reactive species through cross sections several orders of magnitude thicker than the fiber. Also, with such structures the large volume change associated with an approximate 35% weight pickup of acetate would lead to internal stresses which would sharply reduce the strength characteristics. In the case of a 14-μm fiber a 35% weight pickup results in an increase in diameter slightly over 2 μm. To facilitate diffusion of the acetic anhydride into the fiber, a swelling agent such as dimethyl formamide can be used. There appears to be a critical balance among the cross-link density of the precursor fiber, the need for a swelling agent, and the break elongation of the resultant fiber. The acetylated fiber typically displays the same elongation characteristics of the nonacetylated fibers; however, the average tenacity drops from 1.7 g/d to 1.4 g/d.[9] (See Table 3.) The drop in tenacity might be predicted since one is incorporating 35% by weight acetate units into the fibers, and these acetate units act as a diluent to reduce the overall bond density from which the strength properties are derived.

The acetylated fibers are white and display very good color fastness. They can be easily dyed to a range of colors by using any one of a number of different dye systems. The acetylated fibers have a significantly improved thermooxidative stability over the unacetylated fiber, suggesting a potential for use in applications where the fibers would be exposed at elevated temperatures between 150 and 200°C for extended periods. Even higher thermooxidative stabilities may be achieved by incorporating electron-withdrawing groups into the aromatic ring, indicating that the improvement in oxidation stability is related to the degree to which the electron density at the methylene bridge can be reduced.

The flame resistance of the acetylated fibers is not so good as the phenolic fiber as noted by the reduction in LOI values from 0.38 to 0.28. On the other hand, the fibers show very good dimensional stability in a flame, tending to retain their shape and providing excellent protection as a flame barrier. The

TABLE 3
Mechanical Properties of 17-μm
Acetylated Filaments

Tenacity (g/d)	Elongation (%)
1.5	44
1.3	39
1.6	48
1.7	51
1.4	42
1.4	43

latter feature is undoubtedly related to the cross-linked structure present in the phenolic fiber, which acts to stabilize the fiber over a wide temperature range against melting and shrinking even in a hot flame.

3. Properties of Phenolic Fibers

In this section the various properties of the phenolic fibers are summarized under the general headings of flame resistance, mechanical properties, processibility, and thermal and chemical resistance.

3.1. Flame Resistance

A meaningful discussion of the flame resistance of a given material requires that data concerning evolution of smoke and toxic gases also be considered. Since the intrinsic flame resistance of phenolic fibers has been described in an earlier article,[11] emphasis is placed here on the effects of smoke and toxic gases which may be evolved during combustion of these and other fibers. Discussions of the flame-resistant properties of phenolic fiber will be limited to a consideration of the flame barrier performance of the fibers. As shown in Fig. 1, a dense felt of the phenolic fibers when exposed to a flame of 1900°F provides excellent thermal protection.

The flame resistance of phenolic fibers derives from the tendency for the fibers to char in a flame and produce as primary volatile products water and carbon dioxide. The minimal amount of smoke evolved is in part due to the high char yield but more directly to the stability of the cross-linked structure, which prevents volatilization of chain fragments. Since this balance appears unique among fibers, it appears worthwhile to examine more carefully the structural features which are critical for good flame resistance and low smoke evolution. Certainly both char yield and cross-link density appear to be important. For example, uncured novolacs will burn in a flame with evolution of considerable smoke even though char yields of 40% can be obtained during carbonization of the uncured resin. The effect of cross-link density on LOI for phenolics and epoxies is shown in Fig. 2. It is apparent that the flame resistance of the phenolic is greatly increased from an LOI of 0.24 to 0.36, while the value for the epoxy hardly changes with increasing cross-links. Hence the stability of a fiber at the elevated temperatures encountered in a flame depends not only on the presence of cross-links but also on the inherent stability of the structural units comprising the polymer chain. A measure of this stability can thus be obtained by determining the change in char yield with increasing cross-link density. Another example of the role of char yield is afforded by the case of fibers based on pitch.[12] Typically, pitch will yield

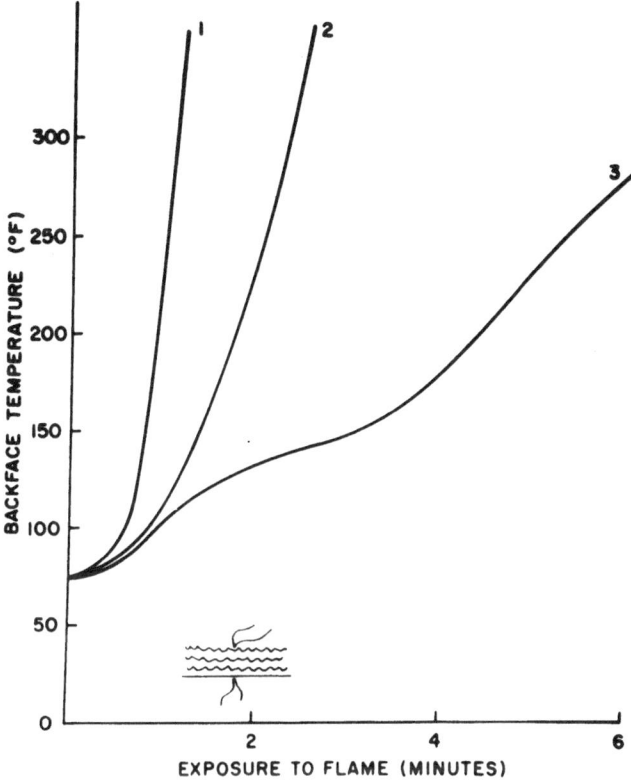

FIGURE 1. Flame barrier performance of Kynol felts. Front face temperature = 1900°F. (1) 36 oz/yd², (2) 52 oz/yd², (3) 79 oz/yd².

a 50% char on carbonization and yet will burn in a flame with evolution of considerable smoke. As shown in Table 4, cross-linking of the pitch yields a fiber which after carbonization will give a char yield of 75% and an LOI of 0.42. The cross-links in the pitch are introduced in the same way as those in the phenolic fiber.[13] In either case the methylene bridges appear to provide excellent short-term stability in a flame, as is also evidenced by the small amount of shrinkage of the fiber (17% in the case of the phenolic). In a flame, the methylene bridge is most likely oxidized to carbonyl, and at higher temperatures of 400–500°C the carbonyl bridge is replaced by an ether bridge derived from the phenolic hydroxyl. At even higher temperatures the aromatic units begin to coalesce into larger condensed polynuclear units which tend to stack in layered structures, leading eventually to a glassy carbon structure. High char yields of 40–45% can also be obtained from acrylic

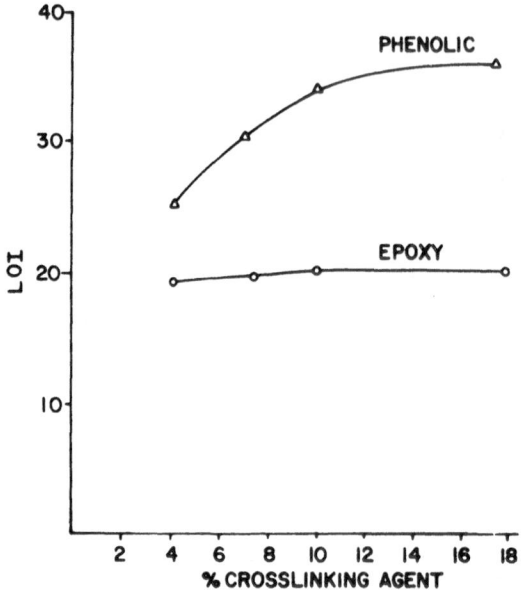

FIGURE 2. Effect of cross-linking on LOI of polymer. (△) Hexamethylene tetramine, (○) diethylenetriamine.

fibers if they have been preoxidized at about 220°C. The structure of this char reportedly consists of condensed polyquinizarine rings which on heating to 500°C begin to display 002 reflections, suggestive of stacking of layered structures.[14] Above 500°C the nitrogen splits out of such structures primarily in the form of HCN. Hence, nitrogen-containing polymers may produce stable chars with the nitrogen directly bound in the aromatic units; however, as temperatures exceed 500°, the nitrogen-carbon bond appears to be relatively unstable and tends to evolve HCN as the primary product.

Structures which tend to form stable chars in a flame might also be expected to produce relatively porous surfaces with high surface areas since

TABLE 4
Char Yield at 800°C

Fiber	% carbon yield
Cured phenolic fiber	60
Uncured phenolic resin	40
Cross-linked pitch fiber	75
Pitch resin	50

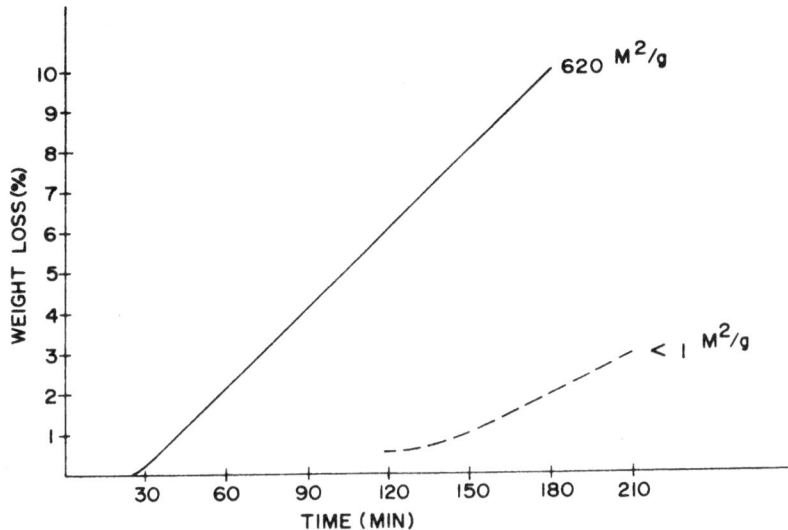

FIGURE 3. Isothermal TGA of carbonized Kynol cloth at 350°C in air.

the surface is being etched by the flame and small molecules and chain fragments are thermally evolved. Increased surface area in the fiber and especially
in the resultant char would tend to facilitate oxidative degradation and thus
lower flame resistance. On the other hand, the tendency of many polymers to
flow and thus fill such voids provides an offsetting process to minimize this
type of problem. A good measure of the latter process for fibers which char is
the shrinkage behavior of a fiber in a flame. Some studies have been carried
out to quantify the effect of surface area on rate of oxidation of carbon
fibers[15] As can be seen in Fig. 3, the carbonized fabric with surface area of
620 m^2/g shows a lower threshold weight loss temperature and a faster rate
of weight loss. Presumably the increased rate of oxidation is related to the
increased surface area available for attack by oxygen and the relatively disordered structure of the carbon, resulting in higher edge concentrations of the
carbon layers. The effect of surface area and nature of the carbon on flame
resistance is illustrated in Table 5. Clearly a highly glassy carbon fiber
derived from the phenolic fiber has the highest LOI value. The much lower
LOI for the carbonized fabric with high surface area is indicative of the role
of surface area on flame resistance. The very low value of 0.19–0.25 for the
activated coconut charcoal suggests that surface area and trace impurities in
the charcoal both contribute to the combustion of carbon. This is also
supported by the value for the carbonized phenolic fabric containing 0.25%
ash where the LOI value is 0.32.

A series of tests was made to determine the relative rates of smoke evolu-

TABLE 5
Effect of Carbon Structure on LOI

Material	LOI
Activated coconut charcoal	19–25
Carbon fabric (0.28% ash)[a]	32
Phenolic fiber	36
Carbon fabric (350–750 m^2/g)[a]	34–41
Carbon fabric[a]	74–81

[a]Carbon fabric obtained from phenolic fabric.

tion and the toxicity of the various combustion gases from Kynol fibers compared to wool, Nomex, and modacrylic fibers.[16] The tests were carried out in an apparatus adapted from a system for testing combustibility of "interior building materials," JIS-A 1321 (see Fig. 4). The ignition was carried out under two sets of conditions, namely with a burner for 3 min and for an additional 12 min using the burner and an electric heater. The amount of smoke produced was optically measured using a CdS photocell. The toxicity of the gases was examined from tests on mice and an analysis made for CO,

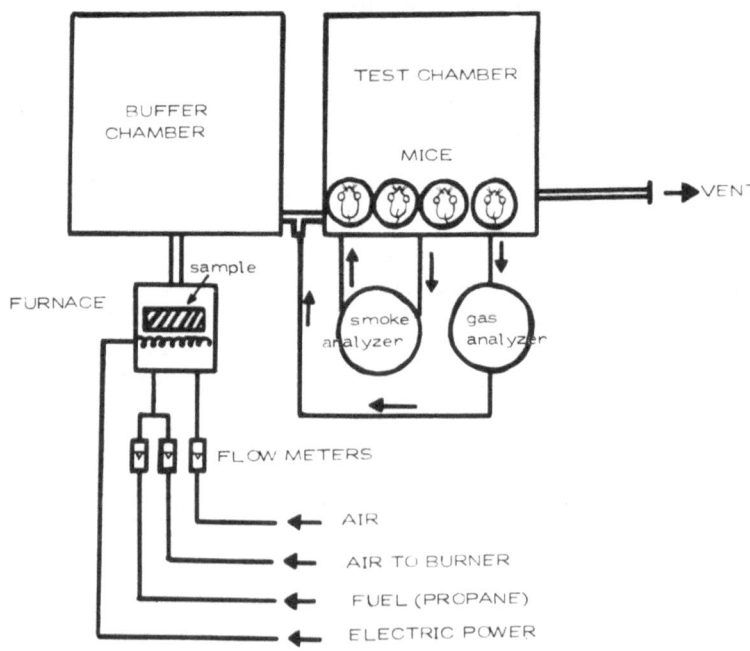

FIGURE 4. Apparatus for testing combustibility.

CO_2, and HCN. The results of these tests are presented in some detail in the following paragraphs, since they appear to provide some real insights into the relative effects of smoke, CO, and HCN.

The smoke density was measured continuously and the peak value recorded as shown in Table 6 for 3- and 15-min ignitions. For the 3-min ignition time, the peak values for modacrylics and wool were grouped closely, although the Nomex showed about one-fifth the smoke density of the modacrylic. The phenolic fibers, on the other hand, displayed a smoke density three orders of magnitude lower. With the additional 12-min ignition, these values changed somewhat to 6.4 for modacrylic, 3.4 for wool, 1.95 for Nomex, and 0.2 for phenolic fibers. Blends of phenolic fiber with wool, modacrylic, or Nomex did not show any synergistic effects, and the smoke evolved was dependent on the percent wool, modacrylic, or Nomex that was present.

Analysis for CO, CO_2, and O_2 in the test chamber showed a considerable variation depending on whether the ignition time was 3 or 15 min (see Table 7). However, for a given ignition time there was little variation in the relative concentration of these gases from fiber to fiber. Thus for the 3-min ignition, the spread of values for all the fibers was small, e.g., CO from 0.05% to 0.12%, CO_2 from 1.60% to 2.14%, and oxygen from 17.6% to 18.3%.

Presence of HCN was determined using infrared analysis (peak at 712 cm^{-1} was used). For the 15-min ignition time the modacrylic fiber yielded

TABLE 6
Smoke Generation (I/M)

Fiber	3-min ignition	15-min ignition
Modacrylic	5.01	6.4
Wool	1.99	3.4
Nomex	0.86	1.95
Kynol	0.002	0.2

TABLE 7
Percent Concentration of Combustion Gases

	CO (max.)		CO_2 (max.)		O_2 (min.)	
	3 min	15 min	3 min	15 min	3 min	15 min
Modacrylic	0.12	0.23	1.67	4.6	18.2	13.0
Wool	0.11	0.21	2.14	4.8	17.6	13.1
Nomex	0.05	0.15	1.62	4.8	18.4	13.0
Kynol	0.08	0.13	1.67	4.1	18.2	15.0

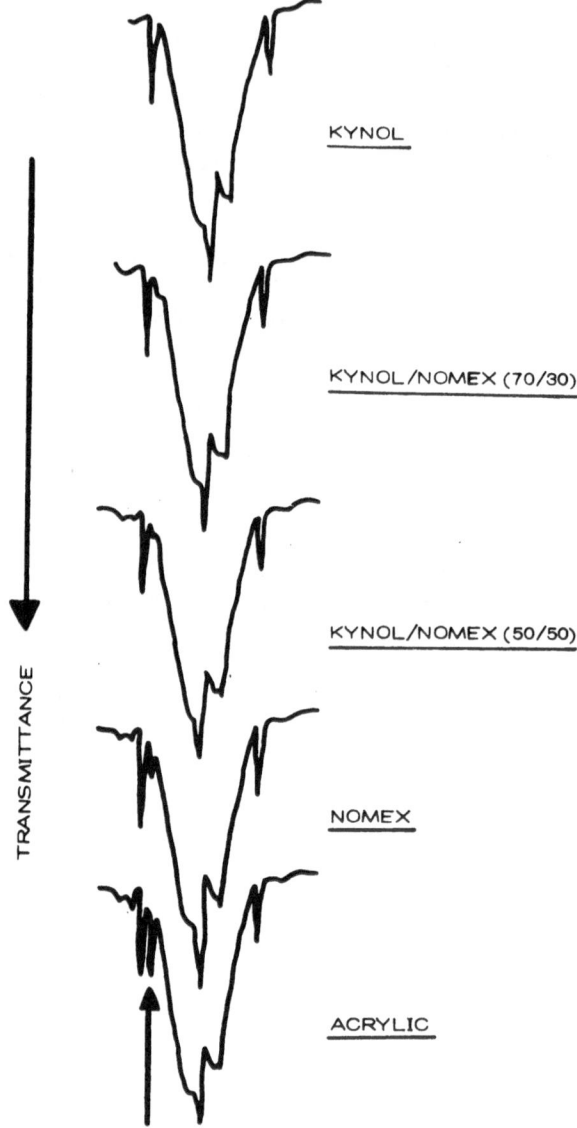

FIGURE 5. Identification of HCN by IR spectra.

$2\frac{1}{2}$ times the HCN compared to Nomex fiber (see Fig. 5). For the 3-min ignition time, wool gave off about 35% the amount of HCN compared to modacrylic fiber.

In terms of actual toxicity, tests based on exposure of live mice to the

TABLE 8
Toxicity of Combustion Gases
(8 mice per test)

	3-min ignition (20-min exposure)		15-min ignition (30-min exposure)	
	Time of death (min)	Expired mice	Time of death (min)	Expired mice
Modacrylic	4–7	8	5–7	8
Wool	6–11	8	7–12	8
Nomex	—	0	10–20	8
Kynol	—	0	—	0

combustion gases produced some interesting differences (see Table 8). Based on the 15-min ignition time all the mice died with the modacrylic, wool, and Nomex. Under the same combustion conditions with phenolic fiber, all the mice survived. With blends of 70:30 phenolics with wool, all the mice survived, while with a 30:70 blend only three survived and five died in from 9 to 13 min. With an 80:20 blend of phenolic–modacrylic, all the mice survived, while three survived at 70:30 (five died in from 9 to 14 min), and for a 30:70 blend, all died (6–9 min).

Several conclusions can be drawn from these toxicity data and from the analyses for CO, CO_2, and HCN evolution. For example, all the fibers appear to produce about the same ratio of $CO/CO_2/O_2$ under a given test condition. For the 3-min ignition time, the ratio of CO to O_2 in the gas is about 0.1%–18%, or just about the level at which hemoglobin takes up equal amounts of O_2 and CO. For the 15-min ignition time, this ratio is closer to 0.20–13%, or well over the level where hemoglobin absorbs CO at twice the rate of O_2. In spite of this somewhat higher CO evolution for the 15-min ignition, all the mice survived the test with phenolic fibers. The results with the phenolic fiber are particularly important since one can isolate the toxic effects of CO, and there is practically no smoke or any other detectable toxic gas. Since none of the mice died even at 30 min, it would seem the mice can tolerate substantial percentages of CO in their system. With humans, a dangerous concentration for a 30-min exposure is 0.4%, or considerably above the amount produced by the fibers tested here. Hence, these results suggest that CO, although considered highly toxic, may not be the primary cause of death in fires even though the amount measured in the blood may range well over 15%.

If the CO is not the primary cause of death, then it is worthwhile to consider more closely the effects of HCN and smoke. Published data[17] indicate that HCN is approximately 30 times more toxic than CO on a volume basis and that a dangerous concentration of HCN for a 30-min exposure is 135 ppm. Clearly, the presence of HCN could account for the toxic effects associated with Nomex, wool, and modacrylic fiber.

In Fig. 5, it was shown that for the 15-min ignition Nomex produced about 30% of the concentration of HCN compared to modacrylic fiber. Since all the mice died with Nomex or modacrylic fiber, there was no way to differentiate the relative toxic effects. When Nomex was blended with 50 and 70% phenolic, the presence of HCN was still detectable from IR, but the concentration was reduced in proportion to the percent of phenolic. With a 70:30 phenolic–Nomex blend, seven out of eight mice survived and the one succumbed at 20 min.

With the 3-min ignition time, it was found that wool produced 35% HCN compared to modacrylic fiber. The evolution of HCN from the blends of Kynol with wool, modacrylic, and Nomex was almost linear when plotted over the range 0–100%, indicating little or no synergistic effect. With the Kynol–modacrylic blends, a sharp difference in toxicity to mice was noted between the 80:20 and 50:50 blends. Thus all the mice survived with the 80:20 blend, while seven of the eight succumbed with the 70:30 blend (in from 9 to 14 min), and all the mice died with the 50:50 blend (5–10 min). In the case of Kynol–wool blends, all the mice survived at the 70:30 level, while six of the eight died with the 50:50 blend in from 9 to 16 min. From these results, there appears to be a direct correlation between the toxicity of HCN and the relative concentrations evolved from the various fibers and fiber blends.

An insight into the cumulative effects of HCN and CO can be realized from an examination of the concentration of CO complexed with the hemoglobin. Thus, the expired mice showed concentrations of 35.5 and 18% CO–hemoglobin in their blood when exposed to combustion gases from wool and modacrylic fibers, respectively. These values were increased to 44.5 and 26.0% respectively, when 50% Kynol was blended with the wool and modacrylic fibers. Hence the Kynol fiber tended to reduce the concentration of HCN by 50% and significantly extended the time of death as measured by the much larger CO content in the blood. In examining these results the role of HCN appears to be far more critical than CO, which is consistent with the earlier report[17] noting the far greater toxic effects of HCN versus CO.

It is of value to now consider the effect of smoke as a contributing cause to the death of mice. From a cursory examination of the data, it would appear that the relative smoke evolution from Nomex, wool, and modacrylic is rather similar to the amount of HCN evolved, which makes it rather difficult to differentiate the role of smoke from that of HCN. It seems reasonably clear that smoke particles can provide a synergistic effect by acting as carriers for toxic gases. How this can happen was dramatically illustrated from the analysis of particles from combustion of gaoline.[18] These particles were shown to be made up of condensed polynuclear structures with a multiplicity of functional groups including phenolic, carboxyl, amines, and amides.

Smoke particles formed in a fire also must contain a multiplicity of functional groups which can form weak bonds with gaseous ingredients. Such particles containing absorbed toxic gases can act as concentrated sources of irritants which are carried directly into the respiratory system. These toxic irritants may act to deaden the cilia which line the bronchial–tracheal system,[19] reducing the capacity for the cilia to provide a mechanism to dilute the potential irritants. Similarly, such particles contacting the eyes may provide a source of irritation which is greatly enhanced by the presence of absorbed toxic gases. It would appear that such particles of smoke could first lead to intense irritation accompanied by excessive coughing and lachrymation and then followed by general disorientation and possibly shock.

3.2. Mechanical Properties

Since the phenolic fibers consist of a cross-linked phenol formaldehyde structure, one might expect a mechanical behavior considerably different from that of typical textile filaments. However, the high elongations (typically 40%) suggest a behavior more similar to natural and synthetic textile fibers rather than the brittle behavior associated with molded phenolic resins. The high elongation values indicate that the cross-link density is much lower than that normally ascribed to phenolic resins. Based on the data from the fibers it would appear that the low elongations of the resins are due to a much higher incidence of defects rather than a high cross-link density. In fact, even with the fibers it has been shown that the stress–strain properties of the fiber are dependent on the diameter. As shown in Table 9, the mechanical properties increase considerably with decreasing fiber diameter.[20] For example, a 1-d fiber displays a tenacity of 2.6 g/d with an elongation of 60% compared to a 2-d fiber with a tenacity of 1.8 g/d and an elongation of 40%. This decrease in stress–strain properties with diameter must be associated

TABLE 9
Dependence of Mechanical Properties on Fiber Diameter

Diameter (μm)	Denier	Tenacity (g/d)	Elongation (%)
9–10	0.7–0.9	2.7	45
10–11	0.9–1.1	2.6	40
11–12	1.1–1.3	2.4	38
12–13	1.3–1.5	2.3	36
13–14	1.5–1.7	2.2	35
14–15	1.7–2.0	2.1	32
15–16	2.0–2.3	2.0	30
16–17	2.3–2.6	1.9	25
17–25	2.6–5.5	1.0	15

with a higher incidence of flaws. The source of these flaws is not immediately evident but could arise from a combination of factors such as a greater potential for nonuniformity in curing with larger-diameter fibers as well as an increased chance for surface flaws. Since most textile equipment has been designed to handle 2-d and larger staple, one cannot take advantage of the better mechanical properties of the finer filaments, at least for standard textile applications.

To better characterize the stress–strain properties of the phenolic fiber, tests have been carried out over a range of strain rates, at different temperatures, and in the presence of water. By increasing the strain rate from 2 to 100%/min, the phenolic filaments show an increase in tenacity of about 25%. At higher strain rates of 1000%/min the elongation at break decreases from 40 to 23%. The effect of temperature and water on the stress–strain behavior of the phenolic fiber is shown in Fig. 6. Exposure of the fibers to water for 16 hr at room temperature does not appear to influence appreciably the shape of the stress–strain curve, although the ultimate stress is slightly reduced. When the fiber is immersed in water at 90°C and then tested at this temperature the yield stress is reduced to 1 g/d, but the elongation is increased by 50%. In a sense the fiber behaves as though it were plasticized by the combination of heat and water. At even higher temperatures of 150°C this effect is magnified further, and elongations approaching 100% can be observed.

The phenolic fibers display a relatively high work to fracture as calculated from the area under the stress–strain curve; however, the abrasion resistance is significantly lower than most textile fibers.[20] This seeming

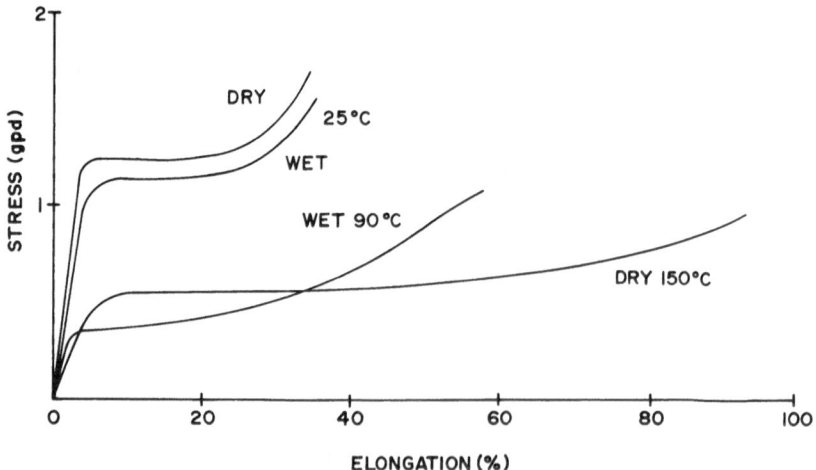

FIGURE 6. Stress–strain of Kynol, dry, wet, and at elevated temperatures.

TABLE 10
Effect of Finishes on Abrasion Resistance

	Neoprene coated (%)	Taber wear index (mg loss/1000°)
Phenolic	0	527
	1.4	220
	1.9	123
Rayon HWM	0	200
Nomex	0	32

contradiction appears to be related to the higher sensitivity of the phenolic fiber surface to formation of defects and the ease with which cracks can then propagate through the amorphous structure. This problem can be minimized through use of appropriate surface finishes.[21] As shown in Table 10, coating a fabric with 1.4% neoprene reduced the abrasion resistance to the level of a high wet modulus rayon, and with 1.9% the abrasion resistance was significantly better than rayon. A somewhat different approach for increasing the abrasion resistance of the phenolic fiber depends on blending 20–30% of a fiber such as wool or Nomex during the textile processing. Fabrics containing such small amounts of such additives show improved wear resistance and can be spun into very fine yarns at acceptable spinning rates.

Another approach for improving the mechanical properties of the phenolic fibers is to introduce molecular orientation by stretching the filament. As shown in Fig. 7, a significant improvement in tenacity and modulus

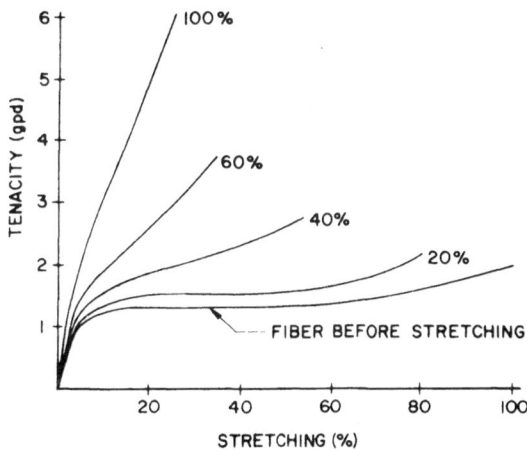

FIGURE 7. Effect of stretching on stress–strain properties of filaments.

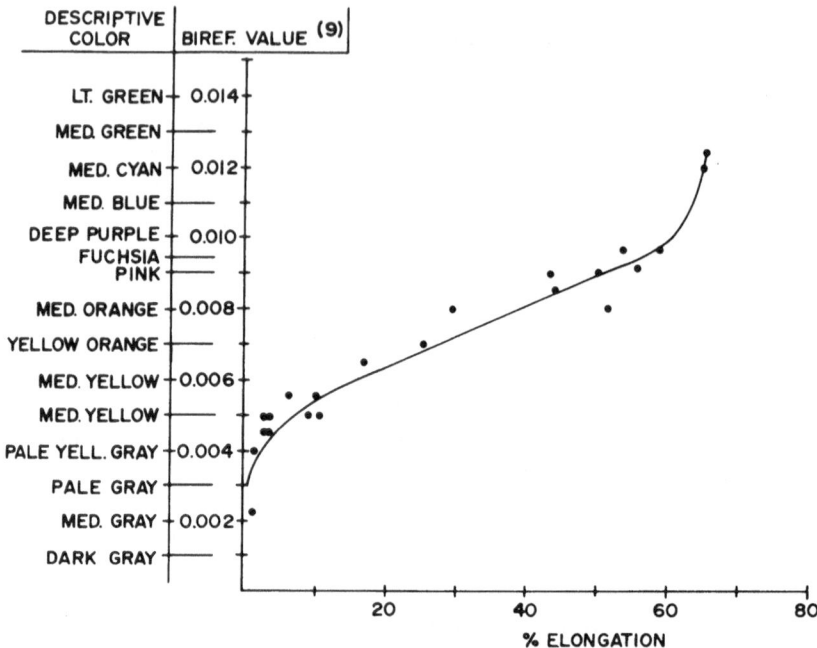

FIGURE 8. Birefringence vs. elongation.

can be obtained by cold-drawing filaments using draw ratios of 1.2 to 2.0. Presumably, the increase in tenacity and modulus is due to an increase in the molecular orientation in the filament direction; however, there is no evidence of crystalline regions in the highly oriented fibers. In the laboratory, filaments with a break stress of 7 g/d have been observed. The increase in molecular orientation with increasing draw ratio can be followed by measuring the birefringence as shown in Fig. 8. The orientation is induced in the filaments only as the yield stress is approached. In the flat part of the stress–strain curve, practically all elongation is reversible. There is a certain time dependence in the reversibility of the elongation. Thus 90% of the reversible elongation is immediately recovered after releasing the load, while an additional minute is required to recover another 9%.

To summarize the above data on mechanical properties of the phenolic fibers, it appears that the stress–strain properties are similar to wool and rayon and can approach acrylics. The tenacity can be significantly enhanced by orientation, although these techniques have not as yet been fully worked out. Optimization of the network structure could also provide significant improvement in the stress–strain behavior. The abrasion resistance, although on the low side, can be improved dramatically by use of finishes, by blending with other fibers, and by minimizing the excessive work on the fibers during carding. it appears safe to predict that phenolic fibers with strengths comparable to acrylics and elongations of 50% should be obtainable in the future.

3.3. Processibility

When Kynol fibers were first made available in experimental quantities in 1969, they were uncrimped and nonuniform in length and diameter so that processing on a cotton system could be carried out only at low rates with relatively high losses. Subsequently, the variation in fiber diameter was brought under control by better design of the melt spinning process, and an exact staple length was obtained by cutting the filaments in the form of a continuous tow. It was also found that the fibers could be crimped readily through use of a stuffer box. A crimp frequency of 8–12/in. can be obtained, which is more than sufficient to provide the necessary cohesive character in carding and spinning. To minimize damage to the fiber, it was found advantageous to carry out the crimping at elevated temperatures in the presence of steam. Most of the crimp is permanent since it arises from an irreversible plastic deformation of the polymer. It was found that carding of the staple fiber was greatly facilitated by the use of an antistatic finish and operating in a controlled environment with a relative humidity of 52–56% and a temperature of 76–82°F. Phenolic fibers have been blended with practically all conventional fibers to provide a range of flame-resistant textile materials.

4. Thermal and Chemical Resistance

The phenolic fibers display an excellent stability at elevated temperatures and especially under inert conditions. For example, at 250°C the fibers show an initial weight loss of 3% with no further change up to 200 hr.[10] The tenacity shows a modest increase from 1.6 to 2.0 g/d after heating at this temperature for 1000 hr; however, the elongation drops to a value of about 10%. When heated at 400°C under nitrogen the fibers show a somewhat higher initial weight loss of 15%, but this value also remains constant up to 200 hr.

In the presence of air at 250°C the fibers show a very sharp loss in tenacity and elongation, as shown in Table 11. This loss in properties appears to be related to oxidative degradation at the methylene bridge, resulting in some chain scission and weight loss. The mechanism of oxidation appears to be related to a transient peroxide formed at the methylene bridge which can break down to form a carbonyl group. This type of reaction may occur at even

TABLE 11

Changes in Mechanical Properties at 250°C in Air

Time (hr)	0	356	476	950
Tenacity (g/d)	1.6	0.6	0.6	0.5
Modulus (g/d)	45	44	42	32
Elongation (%)	3.0	2.9	2.6	1.2

lower temperatures as suggested from DTA, which shows an endotherm at 140°C. The fibers will also form small amounts of carbonyl on standing in sunlight for considerable periods of time, which explains why the fibers darken on standing (reverse fading). Presumably the carbonyl is conjugated with the phenolic hydroxyl, resulting in a chromophoric keto–enol unit. One technique for eliminating the reverse fading is by dope dying or solution dyeing the fibers to a bright red, maroon, brown, or black color. These systems display excellent light stability even after exposure to a fadeometer after 72 hr.

The acetylated fiber shows a significantly improved thermooxidative stability, and no reactions can be detected with air even up to 200°C by DTA. The conversion of the phenolic hydroxyl to an electron-withdrawing acetate ester tends to deactivate the ring and particularly the reactivity of the methylene bridge toward oxygen. An even more dramatic enhancement of the thermooxidative stability of the fiber can be achieved by incorporating halogen atoms into the phenolic structure, and stabilities in air up to 300°C are possible with such systems.

Phenolic fibers display excellent resistance to nonoxidizing acids at room and elevated temperatures.[20] The fibers retain their integrity even when aged in concentrated sulfuric acid at 160°C for a week. In this treatment a weight increase of about 20% occurs due to partial sulfonation of the polymer. Concentrated phosphoric acid at 135°C does not affect the fiber integrity over a period of 5000 hr. The resistance of the fiber against dilute caustic and concentrated ammonia is excellent. The fibers retain their original stress–strain properties when aged for 3 weeks in gaseous NH_3 at 150°C. Kynol fabrics have undergone 50 regular laundering cycles without noticeable change in appearance or hand. Owing to their cross-linked nature, the fibers are insoluble in all organic solvents up to reflux temperature. The chemical resistance of Kynol fibers is summarized in Table 12.

5. Uses

This section is divided into two parts, namely, (1) a summary of the published uses for phenolic fibers as described in various brochures and (2)

TABLE 12
Chemical Resistance of Phenolic Fibers

	Resistance	
Environment	1 week at 25°C	1 week at 100°C
Acids		
HCl, H$_3$PO$_4$, HF, H$_2$SO$_4$	Excellent	Excellent
Dilute HNO$_3$	Excellent	Good
Conc. HNO$_3$	Poor	Poor
Bases		
NH$_3$	Excellent	Excellent
Dilute NaOH, KOH	Excellent	Good
50% NaOH, KOH	Fair	Poor
Organic solvents	Excellent	Excellent

a discussion of the uses of the phenolic fiber as a precursor to prepare other fibers.

5.1. Commercial Uses

Kynol fiber is available in the standard woven and knitted constructions needed for all kinds of clothing, draperies, curtains, bedding, and coverings. Twills, herringbones, satin weaves, tweeds, and plain and double knits of Kynol fiber are currently available from leading textile manufacturers in any desired weight from 3 oz/yd^2 to 10 oz and even heavier for industrial applications. These fabrics can also be coated for water resistance or aluminized for heat reflection. Nonwoven battings and felts of Kynol are available for insulation, upholstery, carpet underlays, wall coverings, and fire barriers. Because they are extremely resistant to acids, alkalis, and organic solvents, nonwoven Kynol materials are also being used for chemical filtration, dust collection bags, and chemical resistance barriers.

For specific end uses, Kynol can also be blended with other fibers to get improved properties. Blends with aramid fibers, for example, provide increased physical strength and abrasion resistance. When Kynol novoloid is mixed with a polyester, one obtains good resistance to corrosive chemicals such as aircraft hydraulic fluids. Blends with modacrylics, FR rayon, and wool and other blends may also be employed to attain specialized properties.

Kynol textiles are available in a number of versatile solid colors: charcoal black, oxford blue, emerald green, crimson red, orange, and natural gold. These solid colors have been interwoven with the natural gold to make attractive patterns.[22]

In limited-risk environments, where people are exposed to the chance of having their clothing catch fire, Kynol fiber offers industry a way to protect workers adequately—and protect itself from compensation claims, replacement costs, and downtime. Because protective shirts, pants, gloves, and coveralls made of Kynol conform to OSHA standards, it is a fiber of choice for public utilities and all kinds of manufacturing plants. And protective clothing made of Kynol fabric is lighter in weight and more comfortable to wear so employees are more willing to use it on a regular basis.

In high-risk environments, where there is a constant threat of fire, Kynol fabrics offer a margin of safety unequaled by any fiber. Metal and chemical industries and industrial fire fighters are now using Kynol novoloid for welder's gloves; aprons; burn, tap, furnace, and proximity coats; underwear; hats; and hoods.

In auto racing Kynol uniforms, liners, and knitted underwear are USAC and NASCAR approved for use by racing drivers, pit crews, and fuelers after extensive testing in actual races. Kynol upholstery fibers are certified by the United States Auto Club for Indianapolis Championship Cars. Kynol batting is also utilized in this special upholstery. Kynol fabrics and nonwoven batting are used exclusively for the new foam-injected suits now being used by racing drivers and are under consideration by the Army, Navy, and Air Force for helicopter and fighter pilots. Kynol is NASCAR approved for safety window nets and is now used on several thousand cars throughout the country.

With respect to military uses, Kynol nonwoven batting meets all requirements for Army, Navy, and Air Force insulation in flight and military clothing. Kynol batting meets the clothing specifications for military fire fighters. Kynol fibers are approved by the Navy for shipboard curtains. Because Kynol fiber will not ignite or give off toxic gases, it is being used on Polaris submarines and many ships where survivability is the critical consideration. Kynol fiber has FAA approval for one-piece multilayered proximity suits used by aircraft fire control personnel at a federally owned facility. Kynol fiber as a carbon precursor is one of the primary candidates for extensive use for composite-structure reinforcement in aerospace applications, including the NASA Space Shuttle.

5.2. Use of Phenolic Fibers as Precursor Fibers

One of the most intriguing features of the Kynol development is the concept of preparing a new fiber by chemical conversion of a suitable precursor since it opens up potential routes to completely new types of fibers. Thus, the Kynol family of flame-resistant fibers was prepared by (1) reacting a novolac filament with formaldehyde and HCl to produce a cross-linked fiber and (2) treating the cured fibers with acetic anhydride to obtain the acetylated version of Kynol. A number of different reactions have also been carried out to prepare completely new types of fibers using a cured phenolic fiber as precursor. Several of these reactions are described in the following paragraphs.

Sulfonation of the phenolic fiber has been studied in some depth with the intent of preparing ion exchange fibers.[23] Typically, the weight of the phenolic fibers can be easily doubled, indicating that the fibers have picked up approximately one sulfonic acid per phenolic unit. The sulfonation process can be carried out directly on the fiber or in a finished textile form. The advantage of an ion exchange fiber lies in the facility with which continuous ion exchange processes can be designed using a fabric in the form of a moving belt. The fabric would be contacted first with the solution to be cleaned and then move directly into a regenerating tank. Anionic exchange fibers have also been prepared using the phenolic fiber as precursor.[24]

Use of the phenolic fibers as a precursor to prepare low-modulus carbon fibers has been studied in some detail.[25] The advantages of a phenolic precursor derives from the high carbon yield of 60% and the simplicity of the carbonization process. For example, the phenolic fibers can be heated directly to 800°C at a rate of 200°C/hr. With rayon the carbonization process must be carefully controlled because of the large weight loss and tendency for the fiber to soften. Carbonization of the phenolic fiber results in a structure which is highly glassy. Fibers heated to 1000°C display an interlayer spacing (d_{002}) of 3.95 Å and crystallite dimensions of about 15–30 Å. On heating to 2800°C the interlayer spacing decreases to 3.75 Å and the crystallites grow to about 20–50 Å. Hence the highly disordered carbon structure formed at 1000°C sharply inhibits the rate of graphitization. Attempts to stress-graphitize these carbon fibers have met with only partial success. Fibers with modulus values of 24.5×10^6 psi could be formed by carbonizing under tension to 800°C followed by further heat treatment to 2750°C under tension. Use of the phenolic fibers to prepare activated carbon fibers has also been successfully carried out and is reported in the literature.[15,26]

6. References

1. R. P. Lavach and A. J. Papa, *Mod. Plast.* **52** (10A), 110 (1975–1976).
2. J. Economy, L. C. Wohrer, and F. J. Frechette, *Text. Res. Inst. Reprints*, 39th Annual Mtg, 73 (1969).
3. J. Economy and R. Clark, U.S. Pat. 3,650,102 (March 2, 1972); J. Economy, L. C. Wohrer, and F. J. Frechette, U.S. Pat. 3,716,521 (Feb. 13, 1973).
4. J. Economy, Chemical Reactions on Organic and Inorganic Fibers, presented at the 7th Akron Summit Polymer Conference, Sept. 7, 1976.
5. R. Y. Lin, J. Economy, H. H. Murty, and R. Ohnsorg, *Appl. Polym. Symp.* **29**, 175 (1976).
6. J. Economy, W. D. Smith, and R. Y. Lin, *Appl. Polym. Symp.* **29**, 105 (1976).
7. W. D. Smith, R. Y. Lin, and J. Economy, *Appl. Polym. Symp.* **29**, 83 (1976).

8. S. Yamada and M. Yamamoto, *Appl. Polym. Symp.* **9**, 339 (1969).

9. J. Economy, L. C. Wohrer, and F. J. Frechette, United States Air Force ML TR-70-72, Part II (March 1972).

10. J. Economy, L. C. Wohrer, F. J. Frechette, and G. Y. Lei, *Appl. Polym. Symp.* **21**, 81 (1973).

11. J. Economy, L. C. Wohrer, and F. J. Frechette, *J. Fire Flammability* **3**, 114, (1972).

12. J. Economy, R. Y. Lin, and H. H. Murty, *Ind. Res.* I.R.-100 Award Announcement (Sept. 1972).

13. J. Economy, F. J. Frechette, R. Y. Lin, and L. C. Wohrer, U.S. Pat. 3,890,262 (June 17, 1975).

14. A. Shindo, Studies on Graphite Fiber, **317** (Gov't Indust. Res. Inst. Osaka, 1961).

15. R. Y. Lin and J. Economy, *Appl. Polym. Symp.* **21**, 143 (1973).

16. J. Economy, in *Proceedings of the Polymer Conference Series, University of Detroit, May 13, 1975*.

17. K. Sumai and Y. Tsuchiya, *J. Fire Flammability* **4**, 15 (1973).

18. *Industrial Research Magazine*, 17 (Feb. 1975).

19. E. J. Kaminski, O. G. Francher, and J. C. Calandro, *Arch. Environ. Health* **16**, 188 (1968); D. J. Tiggleback, *J. Nat. Cancer Inst.* **48**, (6), 1825 (1972).

20. J. Economy and L. C. Wohrer, *Encycl. Polym. Sci. Technol.* **15**, 365 (1971).

21. J. Economy, J. Gardella, L. C. Wohrer, F. J. Frechette, and G. Y. Lei, United States Air Force ML TR-70-72, Part III (Sept. 1973).

22. J. Economy and G. Y. Lei, U.S. Pats. 3,942,947 (March 9, 1976); 3,967,925 (July 6, 1976); 3,993,442 (Nov. 24, 1976).

23. J. Economy, L. C. Wohrer, and F. J. Frechette, Ind. Res. I.R.-100 Award Announcement (Sept. 1972).

24. J. Economy and L. C. Wohrer, U.S. Pat. 3,835,072 (Sept. 10, 1974).

25. J. Economy and R. Y. Lin, *J. Mater. Sci.* **6**, 1151 (1971).

26. J. Economy and R. Y. Lin, *Appl. Polym. Symp.* **29**, 199 (1976).

8

Flame-Resistant Wool and Wool Blends

Mendel Friedman

1. Introduction

Throughout the world, many people die or are injured from burns associated with flammable fabrics. Human suffering and material losses as a result of fires are frightening and staggering. Flammable fabrics were involved annually in about one fourth of the 5.5 million fires which resulted in about 12,000 fatalities and about 250,000 injuries in the United States, which has a population of about 215 million.[5,6,8,9,16,58,59,142,204,205,211,220,230,231,242,257] Burn injuries require long hospitalization and are often fatal. Awareness of this problem is increasing, and the public probably would be willing to pay the added cost of the protection offered by flame-resistant clothing and flame-resistant upholstery, draperies, and related furnishings.

In Sweden, with a population of about 8 million, flammable textiles are responsible for the burns of about 800 patients annually; of those, about 300 die at the accident site or later at the hospital. Injury and fatality rates apparently differ among countries. For example, the United Kingdom, with a much larger population than Sweden,[10] has a much lower fatality rate of 250 per year. An examination of clothing fire statistics by Blum[39] indicates that the mortality rate of patients in Zurich, Switzerland for the years 1957–1966 was 27% for all cases in which clothing played some part.

Mendel Friedman • Western Regional Research Laboratory, Science Education Administration, U.S. Department of Agriculture, Berkeley, California.

Some of the differences may be due to differences in statistical collection of data.

Losses from fires involving clothing run into billions of dollars. The growing concern about this problem has been expressed by new, increasingly restrictive legislation on flammability standards of textiles. In the United States all children's sleepwear must be flame retardant. Moreover, pressures by consumer groups are increasing to extend these requirements to clothing worn by adults, especially the elderly. Ralph Nader[190] reported that old people, along with children, are the chief victims of fabric fires.

These legislative requirements have stimulated interest and research on flammability and fire hazard standards of synthetic and natural fibers and fabrics. I will examine some of the methods and research strategies that have been proposed to impart flame resistance to wool and wool blends.

2. Wool

Wool is a keratin protein composed of amino acids linked by peptide bonds. The peptide units are cross-linked by disulfide bonds and have attached to them reactive groups such as amino, hydroxyl, sulfhydryl, phenolic, carboxyl, imidazole, and guanidino. The concentrations of potentially reactive groups in wool are summarized in Table 1. The disulfide content is variable and can be altered by nutrition of the sheep and weathering of the

TABLE 1
Potential Reactive Groups in Wool[73]

Kind	Concentration (mol/kg)
Peptide (secondary amide)	8.8
Aliphatic hydroxyl	1.47
Half-disulfide	0.86
Total base	0.86
Arginine	0.55
Lysine	0.22
Histidine	0.07
Terminal	0.02
Free carboxyl	0.84
Primary amide	0.76
Phenolic hydroxyl	0.29
Tryptophan	0.04
Methionine	0.04
Sulfhydryl	0.04

wool. The sulfhydryl content can be increased by copper deficiency in the diet of sheep, presumably because of incomplete oxidation of the sulf-hydryl to disulfide bonds during keratin biosynthesis. The content of the carboxyl group can be increased, at the expense of primary amide links, by hydrolysis. The amino acid residues in wool are held together laterally by electrostatic attractions and hydrogen bonding interactions between basic groups (amino, guanidino, imidazole) and carboxyl side chains from aspartic and glutamic acid. The extended polypeptide chains provide fiber strength. Intermolecular disulfide bonds cross-link the peptide units, resulting in a fibrous, resilient, but insoluble keratin material that is very stable to environmental attack. Some special characteristics[73,87] that may give wool special utility are (1) low solubility, (2) steric accessibility to water and to solutes in aqueous media, (3) physical form—as crimped and resilient fibers, and (4) relatively high content of reactive groups, mentioned earlier, that may be chemically modified to provide desirable properties.

Wool and other proteins can be modified in two general ways, by surface deposition or by internal chemical change.[43,44,69–103,144–155,235] The cross-linked nature of the keratin structure is not readily modified. Some reactive compounds, which usually modify soluble proteins, will not modify wool because their molecules are too large or have unsuitable polarity to penetrate the wool structure. Internal structural differences in wool, including crystalline, noncrystalline, and helical regions, further complicate attempts to modify wool chemically.

Although only very small molecules can penetrate the tight wool structure, water and a few other liquids that swell wool permit the internal modification of wool with larger molecules. For reagents that react with water or other protonic solvents, such as alcohols, a nonaqueous, aprotic swelling agent such as DMF is needed. An understanding of wool–nonaqueous solvent interactions is of practical interest and also sheds some light on wool properties. Factors that influence the course of the protein reactions of wool in apolar nonaqueous solvents will be illustrated by a discussion of dimethyl sulfoxide (DMSO).[68,78,88,146,147,154,253,254]

Two main factors influence the course of protein–apolar solvent interactions:

1. Preferential solvation of positive charges by DMSO leaves negative charges and other nucleophilic centers free from destabilizing influences of positive charges and ion pair aggregation.
2. The strong dipole of the sulfoxide group in DMSO acts as a hydrogen bond acceptor and alters acid–base equilibria, nucleophilic and electrophilic reactivities of reactants, and stabilities of transition states.

These factors imply that chemical reactivities of protein functional groups should be subject to strong DMSO–solvent effects for the following reasons:

1. Proteins are polyelectrolytes in which the unequal distribution of positive and negative charges, except at the isoelectric point, depends on the amino acid composition of each particular protein.
2. Hydrogen-bonding and hydrophobic interactions of peptide bonds and side-chain groups are responsible for conformations that proteins assume in solution; in turn, the geometric details, helicity, and folding influence chemical reactivities of functional groups.
3. The degree of ionization of protein functional groups, which determines the concentration of each nucleophilic species at any given pH, and the inherent nucleophilic reactivities of protein functional groups should depend on the dielectric constant, dipole moment, and hydrogen-bonding abilities of their solvent environment.
4. At room temperature, the effects of DMSO on wool are subtle. DMSO primarily swells the wool, enhances reactivities of basic functional groups, and facilitates sulfhydryl–disulfide interchange. In the range 80–120°C, DMSO is a powerful medium for accomplishing chemical modification. At the same time, unless suitable reagents are present to modify side chains in wool, hot DMSO alone can collapse the helical structure, resulting in marked shortening of wool fibers. DMSO can also oxidize protein SH groups to disulfide bonds.[227]

The complex morphology of the wool fiber is illustrated in Figs. 1 through 5.[40,41,167,168,182–184,189] According to Bradbury,[40,41] fine wool fibers contain two types of cells: flattened cuticle cells on the surface and under them long, polyhedral cells aligned in the fiber directions. Coarse wool fibers contain a third type of cell, the medulla, which forms a central core. A cell membrane complex separates the cells. The cuticle cells are made up of epicuticle, exocuticle, and endocuticle, as illustrated in Fig. 1. The cortex is divided into the orthocortex and paracortex. Microfibrillar rods embedded in a disordered protein matrix comprise the substructure of the cortex. The microfibrils may consist of a series of protofibrils arranged in a specific, but presently unknown, manner. Bradbury[41] thinks that the protofibrils probably consist of a two-strand rope of α-helical polypeptide chains.

Figure 2 is a drawing of a cross section of a wool fiber, showing the elongated cortical cells and the protective sheath of scales. Figure 3 is a drawing of the microstructure of wool.[184] About 1000 such microstructures arranged side by side would be equivalent to the diameter of a single

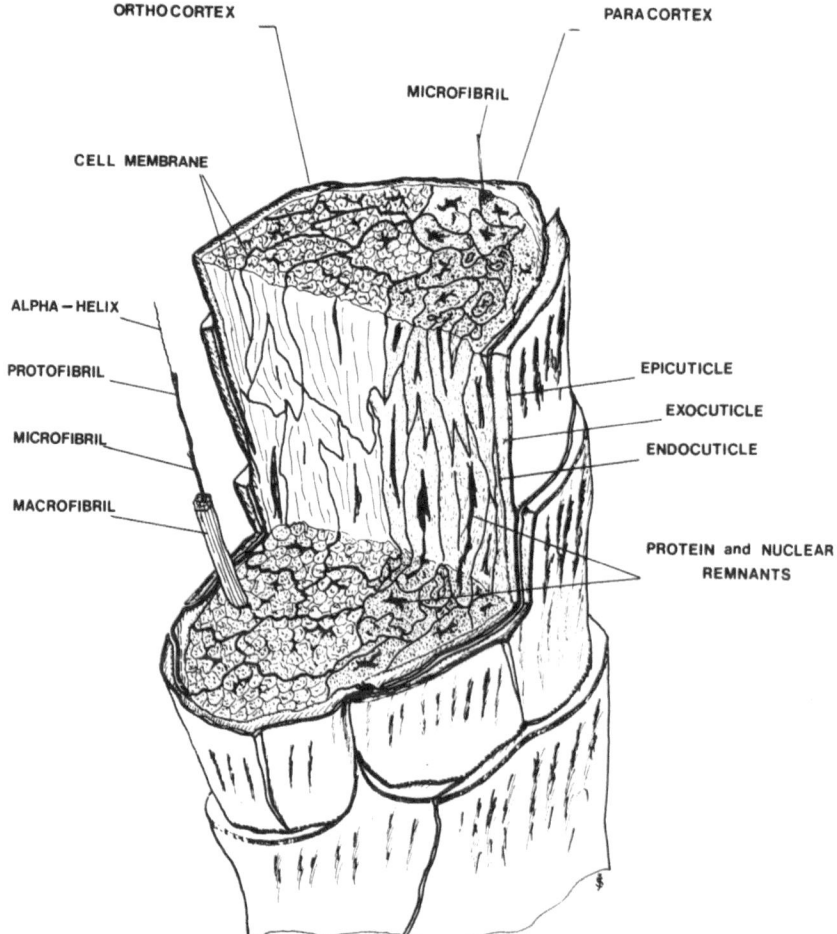

FIGURE 1. Schematic representation of wool fiber.[144] (I thank Professor J. Sikorski, University of Leeds, for this figure.)

fiber. Note the highly oriented structure with regions represented by rods of α-helical chains axially oriented and connected by more random amorphous chains. The whole mass is highly cross-linked by disulfide bonds. The α-helix forms of the protein are illustrated on a still smaller scale in Figs. 4 and 5. If the side chains are ignored, the dipoles of each helix amino acid point in an almost identical direction; thus, the entire fiber, made up of these oriented helices, would have a huge dipole moment. Menefee[183] discovered that by heating slices of keratin between electrodes substantial currents were produced from disorientation of ordered dipoles, first from

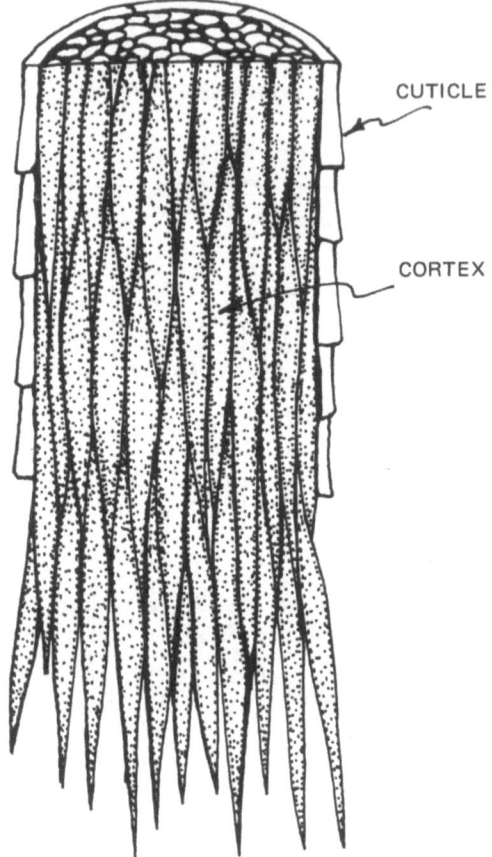

FIGURE 2. Cross section of wool fiber.[183,184]

distortion of ordered water and later from melting of the oriented protein helices. In between, there is a very large, but as yet unexplained, region. Calculations by Menefee[183] show that the amount of charge produced by melting the helix indicates that all the helices are aligned in the same direction.

3. Research Strategies for Flame-Resistant Wool

3.1. Historical

The following details, also taken from the *Ciba Review*,[10] show that in 1735 the Englishman Obadiah Wyld was issued British Patent 551, in which he described a process for making flame-retardant cellulosic fabrics by treating them with aqueous solutions of alum, borax, and iron sulfate. In 1820, Gay-Lussac treated linen and jute with aqueous solutions of ammonium chloride, ammonium sulfate, and borax. Neither of the above treatments was durable to leaching or laundering but did significantly reduce combustibility.

Wool has been considered relatively safe with respect to heat and flame under ordinary conditions. Because of the current emphasis on consumer safety and concern about fabric inflammability, however, major efforts were undertaken in the last decade to devise flame-retardant finishing treatments for wool in its various forms. The various finishing treatments intended to increase the natural flame resistance of wool and wool blends toward combustion will be compared. In all research strategies used to impart flame resistance to wool and wool blends, the maintenance of quality

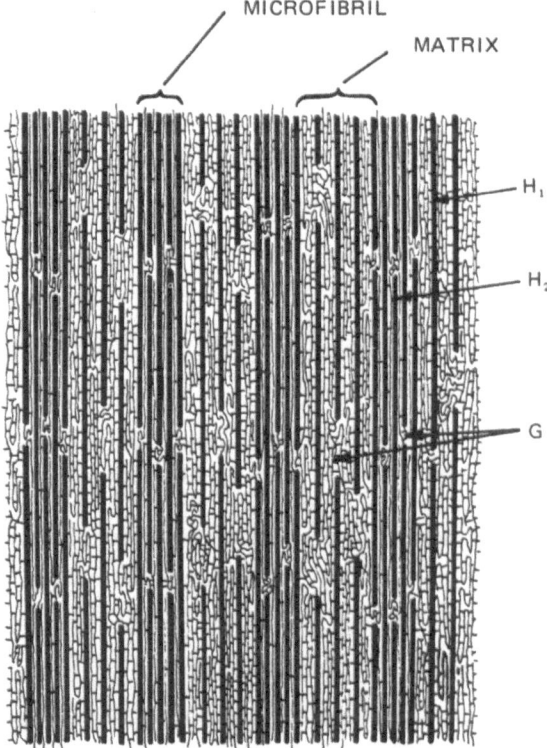

FIGURE 3. Microstructure of wool. Vertical bars are protofibrils containing two or three alpha helices. H_1 is high-melting helical protein in matrix. H_2 is low-melting helical protein in microfibrils. G represents cross-linked, partially oriented amorphous connective protein.[183,184]

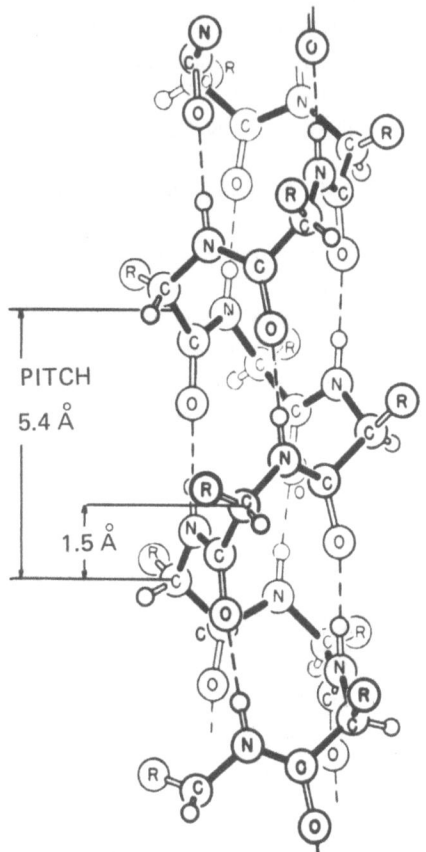

PITCH

5.4 Å

1.5 Å

FIGURE 4. α-Helix structure of a wool protein chain.

throughout manufacturing and practical use, whenever wool may be exposed to heat, was stressed. References 12–14, 17–20, 49, 54, 66, 114, 116, 119–121, 127, 132, 138, 141, 158–160, 163, 164, 188, 194, 195–200, 209, 210, 218, 219, 240, 248, 249, and 251 are an excellent entry into the literature on this subject.

4. Flammability Test Methods

Test methods that have been proposed to evaluate the flammability of textiles in a variety of forms have been discussed by Carroll-Porczynski,[42] Benisek,[22–33] and others[4,11,47,51,104,135,137,161,162,169–171,187,188,216,230,231,237, 238,247] and will not be described here [cf. especially the two papers by G.

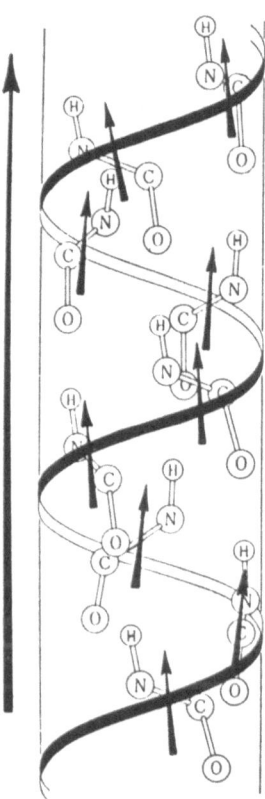

Figure 5. Orientation of polar groups in the α-helix.[183,184]

Stamm,[228,229] who surveyed test methods used by various countries]. Carroll-Porczynski points out that for proper evaluation test equipment is needed to measure the following aspects of textile flammability: ease of ignition, rate of burning, oxygen consumption, heat evolution, smoke density and toxicity. There is apparently no general agreement within the fire research community of the need to measure oxygen consumption [B. N. Hoschke[122]]. Some of these topics will be briefly mentioned later.

Objections are being raised to the severity of the DOC FF-3-71 test, which is a vertical flame test that requires the use of bone-dry textile material. Treated fabrics must retain flame resistance after 50 launderings at 140°F. This is unrealistic, since it does not predict behavior of fabrics in actual use. The standard could be lowered without comprising safety and thus save the consumer money when he purchases flame-retardant textiles. Scientists at the National Bureau of Standards (NBS) who devise flammability tests appear to have deliberately made the standard as rigid as necessary for extreme situations, since otherwise a large number of tests would be needed to meet the different situations to which garments are exposed to during their lifetime. Moreover, it is not practical to restrict use of a particular fabric to those situations to which it is safe.

The NBS is exploring ways to revise the standards and has proposed the Mushroom Apparel Flammability Test (MAFT). The test depends on the use of a cylinder of textile fabric wrapped around a copper core which contains heat sensors. The fabric is ignited on the surface, and the instruments then measure the rate of heat transfer from the ignited fabric to the heat sensors. The fabrics are then rated as follows from least to most flammable: Class A, maximum rate of heat transfer of 0.1 cal/cm^2/sec; Class B, maximum rate of heat transfer greater than 0.1 cal/cm^2/sec and igniting in more than 1 sec; and Class C, ignition in 1 sec or less. This test is considered a good indication of the danger to the body resulting from heat transfer from burning fabrics. It remains to be established whether the MAFT test will correlate with real burning situations.

The State of California has adopted standards for flammability that are stricter than any of the federal standards. Those standards require an average char length of 3.5 in. with no specimen to exceed 6 in. compared to an average char length of 7 in. for the DOC FF-3-71 test.[11]

Although bone-dry requirements were developed in the apparent belief that wool had no share in the children's sleep-wear market, I think that the oven-dry requirements should be eliminated for natural fibers, such as wool and cotton, that ordinarily have a high natural moisture content. All fibers should be tested at their normal moisture contents except for special uses which require low humidity. The present standard unreasonably favors certain synthetics, such as polyester, that have low normal moisture contents.

5. Inorganic Compounds

5.1. Sodium Hydrosulfite–Formaldehyde–Borax Treatment

Gilbert and Liepins[105] flameproofed wool shag carpets by immersion in a solution of 7.7% formaldehyde, 7.7% borax, and 1.5% sodium hydrosulfite for 20 min at 90°C and then for 20 min at 100°C. This resulted in a 300% wet pickup. The carpet was then rinsed for 15 min with tap water and for 3 min with trichloroethylene and then dried for 2 hr at 100°C in a forced-draft oven. The boron content of the treated carpet was about 0.5%. The treatment caused about 10% shrinkage of the pile yarn. Treatment to the same level with borax only marginally improved flame retardancy. The treatment was effective after the seventh but not after the eighth shampoo. Similar treatment of wool fabric resulted in 15–22% shrinkage.

The authors attributed the flame-retardant effect to the interaction of borax with sulfhydryl groups formed on reduction of disulfide by sodium hydrosulfite. The resulting thioborate derivative might be further stabilized by interaction with amino groups present in wool. The role of formaldehyde is not immediately apparent, but it presumably stabilizes the yarn by cross-linking some of the generated SH groups as well as other functional groups in wool.

Disadvantages of this treatment include (1) weakening of the fibers by reduction of disulfide bonds, (2) the possibility of formaldehyde odor, (3) the possibility of yarn stiffening due to cross-linking by formaldehyde, and (4) shrinkage of fabric.

5.2. Phosphoric, Sulfuric, and Sulfamic Acids and Their Salts

O'Brien and Weyker[194] evaluated combinations of cyanamide and phosphoric acid for flameproofing wool shag carpets. They found that add-on levels of 6–10% impart flame retardancy that is durable to shampooing, provided the wash procedure is not too drastic and does not alter the physical characteristics of the yarn. The Pyroset CP solution may be applied by either padding or spraying. In contrast to cellulosic materials, wool shag carpet does not require steaming and fuming to cure the reagents.

The chemistry of the treatment was not described. We may surmise that the phosphoric acid phosphorylates various functional groups in wool including new functional groups formed from wool and cyanamide. It remains to be shown whether this treatment also works for wool fabrics and whether physical properties are impaired.

Phosphoric acid alone is an excellent flame retardant for wool, but the treatment is not durable to laundering. Simpson[224,225] also showed that sulfuric and sulfamic (but not nitric or hydrochloric) acids, ammonium phosphate, ammonium sulfamate, ammonium borate, potassium hydroxide, and potassium carbonate impart flame resistance to wool.

Lewin and associates[165,166] discovered that sulfation of wool imparts durable flame resistance to wool and wool–cotton blends. The degree of flame resistance appears to depend on the amount of introduced sulfur. About 4% of sulfur is required to impart flame resistance that is durable to at least 50 launderings according to the DOC FF-3-71 Children's Sleepwear Standard. The treatment did not significantly change the brightness, hand, and bursting strength of all-wool knitted fabrics. The treatment should be carried out after dyeing since sulfation decreases the dyeability of wool with acid dyes. The authors speculated about the nature of the sulfation process of proteins. In a related study, Kirkpatrick and Maclaren[142a] report that sulfated wool, formed by the action of sulfuric acid on wool, is flame-resistant.

5.3. Tris(hydroxymethyl)phosphonium Derivatives

Beninate and collaborators[21] imparted flame resistance to cotton and wool by impregnating fabrics with a pH 7.5–7.9 aqueous solution containing 10–40% tris(hydroxymethyl) phosphine. The fabric was then exposed to ammonia for 2–6 min.

Dupraz[60] and Crawshaw et al.[48] describe the application of the Proban flameproofing treatment to wool. At pH 4–5 a condensate of tetrakis (hydroxymethyl)phosphonium chloride and urea and a melamine resin (Proban 420A and 420B, respectively) is applied in a 2:1 weight ratio for a total deposit on wool of about 15%. The wet pickup should be about 60–80%. The fabric is then dried, cured, and subjected to an oxidizing afterwash in which the fabric is scoured with a hydrogen peroxide solution. This treatment is designed to counteract the reducing action of the THPC on wool. The effectiveness of THPC and related phosphorus-containing wool flame retardants is still being tested.[15,136,245,246,255]

5.4. Zirconium and Titanium Compounds

Benisek[22–37] and Gordon and associates[107–112] systematically studied the reaction of wool and chemically modified wool with potassium hexa-fluorozirconate and potassium hexafluorotitanate. They report a number of observations which are helpful in understanding the mechanism of flame protection. (1) The amount of fluorozirconate needed to impart durable flame resistance depends on the type of fabric, fabric construction, and fabric history. (2) A zirconium content of more than 0.8% is required for fabrics to pass the stringent Federal Aviation Administration test (FAA Test FAR 25.853). (3) Chemically modified wools show a variable uptake (Table 2) which permits conclusions about the nature of the binding sites involved in zirconium binding. The results in Table 2 suggest that negatively charged groups, such as carboxyl and sulfhydryl, do not interact with the fluoro-zirconate but that positively charged basic groups, such as amino groups of lysine, guanidino groups of arginine, and imidazole groups of histidine, have a strong affinity for fluorozirconate anion. The authors suggested that the increased uptake of zirconium in esterified wool is probably due to the release of additional amino groups from salt bridges which can now form as salt with a fluorozirconate anion. This suggestion is also consistent with the observed increase in uptake by S-aminoethyl wool (thialysine wool) which contains additional amino groups ($W-S-CH_2-CH_2-NH_2$). Sulfated wool shows a zero-level uptake. The negative charges on the introduced OSO_3- (sulfate) residues evidently electrostatically repel the negatively

TABLE 2
Uptake of Fluorotitanate and Fluorozirconate (as % titanium
and zirconium) by Chemically Modified Wools[108,110,240]

Sample	Titanium, content (%)	Zirconium content (%)
Control	0.43	
Acetylated	0.25	1.03
Esterified (methyl ester)	0.38	1.76
Esterified and acetylated	0.38	1.28
Reduced	0.45	1.48
S-aminoethyl (reduce + 2-bromoethylamine)	0.50	1.82
Sulfated	0.01	0.00

charged fluorozirconate anions or preferentially form salt bridges with the available basic groups, thus preventing any chelation by the inorganic anions. (4) Pyrolysis of treated fibers left a residue of zirconium dioxide tubules which retained the physical appearance of the original fibers. These results suggest that the fluorozirconate is situated as a ring in the cuticle and outer cortex. (5) A kinetic study of the rate of exhaustion shows that there is an initial rapid uptake followed by a much slower step. An equilibrium appears to be established since only about 75% of the hexafluorozirconate is exhausted from the bath even if the initial concentration is varied from 4 to 14% of weight of fabric. However, a rate study showed that the exhaustion of potassium hexafluorozirconate increased from 75 to 98% when sodium tungstate and citric acid (citratotungstate complex) were added to the bath.[22] The maximum amount sorbed is about 1.8% zirconium as fluorozirconate. This amount is not changed by increasing bath temperature up to 80°C. (6) Treatments frequently are uneven when applied at low temperature.[107] This problem can be avoided by progressively decreasing the pH during the exhaustion process by slow addition of HCl to the bath. (7) Although the treatment is durable to laundering, the potassium hexafluorozirconate appears to hydrolyze, during alkaline laundering conditions, to zirconium dioxide. The washfastness appears to be due to the insolubility of the dioxide formed within the fiber. (8) Fabrics treated with hexafluorozirconate after a dichloroisocyanuric acid shrinkproofing treatment lost less fluorozirconate during laundering than fabrics treated in the reverse order. Less fluorozirconate is exhausted on DCCA-treated wool than on native wool. Fluorozirconate is lost from flame-resistant wool when the fabrics are treated with a DCAA shrinkproofing process. The authors suggest that DCAA aftertreatment is less desirable and that compatibility with chlorine-mediated shrinkproofing treatments is accomplished when fluorozirconate exhaustion follows the chlorination step.

Several additional facts are noteworthy: (1) We found that potassium hexafluorozirconate did not flameproof wool which had been shrinkproofed by the Wurset (polyurea) process. (2) Since only 75% of the hexafluorozirconate is exhausted from the dyebath on the wool, the remainder could create an effluent (pollution) problem. There appears to be disagreement about the extent of hydrolysis of the hexafluorozirconate on wool during laundering,[19,130,134] but it is nevertheless clear that fluoride ions are liberated during the alkaline conditions. Although the concentration of fluoride ions may be low in laundry water, these fluoride ions could wind up in the sewage system as part of the laundry water and could constitute an additional problem. I feel that studies are needed to determine whether this constitutes a health hazard. Basic information is also needed about the extent of hydrolysis of the various hexafluorometallo salts as a function of pH both in solution and on wool (cf. also Refs. 52, 53 and 217). (3) Since hexafluorozirconate apparently does not penetrate the complete wool structure, but appears to localize in the cuticle and outer cortex, it would be worthwhile to evaluate the effectiveness of swelling agents that improve the penetration of the inorganic salts into wool. (4) Wool treated with hexafluorotitanate shows a pronounced yellowness which could not be eliminated by EDTA or by several reducing agents (Table 3). (5) With the potassium hexafluoro-

TABLE 3

Yellowness Index of Wool after Various Fluorotitanate Treatments[37,111]

Treatment	Yellowness index	Titanium content (mol g^{-1} of wool $\times 10^{-4}$)
Untreated	6.6	
TiF_6^{-2}	19.4	1.0
TiF_6^{-2} + 1% EDTA	19.0	1.1
TiF_6^{-6} + 5% Zn	14.7	0.06
TiF_6^{-2} + 2% $Na_2S_2O_4$	19.7	0.72
TiF_6^{-2} + 2% $NaHSO_3$	19.0	0.83
TiF_5^{-2} + 5% Zn	11.2	0.09
TiF_5^{-2} + 2% $Na_2S_2O_4$	18.7	0.40
TiF_5^{-2} + 2% $NaHSO_3$	17.2	0.36

titanate treatment the fastness to washing appears to be due to the formation of titanium dioxide from the hydrolysis of the fluorotitanate. The dioxide is trapped within the fibers because of its insolubility and appears to be the active flame retardant. (6) Although the fluorozirconate and the related zirconium–tungsten treatments are durable to multiple washing at 50°C, they do not appear to pass the DOC FF-3-71 specifications, which require laundering at 60°C.

Durability to washing, however, can be increased by treating wool simultaneously with potassium hexafluorozirconate and sodium meta-vanadate.[22] Unfortunately, this treatment turns the wool yellow-green. Possibly zirconium and titanium salts form salt or covalent chemical bridges with amino or other basic groups in wool. Such cross-links are a well-recognized phenomenon in protein–metallo chemistry.[68,87,91,150,155,173,178,208]

5.5. Tungsten, Molybdenum, and Vanadium Compounds

Further studies by Benisek[22] showed that negatively charged isopoly-molybdates, isopolyvanadates, and isopolytungstates are readily sorbed by wool at pH 3 under dyebath conditions. The resulting wool is flame resistant but discolored. The degree of discoloration decreases in the order $V > M > W$. The blue-gray tungsten discoloration is light catalyzed and reversible. A combined treatment of fluorozirconate or fluorotitanate with vanadate, molybdate, and tungstate derivatives results in an additive improvement in flame resistance and washfastness of the treatment. For example, the LOI value of untreated wool, 24.5, is increased to 28.9 on treatment with 8% K_2ZrF_6 and to 31.5 on treatment with K_2ZrF_6, 5% Na_6WO_4, and 4% citric acid. Benisek[22] suggests that tungstates probably form complexes with zirconium and titanium salts and act as a carrier for zirconium and titanium salts to improve penetration of the zirconium and titanium into the wool fiber.

Wool treated with complexes of sodium tungstate with citric, oxalic, malic, tartaric, and phosphoric acids is flame resistant and lightfast.

5.6. Tin Compounds

Ingham and associates[129–134] devised a process for imparting flame resistance to woolly sheepskins. Because the wool surface of sheepskins is highly permeable to air, flame is rapidly propagated on the fiber tips. A number of inorganic compounds, including ammonium dihydrogen phos-phate, phosphoric acid, ammonium sulfamate, and borax–boric acid, were effective to various degrees. However, since all of these compounds are water-soluble, they could be leached out by children sucking the wool tips of the sheepskins. Toxicity considerations[140] therefore precluded their commercial use on sheepskins. Studies with potassium hexafluorotitanate showed that the white sheepskins became a light creamy yellow which intensified on exposure to light. Reductive treatment with bisulfite or hydrosulfite did not decrease the discoloration, and oxidative treatments with hydrogen peroxide changed the yellow to orange. Similar studies with potassium hexafluorozirconate revealed that although the wool retained an acceptable color and hand after treatment, the leather part of the sheepskins became stiff and harsh. Further studies showed that zirconium salts have a strong affinity for leather. This is not surprising since both wool and leather are protein in nature. Further-more, when the treatment was carried out before tanning, most of the zirco-nium was lost during tanning. Zirconium salts are widely used in leather tanning.[116a]

Because of the cited problems with some of the known flameproofing methods to treat sheepskins, Ingham[130] successfully devised a process based on the use of stannic chloride and ammonium bifluoride in isopropanol dissolved in water at a final pH of 1.1. The flameproofing solution can be applied either by spray or brush. Treated samples passed the specifications of the DOC FF-1-70 U.S. Federal Standard Flame Test (methenamine tablet). Representative specimens are subjected to ten standard washing cycles. The specimens are then heated at 105°C for 2 hr and cooled for at least 1 hr in a desiccated cabinet. A methenamine tablet is then ignited on the surface of a specimen under controlled conditions and burns for 3 min. The flame spread has to be less than 1 in. (2.54 cm) of an 8-in. circle specimen centered on the tablet. The specifications state that only one of eight specimens can fail this test. The treatment was durable to ten launderings if the sheepskins were uniformly sprayed with the flameproofing solution.

Because of possible toxicity and effluent (pollution) problems, the amount of extractable fluoride was determined in treated wool sheepskins. Wool shavings from sheepskins were shaken for 20 min in distilled water. The concentrated extracts were analyzed for fluoride content by a gravimetric procedure as calcium fluoride. The extractable fluoride level was 1.4 g (0.17%) for the weight (800–850 g) of wool from an average sheepskin. A similar determination of sheepskins treated with K_2ZrF_6 by the exhaustion method gave an extractable fluoride content of 3.7 g (0.40–0.50%). Fluoride was extracted with distilled water, so the procedure does not approximate laundering conditions. Extractable fluoride should be determined after mul-tiple launderings. In a related study, Gordon[108] found that the zirconium-to-fluoride ratio remains almost constant. Inorganic flame retardants for wool are summarized in Table 4.

TABLE 4
Inorganic Flame Retardants for Wool

Compound and references	Remarks
Ammonium dihydrogen phosphate (224, 225)	Dusty surface, not durable
Ammonium sulfamate, sulfamic acid (165, 166, 224, 225)	Some resistance to dry cleaning; used to flame-proof airplane upholstery
Antimony oxide (224, 225)	Effective for wool–cotton blends
Borax–boric acid (224, 225)	Forms glassy dustless layer of solids; not durable
Inorganic acids (224, 225), boric, phosphoric, sulfamic, sulfuric	Order of effectiveness: free acid, ammonium salt, sodium salt for strong acids; reverse order for weak acids; phosphoric acid highly effective
Sb_2O_3 (224, 225)	Wool containing Sb_2O_3 becomes flameproof when soaked in dilute HCl; not durable to laundering
Dicyanamide plus H_3PO_4 and urea (194)	Variable washfastness; effectiveness increased by melamine, urea, or ammonia; wool degraded and stiff; some yellowing; good flameproofing
HBF_4 (215)	Yarns immersed for 30 min at 60°C in aqueous HBF_4 solution; passed Japan Transportation Ministry Flammability Test Standards
Tris(1-azirdinyl) phosphine oxide (APO) (224)	Decomposes with evolution of pungent fumes when treated wool is heated over 100°C; BF_3 synergizes flameproofing effect
Phosphorus oxychloride in DMSO and H_3PO_4 (224, 225)	Only limited phosphorylation needed to achieve flame resistance
Stanic chloride plus ammonium bi-fluoride (129, 130)	Effective for sheepskins
K_2ZrF_6 (22–38, 61, 108–112, 116, 123–125, 130, 240)	Effective for wool; high affinity for leather; see text
Tetrakis (hydroxymethyl) phos-phonium chloride (15, 21, 48, 60, 181, 213, 214, 223–225, 232, 233, 243, 245, 246)	
$TiCl_4$ (22, 174)	Complexation with citric or other acids improves effectiveness; can be applied from DMF
$TiCl_4$ plus NH_4F (22, 91, 108, 129, 130)	Yellowing of wool; harsh hand; stiffens leather of sheepskins
$ZrOCl_2$ (28)	Complexation with organic acids improves effectiveness
$SbCl_3$ (174)	As above
$BiCl_3$ (174)	Flameproof and mothproof when applied from DMF at 110°C for 1 hr
$ZnCl_2$ (174)	As above
$ZrCl_4$ (174)	As above
Sodium tungstate (22)	Enhances effectiveness of hexafluorozirconate to alkaline aftertreatment and Hercosett shrink-resist treatment; flameproofing and washfast effect
Vanadates, tungstates, and molyb-dates (22)	Can be exhausted as isopoly salts; additive effect with hexafluorozirconates and hexafluoroti-tanates; discoloration

6. Organic Compounds

6.1. Tetrabromophthalic Anhydride and Related Compounds

6.1.1. Reaction in Dimethylformamide

Whitfield and Friedman[253,254] showed that the flame resistance of wool fabrics is usefully increased by treatment for 15–30 min at 50°C in DMF with haloorganic acid anhydrides and halides (Figs. 6 and 7). Chlorendic anhydride imparted flame resistance at uptakes of 3% or more. The effect was durable to both dry cleaning and laundering. Evidently, the anhydride acylates functional groups in wool, so it falls under the category of reactive flameproofing compounds. Additional studies showed that related acid anhydrides and halides are similarly effective. Such compounds include chloroacetic, dichloroacetic, dichloromaleic, tetrachlorophthalic, and bromoacetic anhydrides and a comparable group of chloro- and bromo-substituted acid halides. Moreover, polyurea-shrinkproofed wool may be treated by this process.

6.1.2. Two-Solvent Process

Chemical modification of wool requires reagent penetration into a cross-linked polypeptide structure. Such internal modification can occur in dipolar, aprotic swelling agents such as DMF, dimethyl sulfoxide (DMSO), and 1-methyl-2-pyrrolidinone. From a commercial standpoint, these solvents are relatively expensive and are difficult to recover because they have high boiling points.

An attempt was made to solve this problem by carrying out the reaction in a small amount of DMF plus a large amount of an inert solvent, such as perchloroethylene, but reaction did not occur. The idea then occurred to us[88,89] that a low-boiling inert solvent could be evaporated (recycled). The residual solvent would become enriched in DMF, and reaction should then proceed to give durable, flame-resistant wool. We reasoned that evaporation of the mixed solvent should enhance the reaction because evaporation makes the reaction medium more polar and enhances the desired acylation of wool by the flameproofing compound.

A volatile solvent, dichloromethane (DCM), was added to aid even distribution of TBPA and DMF and decrease the amount of DMF (or other reaction medium) needed. When the system is heated DCM evaporates and the anhydride presumably reacts covalently to form wool–amide and wool–ester derivatives.

The treated wool is flame resistant and shrinks less than native wool when laundered. Possible advantages of such a two-solvent treatment include

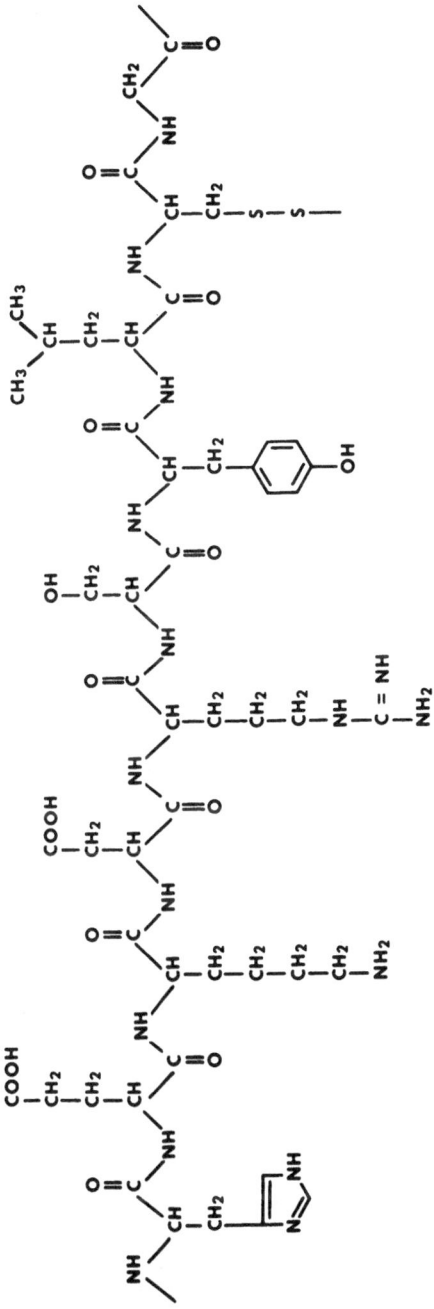

FIGURE 6. Potential sites in a protein for covalent and noncovalent binding of flame retardants.

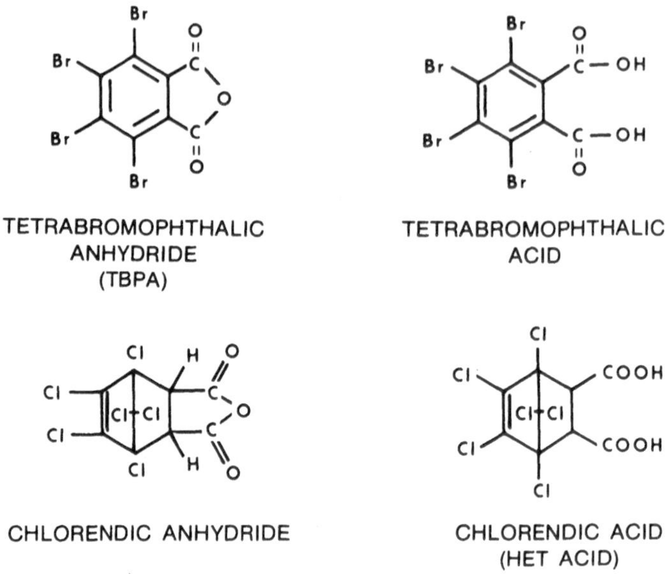

TETRABROMOPHTHALIC
ANHYDRIDE
(TBPA)

TETRABROMOPHTHALIC
ACID

CHLORENDIC ANHYDRIDE

CHLORENDIC ACID
(HET ACID)

FIGURE 7. Structures of halogenated flame retardants.

continuous application instead of in a batch process and lower cost. The system is economical because use of a low-boiling solvent, dichloromethane, reduces heat energy cost and the solvent can be recycled. The concept of the two-solvent process, in which one of the solvents (in this case DMF) enhances the reaction and a more volatile, second solvent (e.g., DCM) serves as a temporary diluent, should be generally useful to modify wool and other textiles with various reactive compounds such as acid halides and isocyanates.

6.1.3. Dyebath Applications

Because the probability of commercial acceptance of a flameproofing treatment would be increased if it could be combined with a finishing operation such as dyeing, we are testing the possibility of incorporating flame-retarding dye analogs into standard dyeing operations. We found that chlorendic acid (HET acid) can be conveniently applied to wool during standard dyeing.[84,85]

The following treatment is typical: One liter of dye solution was made up at 120°F (49°C) with 2% of representative dyes of weight-of-fabric (owf), 10% (owf) Glauber's salt, 4% (owf) H_2SO_4, and 1% (owf) Albegal B (dispersing agent) heated to boiling over a period of 20 min, and the dye solution was boiled for about 10 min with a 10 × 20 in. wool swatch (enough for four burning tests). HET acid was then added to the boiling dyebath. After 20 min of further boiling, the fabric was washed with water, air-dried, and steam-decated. Simultaneous application of HET acid at the start of dyeing or at room temperature was unsatisfactory. Flame tests on fabrics were carried out according to AATCC 34-1949. The treatment is judged effective when the average char length of five specimens ($3\frac{1}{2} \times$ in.) is less than 7 in. and the afterflame time of none of the specimens exceeds 10 sec.

Ultraviolet spectrophotometry proved convenient for measuring exhaustion of HET onto wool. The compound has a λ_{max} at 213 nm at pH 2.7 with $\epsilon = 6140$. Measurement of the exhausted dyebath solutions showed that 95% of HET was adsorbed when initial concentrations ranged from 2 to 12% of weight of fabric. Evidently, the extent of uptake was independent of initial HET concentration in the range studied.

Satisfactory flame resistance was achieved with HET concentration of about 8% owf. Treated fabrics retained effective flame resistance after ten dry cleanings and after ten washings in Woolite (Boyle-Midway Co.) at room temperature acording to label specifications. However, with shrink-resist fabrics, the HET treatment was not fast to normal laundering. Knit fabric was effectively treated. For example, 4/8's worsted knitting yarn treated with 12% HET (owf) after dyeing with 2% (owf) acid leveling dye and knitted after treatment had an average afterflame time of 0.7 sec (control: 75 sec) and an average char length of 0.8 in. (control: completely burned). Polyurea-treated shrink-resistant wool was also effectively protected without impairing shrink resistance. A woven shrink-resistant fabric, treated with 8% HET (owf) in a paddle wheel dyer after dyeing with 2% (owf) Chrome Brown RLL, had an average afterflame time 1.9 sec (control: 29 sec) and an average char length of 5.0 in. (control: completely burned).

Although the chemistry of interaction between the HET acid and wool has not been studied, we .presume that its hydrophobic ring system and acid groups promote binding to wool somewhat analogously to dyeing with an acid dyestuff. Thus the carboxyl groups may form salt linkages with amino, guanidino, and imidazole groups in wool assisted by hydrophobic interactions between the chlorinated ring system of HET acid and aromatic and aliphatic hydrophobic side chains in wool. It is also possible that HET acid may interact or complex with the metal-dye molecules. Covalent bonding with the protein is unlikely under dyeing conditions. Although such processes imply competition between HET acid and dye for binding sites on wool, no competitive effect of HET acid and dye is apparent; i.e., the treatment did not noticeably affect the shades produced by dyeing. Also, the treatment had little or not effect on fabric hand. Chlorendic anhydride turned wool brown on heating (Table 5) and should therefore be used only with dark-shaded fabrics.

TABLE 5
Color Changes on Heating Native and Treated Wool Fabrics[85]ᵃ

Fabric	Temp. (°C)	Time (min)	R_D	a	b
Native wool	Before heating	5	65.9	−1.2	15.5
	125	5	64.8	−0.8	15.3
	125	15	65.2	−0.8	16.3
	150	5	64.4	−0.7	16.7
	150	15	64.4	−0.9	18.4
Wool plus	Before heating		66.2	+1.7	14.8
chlorendic acid	125	5	62.7	+3.7	17.7
(12% owf)	125	15	56.1	+6.7	20.4
	150	5	50.9	+8.2	21.8
	150	15	31.8	+9.5	22.6
Wool plus tetra-	Before heating		69.3	−0.6	14.9
bromophthalic	125	5	68.8	−0.6	14.9
anhydride	125	15	69.5	−1.0	16.9
(12% owf)	150	5	68.4	−1.1	18.3
	150	15	67.4	−2.1	21.8

ᵃ Fabric color changes are defined objectively by reflectance and color parameters obtained with a Gardner Model 7946 Automatic Color Difference Meter and shown in the table. R_D is the overall reflectance of visible light by the fabric. Parameter a measures redness (positive values) *vs* greenness (negative values). Similarly, b measures yellowness (positive values) *vs* blueness (negative values). Native wool yellows when heated. The effect is slow at 125°C. The slight differences between native wool and wool treated with tetrabromophthalic anhydride are of doubtful significance, while the rapid browning of wool with 12% chlorendic acid is clearly measurable (even in minutes) at 125°C. The major advantage of TBPA over chlorendic acid results from the elimination of fabric browning during heating. Prolonged heating of TBPA-treated fabric can cause slight yellowing. This does not, however, appear to be a serious problem. See R. S. Hunter, Photoelectric Tristimulus Colorimetry with Three Filters, National Bureau of Standards Circular C429, U.S. Government Printing Office, Washington, D.C. (1942).

To avoid the browning problem we evaluated several other halogen derivatives.[56,79,85] Tetrabromophthalic anhydride (TBPA) is the best we have found (Tables 6-8, Figs. 7 and 8). Many kinds of wool yarn and fabric, including airplane upholstery, blankets, and carpeting, have been effectively protected. We also treated wool blends with cotton, rayon, nylon, and dacron (polyester) as well as polyurea-treated shrinkproof wool.[199] The treated fabrics do not turn brown at normal ironing temperatures.

For satisfactory protection of wool–polyester fabric and wool–rayon felt (Table 8), we had to add 1% formic acid to the dyebath (pH 2.6). The formic acid may act as a dyeing assistant by swelling the fabric; other assistants tried, including acetic acid and urea, were less effective. We are testing other swelling and suspending agents.

The flame resistance of TBPA-treated wool and wool blends remains satisfactory after dry cleaning with perchloroethylene and washing in warm water and Woolite (a mild detergent). TBPA can be used effectively with

TABLE 6

Flame-Retardant All-Wool Materials Prepared with Tetrabromophthalic
Anhydride (TBPA) under Dyebath Conditions at pH 2.0–2.2[81,82]

Material	TBPA in bath (% owf)	Flammability[a] after flame (sec)	Char length (in.)	Remarks
Wool fabric,	2	12.6	5.6	
7 oz/yd²,	4	7.2	3.1	
2/1 twill	6	2.3	2.1	
	8	1.8	1.5	
	10	0.4	1.6	
	12	0.2	1.3	
	Control	33.6	Total	
Wool Fontana red commercial fabric, 12 oz/yd²	12	0.6	0.0	
Wool serge,	18	2.0	2.7	As is
type 1, 12 oz/yd²		0.3	3.3	After 15-min wash and after dry cleaning
Wool double knit, 7 oz/yd², 2/1 twill	15	0.0–0.3	1.8–2.5	
Ozone-treated	8	0.8	3.8	
shrinkproof wool, 7 oz/yd², 2/1 twill	12	0.5	2.9	
Wurset shrinkproof wool fabric, 7 oz/yd², 2/1 twill	16	0.0	3.3	
Wool knitting yarn,	12	0.4	2.3	Flame tests on yarn
1/30's	Control	40.1	Total	knitted into fabric
All-wool blanket				Oxygen index:
	4	31.0	Total	29.0
	8	0.0	2.3	31.8
	12	0.0	2.3	34.6
	Control	32.5	40.1	25.4

[a] Numbers listed are averages of five to ten determinations.

TABLE 7

Flammability Characteristics of Knitted Samples from Wool Carpet Yarn
Treated with Tetrabromophthalic Anhydride[81]

Sample	After flame (sec)	Char length (in.)	Oxygen index[a]	ASTM E8 Tunnel test rating[b]
Untreated	50+	Total	25.3	275
Treated	0.5	1.5	31.3	35

[a] % O₂ needed to sustain combustion (ASTM D2863-70).
[b] Hospitals receiving funds through the Hill-Burton Act must have a flame spread of 75 or less.

TABLE 8

Flame-Retardant Wool Blend Fabrics Prepared with Tetrabromophthalic
Anhydride (TBPA) under Dyebath Conditions at pH 2.0–2.2[81]

Material	TBPA in bath (% owf)	Flammability[a] after flame (sec)	Char length (in.)	Remarks
90:10 wool–nylon	8	0.7	1.3	Oxygen index: 12%:
airplane uphol-	12	0.5	1.2	37.1
stery, 12 oz/yd²	16	0.1	1.2	
	Control	70.0	Total	
85:15 wool–nylon	16	0.2–0.7	2.7–3.0	In commercial use
blankets				
80:20 wool–dacron	16	0.2–0.7	2.8–3.0	10% urea (owf) added
fabric (red), 11½				to dyebath
oz yd², knitted				
from 1/30's yarn				
75:25 wool–dacron	16	0.0	2.9–3.2	1% (owf) formic acid
fabric, 11½				is needed to achieve
oz/yd², knitted				flame resistance
from 1/30's yarn				
65:35 wool–nylon	12	2.2	2.5	
fabric, 11 oz/yd²,	Control	28.2	8.3	
knitted from				
1/30's yarn				
50:50 wool–cotton,	16	0.15–1.1	5.8–6.2	Fabric should be
12 oz/yd²	Control	46.3–52.1	Total	predyed
50:50 wool–rayon	16	0.4–0.7	1.7–3.3	pH study with all-wool
felt				felts shows treat-
				ment effective up to
				about pH 4.0

[a] Numbers listed are averages of five to ten determinations.

various types of dye. Treated fabrics showed no irritating effect when tested
for dermal toxicity by the rabbit ear test (see below).

We received a report from a commercial carpet company on a trial in
which 100% wool yarn, treated by us for flame resistance with TBPA under
dyebath conditions, was made into an institutional carpet. Samples were then
tested by the drastic building materials Tunnel Test ASTM E84. The test
sample came through with an outstanding rating of 35, against 275 for the
untreated wool (Table 7). United States Department of Health, Education
and Welfare rulings stipulate that contract-grade carpets for certain critical
installations, such as hospitals, rest homes, etc., have this test. Carpets for
hospitals receiving funds through the Hill-Burton Act must have a flame-
spread rating of 75 or less.

FIGURE 8. Vertical flame test results for 90:10 wool–nylon airplane upholstery; left side, control; right side, treated with 12% (owf) of TBPA under dyebath conditions.

We have also treated commercial wool–nylon blanket material in a paddle wheel dryer at 200°F for 20 min in the presence of 16% TBPA (of weight of fabric). We then tested the following swatches for flammability according to DOC PFF 5-73 (vertical flame test): (1) pressed then steamed to give an "as-is" appearance, (2) washed in a mild detergent (Woolite), (3) washed for 15 min under normal laundry conditions, and (4) leached for 1 hr in water at 190°F to stimulate sterilization conditions. All swatches were flame resistant.

Although the chemistry of the interaction of wool with chlorendic acid or tetrabromophthalic acid (presumably formed by hydrolysis of TBPA in the dyebath) had not been studied, we presume that the hydrophobic ring system and acid groups promote binding about as effectively as an acid dye. With the help of hydrophobic interaction between the halogenated ring and hydrophobic groups in wool, the carboxyl groups may form salt linkages with basic groups in the wool. Since tetrabromophthalic acid is not efficiently sorbed on wool above pH 4, where the carboxyl groups are completely ionized and where the hydrophobic nature of the molecule is decreased, hyrophobic interactions between halogenated aromatic diacids and related compounds and wool are probably important.

Recent studies in our laboratory also suggest that it is possible to prepare flame-resistant wool durable to laundering by applying silicone-type shrink-proofing treatments[7,79,80] to wool which has been previously made flame resistant with TBPA under dyebath conditions or treated with TBPA in hot ethylene glycol.[203] Additional studies are needed to optimize these processes.

Because of these and other favorable tests, we believe that TBPA, a commercially available, inexpensive compound, should be widely useful for commercial flame-retardant treatment of wool and wool blends.

The excellent results were confirmed by Thorsen,[241] who also tested the following additional aromatic bromine derivatives for their ability to flame-resist wool: 2,4,6-tribromo-3-aminobenzoic acid, tetrabromophthalimide, pentabromophenol, 2,4,6-tribromo-3-aminobenzenesulfonic (metanilic) acid, and the indicator sulfobromophthalein. The first three compounds were ineffective and the tribromometanilic acid turned the wool brown. None of the compounds were washfast. Since in addition to TBPA, tribromometanilic acid showed the most promise as a potential flameproofing compound for wool, an effort was made to increase its durability by anchoring it to wool with several reactive compounds.[72] Preliminary studies show that it is possible to successfully anchor flame retardants, such as tribromometanilic acid, 2,5-dibromosalicylic acid, 3,5-dichloroanthranilic acid, 3-amino-2,5-dichlorobenzoic acid, as well as potassium hexafluorozirconate, with the aid of suitable cross-linking agents, e.g., formaldehyde, glutaraldehyde, and chromium acetate, to achieve a certain degree of durability to laundering. Other cross-linking agents, including divinyl sulfone, N-methylolacrylamide, and tetramethylene-bis-acrylate, did not anchor 2,4,6-tribromometanilic acid to wool because the treated fabrics were not washfast. More work is needed to optimize these reactions.

6.1.4. Multipurpose Finishes

The excellent textile properties of wool are well known. However, wool has certain limitations. If it is not properly cared for, it is subject to laundering shrinkage and insect damage. Although, compared to other fabrics, it is relatively resistant to ignition, it will burn. Thus, it would be more useful as a garment material if it could be made resistant to (1) laundering shrinkage, (2) insects, and (3) fire. Studies intended to develop finishing treatments for wool and wool blends to overcome these limitations[79,80] clearly show that the combined, multipurpose treatment applied in a normally used process is most economically feasible. Dyeing is a particularly logical finishing step with which other wet finishing treatments can be combined. Our current experiments include the following combinations: (1) *Insectproofing and dyeing:* The highly effective synthetic pyrethroid mothproofing agent, permethrin (*m*-phenoxybenzyl-*cis-trans*-(±)-3-(2,2-dichlorovinyl)2,2-dimethylcyclopropanecarboxylate), was found compatible with several dye types (Tables 9,10). (2) *Insect- and flame-proofing:* Permethrin can be applied to tetrabromophthalic anhydride flame-resistant wool to impart both insect- and flame-resistance to wool (Table 11). (3) *Insect-, flame-, and shrink-proofing:* Certain shrinkproofing treatments can be applied to in-

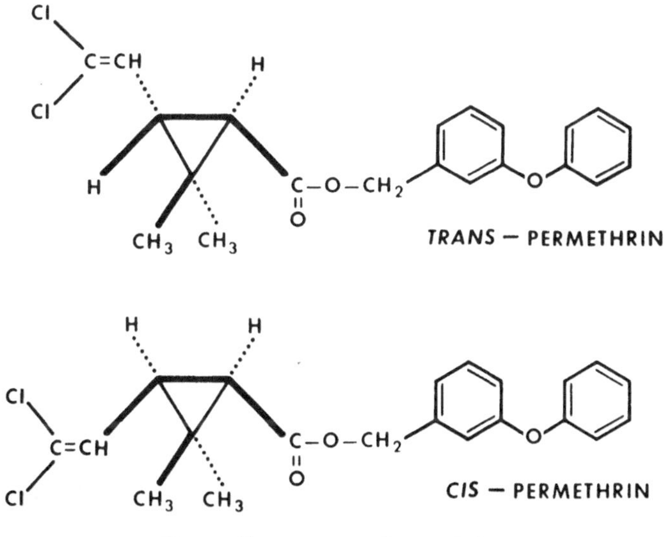

FIGURE 9. Structures of permethrin.

TABLE 9

Mothproofing of Wool Fabrics, Wool Carpet Yarn, and Wool
Hand-Knitting Yarn During Dyeing with Permethrin (0.005%
of weight of fabric)[79,80]

Dye type[a]	Conditions of treatment[b]	Excrement per larva (mg)[c]	
AL—Alizarine Sky Blue 6 GLW	A	0.13	
	B	0.11	(0.09)[d]
	C	2.90	(0.27)
C—Chrome Fast Blue GBX Conc	A	0.12	
	B	0.11	
	C	2.29	
FR—Lanasol Scarlet 2R	A	0.12	
	B	0.10	(0.10)[e]
	C	2.36	(2.57)
NM—Supralan Yellow NR	A	0.12	
	B	0.16	
	C	1.49	
AM—Vitrolan Yellow BE	A	0.13	
	B	0.14	
	C	1.63	
None	—	2.88	

[a]AL, acid leveling; C, chrome; FR, fiber reactive; NM, neutral metalized; AM, acid metalized.
[b]A, permethrin first, then dye; B, dyed and treated with permethrin simultaneously; C, dyed controls.
[c]Black carpet beetle larvae were allowed to feed on the samples at the Stored Products Insects Research and Development Division, Agricultural Research Service, U.S. Department of Agriculture, Savannah, Georgia. Treated and untreated samples (0.5 g) were exposed to the larvae and excrement weight was measured. Wool is considered satisfactorily resistant if the average quantity of excrement per larva is not over 0.5 mg, provided no single value is over 0.6 mg, and the untreated control is over 1.5 mg. (Cf., N. H. Koenig and M. Friedman, in *Protein Crosslinking Biochemical and Molecular Aspects* (M. Friedman, ed.), Plenum Press, New York, pp. 355–382 (1977); R. E. Bry, R. A. Simonaitis, J. H. Lang, and R. E. Boatright, Soap Cosmetics Chemical Specialties, July 1976, 4 pages; and Soap and Chemical Specialties, Textile Resistance Test, *Blue Book*, **47** (4A), 168–171 (1971).
[d]Values for carpet yarn.
[e]Values for hand-knitting yarn.

sect- and flame-resistant wool to impart a triple benefit (Table 12). The described multipurpose finishes merit evaluation for commercial adoption.

It is also possible that the two chlorine atoms in permethrin (Fig. (9) may impart a certain degree of flame-resistance to wool and/or synergize and complement the flame-retardant effect of TBPA or other flame-retardants.

TABLE 10

Results of 14-Day Chemical and Soap Manufacturer's Association (CSMA) Carpet Beetle Feeding Test Conducted with Permethrin-Treated Wool/Blend Fabrics

Fabric	Bath concentration of permethrin (%/wt. of fabric)	Excrement per larva (mg)	
		Treated	Untreated
76/24 wool/acrylic	0.003	0.14	
grey plaid	0.005	0.15	2.67
55/35/10 wool/dacron/mohair	0.003	0.15	
red plaid	0.005	0.16	4.78
65/35 polyester/wool	0.003	0.13	
green plaid	0.005	0.14	1.15
70/30 wool/nylon	0.003	0.13	
maroon plaid	0.005	0.14	3.03
55/45 dacron polyester/wool	0.003	0.15	
light yellow	0.005	0.14	1.42
50/50 wool/nylon	0.003	0.16	
dark red upholstery	0.005	0.14	0.99
65/35 wool/nylon	0.005	0.12	3.12
50/50 wool/dacron	0.005	0.11	3.99
50/50 wool/cotton	0.005	0.11	2.53
55/45 wool/cordelan	0.005	0.11	2.58
100% wool	0.005	0.11	2.95; 3.41

TABLE 11

Compatibility of TBPA-Treated Flame-Resistant Wool with Permethrin[79,80]

Treatment	Excrement per larva[a] (mg)
Untreated control	2.32
Dyed control	1.20
8% (owf) TBPA control	0.82
16% (owf) TBPA control	0.77
Acid-premetalized dye + 8% (owf) TBPA	0.64
Acid-premetalized dye + permethrin (0.005%)	0.11
TBPA + permethrin (0.005%)	0.10

[a] Results of a 14-day Chemical and Soap Manufacturer's Association (CSMA) feeding test with black carpet beetle larvae of TBPA-treated flame-resistant wool.

TABLE 12
Flame–Moth–Shrink-Resistant Wool

Treatment	Excrement per larva[a] (mg)
Ozone shrinkproof + TBPA (8% owf) + acid-leveling dye	0.56
Ozone shrinkproof + TBPA (8% owf) + permethrin (0.005%)	0.10
Wurset shrinkproof + acid-leveling dye + TBPA	0.43
Wurset shrinkproof + acid-leveling dye + TBPA + permethrin (0.005% owf)	0.13
ZM-799-polysiloxane shrinkproof	2.34
Permethrin (0.005%) on ZM-799-polysiloxane shrinkproof, dry clean	0.06
Permethrin (0.005%) on ZM-799-polysiloxane shrinkproof, 75 min accelerator wash	0.06
Untreated control	2.11

[a] Results of a 14-day CSMA feeding test with black carpet beetle larvae of shrinkproof and TBPA-treated flameproof wools.[79,80] Permethrin was applied under dye-bath conditions to wool fabric previously treated with flame- and shrink-proofing compounds, as indicated.

6.1.5. Bromine Analysis

The applicability of X-ray fluorescence analysis to the determination of bromine in textiles has been demonstrated by Nelson *et al.*[192] The X-ray method, in contrast to the chemical method, is fast and nondestructive and does not suffer from interferences by chlorine or iodine; it also eliminates the possibility of low results from incomplete combustion during analysis of flame-resistant textiles. Cut or whole fabric is measured directly without ashing or grinding.

We developed a method in which bromine is measured in flame-resistant wool fabric by X-ray fluorescence spectrometry with a relative precision of 3–8% and relative accuracy of better than 10%.[207,208] The method computes the bromine concentration from fluorescence measurements of the sample. X-ray fluorescence is generally useful for routine analyses of bromine in textiles and has advantages over wet chemical analysis (Table 13).

TABLE 13
Percent Bromine in TBPA-Treated Wool[207]

	Undyed			Dyed	
Sample	X-ray (%)	Chemical (%)	Sample	X-ray (%)	Chemical (%)
1	0.0 ± 0.0	0.0	1	1.19 ± 0.03	1.21
2	1.20 ± 0.03	1.31	2	2.13 ± 0.15	2.26
3	3.16 ± 0.05	3.40	3	2.92 ± 0.23	3.13
4	5.07 ± 0.14	5.67	4	3.79 ± 0.12	3.93
5	5.78 ± 0.13	6.38			

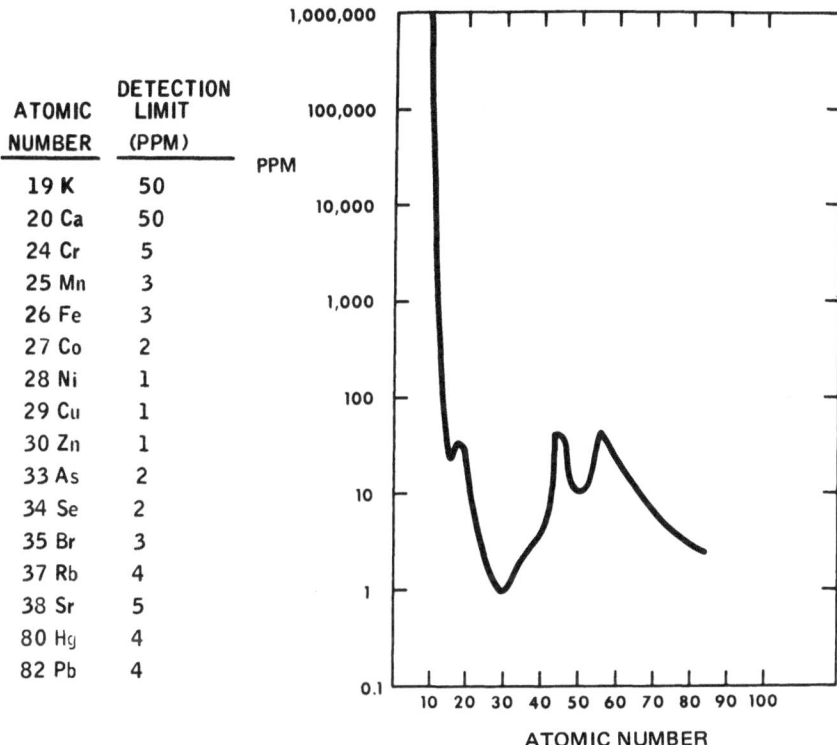

ATOMIC NUMBER	DETECTION LIMIT (PPM)
19 K	50
20 Ca	50
24 Cr	5
25 Mn	3
26 Fe	3
27 Co	2
28 Ni	1
29 Cu	1
30 Zn	1
33 As	2
34 Se	2
35 Br	3
37 Rb	4
38 Sr	5
80 Hg	4
82 Pb	4

FIGURE 10. Detection limits in biological materials including wool by X-ray fluorescence spectrometry.[207]

On the basis of our experience with this technique, the described quantitation procedure for bromine should be applicable to any textile chemical treatment that introduces compounds which contain elements that can be measured by X-ray fluorescence (Fig. 10). However, light elements such as chlorine, sulfur, and phosphorus (atomic no. 17 or less) are possible exceptions to this generalization because low-energy X-rays from these light elements can be highly attenuated by textile fabric samples.

Atomic absorption is also used extensively to determine flame-retardant elements in textiles.[115,127] X-ray photoelectron spectroscopy is finding similar applications in wool research.[185,186]

The question arises about the amount of residual tetrabromophthalic anhydride remaining in the dyebath after the dyebath application of the flame retardant is completed. Possible answers may come from the following experiments:

1. Analysis of the treated fabric for bromine should indicate how much tetrabromophthalic anhydride is left in the dyebath following treat-

ment. This should be confirmed by analysis of the spent dyebath and effluent from subsequent finishing treatments.

2. Treatment of new fabric or of waste wool first with spent dye liquor to absorb any residual tetrabromophthalic anhydride and then in a new dyebath solution, i.e., use of the countercurrent principle. The bath may be replenished a number of times and used over and over again before discarding.

3. Neutralization of dyebath with sodium hydroxide may precipitate most of the tetrabromophthalic acid left in solution.

4. Treatment of the effluent with an anion exchange resin to adsorb any tetrabromophthalic acid left in the dyebath.

5. Use of ultraviolet spectroscopy, as described above for chlorendic anhydride, when tetrabromophthalic anhydride is applied under dyebath conditions but in the absence of any dye. Note that the absorbance of a dye would interfere with the absorbances of both anhydrides.

6.1.6. NMR Studies

The use of nuclear magnetic resonance (NMR) spectroscopy to characterize newly synthesized halogenated aromatic compounds that had potential utility as flameproofing compounds[90] is illustrated with the brominated product of metanilic acid.

The NMR spectrum for metanilic acid in d_6-DMSO at 12°C referenced to acetonitrile permitted the following assignments to the various protons: H_4, 5.21 ppm; H_5, 5.38 ppm; H_6, 5.62 ppm; and H_2, 5.67 ppm. The corresponding spectrum for tribromometanilic acid has a single aromatic peak at 5.67 ppm and labile proton peaks at 3.55 and 3.57 ppm, presumably due to the amino and sulfonic acid protons. Using ortho and meta shift values obtained from disubstituted benzene derivative[226] in organic solvents, we calculated a theoretical value of 5.70 ppm for 2,4,6-tribromometanilic acid and of 5.45 ppm for the 4,5,6-tribromo isomer. These results strongly suggest that the bromination product of metanilic acid is 2,4,6-tribromometanilic acid and not one of the other possible isomers.

The NMR spectrum also shows the presence of 1.5–2.0 moles of H_2O per mole of tribromometanilic acid. See Fig. 11.

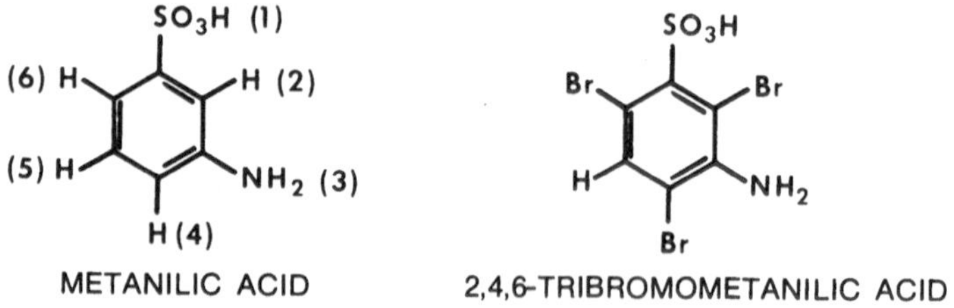

FIGURE 11. Structure of metanilic acid and brominated derivative.

6.2. Iodinated Wool

Since chlorine and bromine derivatives can impart flame resistance to wool, I wanted to find out whether iodinated wool would be flame resistant. I placed a 3 × 10 in. wool swatch into 600 ml of borax (pH 9.1) which was made $0.05M$ in I_2 and $0.25M$ in KI. The wool swatch was shaken for 24 hr and then extensively rinsed with water and $0.1N$ sodium thiosulfate. Elemental analysis showed that it contained 10% iodine by weight. The iodinated wool was flame-resistant. Iodinated wool is brown, and the extent of discoloration was related to the percent of iodine present in wool.

6.3. Tris(2,3-dibromopropyl) Phosphate and Related Compounds

A major problem in flame-retardant studies of wool is that, unless fabric shrinkage is prevented, it is hard to assess the fastness of the treatment to laundering, since the flammability characteristics of wool change greatly as a result of shrinkage and felting. Because of its compactness, felted wool is more resistant to fire than unfelted wool. For this reason evaluations of the stability of wool flameproofing treatments to laundering are questionable unless the fabric shrinkage has been simultaneously or independently controlled.

To avoid this difficulty, we[239] explored the feasibility of introducing flame retardants into polymer films deposited interfacially on the fiber surface. We reasoned that such an agent locked onto the film should provide permanent flame and shrink resistance without impairing fabric hand significantly more than the polymer finish alone. An inherent advantage of this treatment is that flame retardants that do not react chemically with the wool can be used.

The fabric was padded through a perchloroethylene solution containing 15–25% (w/v) tris(2,3-dibromopropyl) phosphate (tris) and 3–5% (w/v) polyurethane prepolymer (derived from a polyether triol, molecular weight about 3000, reacted with toluene diisocyanate) and tumble-dried at 160°F for 10 min or until dry. The fabric was then padded through an aqueous solution of 2% (w/v) triethylenetetramine, lightly scoured, and again tumble-dried to dryness at 160°F for 25 min. The amount of flame retardant deposited was calculated from weight increase and elemental analysis of phosphorus and bromine. The phosphorus content of the treated fabrics was about 1%, and the bromine values ranged from 12 to 15%.

Results indicate that application of either 20 or 25% tris solution with either 3 or 5% prepolymer imparted satisfactory flame retardance before laundering. A 25% solution is needed for complete effectiveness after laundering. The treatment is not durable to dry cleaning. Our observations were confirmed by Fincher and collaborators,[65] who also successfully incorporated the related dichloropropyl phosphate and hexabromodiphenyl but not triphenyl phosphate into a polyurea shrink-resist coating for wool (cf. also Ref. 139). Although the use of tris in flameproofing formulations is inadvisable for safety reasons (see below), the described strategy should work with numerous other nonreactive flame retardants.

The same authors also reported that zirconium mordanting or THPC flame-resist applications were completely lost if they were followed by an application of polyurea, polyacrylate, or polythiol coatings for shrinkproofing treatments. (However, the zirconium treatment seems to be effective when the polymer is applied first.) They explained these results by noting that these shrink-resist polymers have limiting oxygen indices that are about 30% lower than the index for chemically treated flame-resistant wool. We observed similar results in our attempts to impart flame resistance to previously shrinkproofed wool. For flame resistance polyurea-shrinkproofed wool required much higher add-on of the flameproofing compound than native wool. Additional flameproofing compound is therefore required to overcome the high flammability of the polymeric coating used to shrink-resist wool.

Retention of the flame-resistant effect of zirconium mordanting or related treatments requires the use of a shrink-resist resin with a limiting oxygen index which is comparable to or greater than that of wool. Certain chlorinated polymers, such as vinyl chloride polymers, appear to meet this requirement.[3,65] These strategies nevertheless merit further exploration because they impart to wool a double benefit, flameproofing and shrinkproofing, in a single treatment. Further studies may disclose that bulky and reactive flameproofing compounds can be successfully incorporated into shrinkproofing resins to achieve a double treatment that is durable to laundering and to dry cleaning.

6.4. Vinyl Phosphonate Esters

Friedman and Tillin[96] studied effects of bis(β-chloroethyl)vinyl phosphonate (VP, I) on combustion of wool. This compound was known to combine with protein functional groups in nucleophilic addition reactions and was also expected to chemically modify wool.[95] Although treatment of native wool under ionic conditions did not result in a sufficient uptake to achieve flame resistance, modification of reduced wool at pH 8 resulted in a weight uptake of about 6%. This modified wool was flame resistant in a vertical flame test. Amino acid analysis of the hydrolysates of the modified wool showed that VP alkylated nearly all of the sulfhydryl groups generated by reduction.

Since reduction of wool decreases fiber strength, we tried grafting VP to native wool fabrics under various free-radical conditions, including the use of aqueous ceric ion to initiate the polymerization. Such treated fabrics were flame resistant, and the treatment was fast to dry cleaning but not to laundering.

The flame resistance of the vinyl-phosphonate-treated, 50:50 wool-cotton blend fabric is, presumably, due to a combination of this compound with the wool component. We, therefore, wanted to determine whether a flameproofing treatment that was effective for cotton would also protect wool–cotton blends. We tested the flame-retarding effectiveness of a commercially available cotton flame-retarding compound in wool–cotton blends.[98] This compound, which contains 22.5% phosphorus, is a water-soluble, oligomeric vinyl phosphonate with an average of more than one vinyl unit per molecule in which the repeating units have the structure shown in (2). It is formulated with *N*-methylolacrylamide and the potassium persulfate catalyst.

$$
\begin{array}{cc}
\underset{\substack{\|\\ \text{P(OCH}_2\text{CH}_2\text{Cl})_2 \\ |\\ \text{CH}=\text{CH}_2}}{\text{O}} & \underset{\substack{\|\\ (-\text{O}-\text{P}-\text{O}-\text{R})_x \\ |\\ \text{CH}=\text{CH}_2}}{\text{O}} \\
(1) & (2)
\end{array}
$$

The Fyrol E-76 formulation remains stable for several days if it is kept under mechanical or air agitation. However, we prepared a fresh batch for each treatment. Fabrics were immersed in the solution and padded to about 85–100% wet pickup. From the curing conditions recommended,[64] 15 min at 250°F without predrying was selected. The aftertreatment wash was 1% sodium perborate solution at 80°F for 2 min. The wash is recommended to neutralize the fabrics and to remove any unreacted chemicals or colored products.

Since the free-radical curing may be somewhat inhibited by disulfide or other groups in wool, raising the persulfate ($K_2S_2O_8$) level and decreasing the *N*-methylolacrylamide level (as was found useful for sulfur-dyed cotton) might improve results with respect to durability.[64]

We found that wool–cotton and wool–cotton–polyester blends with at least $32\frac{1}{2}\%$ cotton become flame resistant when treated with an oligomeric, water-soluble vinyl phosphonate (Table 14). Fabric hand is excellent. The treatment is durable to dry cleaning and three 15-min launderings. Flame resistance is good in these fabrics, with a limiting oxygen index of 26.6–27.4. The cotton portion of the wool–cotton blend evidently acts as a carrier for the polymerized phosphorus compounds. Blend flammability is not predictable on the bases of flammability of the individual components.[87]

Needles[191] has shown that Fyrol 76 can be successfully and rapidly grafted to wool when exposed to an electron beam (1–10 mrad). The percentage Fyrol reacted can be as high as 80% and is highest for dry wool under a blanket of nitrogen at dose rates as low as 1 mrad. Graft uptakes of as little

TABLE 14

Flammability of Wool, Cotton, and Wool–Cotton–Polyester Blend Fabrics Tested with Fyrol® E-76[98]

Fabric	Uptake (%)	Uptake after rinse (%)	After-flame (sec)	Char length (in.)	LOI
Wool control				Total	25.6
Wool	18	7.5		Total	
Wool[a,b]		5.0		Total	
Wool	25		0	4.3	
Wool	22[c]		0	4.7	
Woven cotton[b]	35	26	0	5.2	
Knitted cotton control				Total	20.4
Knitted cotton[b]		18.9	0	5.0	27.4
50:50 wool–cotton control				Total	22.0
50:50 wool–cotton[b]	22.4	18.5	0	4.7	26.6
50:50 wool–cotton[b,d]		18.8	0	5.8	
50:50 wool–cotton[b]		11.0	0	6.2	
37.5:37.5:25 wool–cotton–polyester control				Total	21.4
37.5:37.5:25 wool–cotton–polyester		17.4	0	5.5	27.2

[a] Presoak in 5% $Na_2S_2O_2$.
[b] Borate rinse.
[c] Borate rinse after dry cleaning.
[d] Three 15-min washes.

as 6% Fyrol 76 applied to slack weight wool fabric are sufficient to raise the limiting oxygen index of the wool from 24.5 to 28.5 and to render the fabric flame retardant by DOC 3-71 standards without significant loss in the physical properties of esthetic characteristics of the wool.

6.5. Tris(1-aziridinyl) Phosphine Oxide (APO)

Simpson developed a procedure for flameproofing by polymerizing APO by BF_3 in the presence of wool.[224] As with cotton, the reactive compound could probably be applied from chlorinated solvents which are commercially practical for finishing.

Organic flame retardants for wool and wool blends are summarized in Table 15.

TABLE 15
Organic Flame Retardants for Wool and Wool Blends

Compound and references	Remarks
3-Amino-2,5-dichlorobenzoic acid (67, 72)	Ineffective when applied from dilute HCl; effective and durable to one-15 min wash when treated in the presence of formaldehyde
Bis(β-chloroethyl)vinyl phosphonate (76, 97)	Durable when applied to reduced wool
Brominated alcohols and phenols (236)	
Brominated and chlorinated acid anhydrides and halides (98, 101, 253, 254)	Durable when applied from DMF
Brominated triallyl phosphoric triamide (250)	
p-Bromobenzenesulfonic acid (67, 72)	Ineffective when applied form dilute HCl
p-Bromobenzoic acid (67)	Ineffective when applied from dilute HCl
Chlorendic acid and anhydride (56, 84, 85, 101, 253)	See text
Chlorinated polymers (3, 65)	Treatment is durable to laundering; antimony oxide acts as synergist
Dibromosalicylic acid (67, 72)	Treatment durable to dry cleaning and washing when done in dilute HCl at 70° C in the presence of formaldehyde
Dichloropropyl phosphate (65)	Can be coapplied with shrinkproofing treatments
Dimethylaminoethyltetrabromo acid phthalate (56)	
Diphenyl ether-4,4'-disulfonyl chloride (144, 145)	Treatment imparts both moth and flame resistance
Hexabromodiphenyl (65)	Can be applied with shrinkproofing treatments
Novolac wool fireproof phenolic resin [*Chem. Abs.* **84** (24), 166, 196 (1976)]	
Phosphazene derivatives [*Chem. Abs.* **87** (6), 40, 702 (1977)]	
Phosphonate esters (Br. Pat. 1,429,004 to Hooker Chemical)	
Phosphorus-containing vinyl compounds (64, 96–98, 156, 243)	Grafting to wool changed character of thermal decomposition and decreased the amount and rate of gas evolution
Polyamino polyphosphonate as coreactant with long-chain aliphatic diisocyanate (222)	Imparts both flame and shrink resistance
Polyethylene glycol phosphonium chloride [*Chem. Abs.* **84** (20), 137, 166 (1976)]	Effective for blends
Propylene phosphate derivatives (244)	Effective for nylon–wool carpets
Sulfonamidoethylphosphonates (106)	
Tetrabromofluorescein sodium salt (67)	Ineffective when applied from water or dilute HCl

TABLE 15 (*Continued*)

Compound and references	Remarks
Tetrabromophthalic acid and anhydride (56, 79–85, 241, 250)	See text; highly effective for wool–nylon blends
Tetrabromophthalein (67)	Ineffective when applied from water or dilute HCl
Tribromometanilic acid (2, 67, 72, 241)	Treatment is durable to three 15-min washes when done in presence of formaldehyde or glutaraldehyde in dilute HCl at 70°C for 30 min
Triethyl- and hexamethyl phosphate (50)	
Tetrabromo-*o*-cresol (67)	Not sufficiently soluble to be applied from aqueous solution
Tris(1-aziridinyl) phosphine oxide (APO) (224)	See text
Tris(2,3-dibromopropyl) phosphate (65, 101, 239)	See text
Vinyl phosphonate oligomer (64, 97, 98, 252)	See text; effective for wool–cotton and wool–polyester blends
p-Vinylbenzenephosphonic acid [*Chem. Abs.* **58**, 1419f (1963)]	
Wool–synthetic fiber blends (28, 30, 61, 79–81, 124, 125, 187, 188, 212, 237, 245, 246)	
Zinc chloride (bromide) vinyl pyridine (145)	Treatment imparts both flame (unpublished tests) and shrink resistance

7. Mechanisms of Flame Retardation

Many mechanisms have been proposed to rationalize the observed flame-proofing effects of certain compounds.[55,113,159,169] None of the proposed mechanisms can be considered as proven, but some of the rationalizations seem plausible. An understanding of the mechanisms of flammability and flame retardance is not only of theoretical interest and serves not only to satisfy our intellectual curiosity but might lead to the intelligent design of simple flameproofing treatments to achieve specific objectives.

Pearce and Liepins[202] note that four fundamental stages characterize polymer flammability: preheating, decomposition, ignition, and combustion followed by flame propagation. They further note that a flame retardant may act by suppressing or eliminating any of these processes. Flame retardants may therefore act at different stages of the combustion process. The mechanism of flame retardancy presumably differs for each of the four stages, although considerable overlap among them probably occurs in a real situation.

According to Pearce and Liepins,[202] some of the mechanisms by which flame retardants could affect the stages of flammability are the following: (1) Flame retardance may result from formation of a surface film of low thermal conductivity and/or high reflectivity that reduces the rate of heating. (2) Flame retardants may decompose preferentially at low temperature and serve as a heat sink. They can change the products of combustion so that they are less combustible. (3) Flame retardants may form a coating that intumesces when heated into a foamed surface layer with low thermal conductivity.

Compounds that are used to impart flame resistance to textiles usually contain one or more of the following elements: halogen (bromine, chlorine), phosphorus, nitrogen, and metals (antimony, titanium, zirconium, tin, bismuth, and zinc). Frequently, compounds or formulations contain two or more of these elements, which act synergistically; i.e., the combined treatment is more effective than the added sum of individual treatments.

Fiber penetration seems to play a key role in influencing the degree of flame resistance imparted by inorganic metal salts.[22] Thus, because titanium has a smaller atomic radius than zirconium, it is a better flame retardant. Similarly, the flame resistance and washfastness for vanadium, molybdenum, and tungstate salts decreased in the order of increasing atomic radii of the three metals, $V > Mo > W$. However, this conclusion may not be valid or be generally true because it is the complete anion molecule (ZrF_6^{2-}) and not the metal which has to penetrate the fiber structure to exert its flameproofing effect. The molecular radii, not atomic radii, should be compared. Moreover, studies by Gordon and associates,[107–112] cited earlier, seem to indicate that surface penetration is sufficient to impart flame resistance.

Benisek[22] suggests that although the mechanism of flame retardancy by metal compounds is not well understood, the metal derivatives probably increase formation of char and/or acid fragments on mild heating of treated wool. The char or the acid fragments then might trap flammable volatile compounds in the solid or vapor phase and suppress flame propagation.

Beck and associates[17,18] recently evaluated the pyrolysis products of wool and the mechanism of flame retardation by using thermogravimetric analysis, limiting oxygen index, and limiting nitrous oxide index measurements (cf. also Ref. 45). They found no substantial differences in volatile gas emission between zirconium-treated fabrics and untreated wool. They suggest that both zirconium and titanium complexes exert their flame-retarding influence by operating in the solid phase to catalyze wool decomposition. These results contrast with the observed increase in char formation and decrease in weight loss of wool in the presence of phosphoric acid, ammonium dihydrogen phosphate, sulfamic acid, and boric acid–borax.

Antimony exerts a synergistic effect with halogen for wool. For example, antimony oxide synergizes the flame-retardant effect of halogen-containing compounds.[70] Phosphorus acts synergistically with nitrogen-containing compounds, especially for cotton. This synergism does not appear to operate for wool presumably because wool contains nitrogen.

One possible pathway by which halogen-containing flame retardants might exert their effects is by preventing the free-radical chain propagation process that is thought to take place during combustion. This pathway may be illustrated with the following equations:

$$R—H + O_2 \longrightarrow HO_2· + R· \tag{1}$$

$$R—CH_3 + ·OH \rightleftharpoons R—CH_2· + H_2O \tag{2}$$

$$R—CH_2 + \tfrac{1}{2}O_2 \rightleftharpoons R—CHO + H· \tag{3}$$

$$Br_2 \rightleftharpoons Br· + Br· \tag{4}$$

$$H—Br \rightleftharpoons H· + Br· \tag{4a}$$

$$R—CH_2· + ·Br \rightleftharpoons R—CH_2—Br \tag{5}$$

$$W—S—S—R + heat \rightleftharpoons 2W—S· \tag{6}$$

$$R—CH_2· + ·S—W \rightleftharpoons R—CH_2—S—W \tag{7}$$

$$W—NH_2 + R—CHO \rightleftharpoons W—N{=}CH—R + H_2O \tag{8}$$

During combustion substrate (R—H) is decomposed to R· and HO_2· [Eq. (1)]. The latter can extract a proton from a methyl group of the substrate to give an unstable free radical [Eq. (2)], which then combines with oxygen to form combustible aldehydes [Eq. (3)], which can be further oxidized to other compounds by a similar free-radical oxidation. The flame retardants also liberate Br_2 or HBr, which decomposes to a free radical [Eq. (4)]. The bromine free radical can stop the chain propagation by combining with $R—CH_2·$ to terminate the chain and thus retard flame propagation. Generally, aromatic chlorine-containing compounds are less efficient than aliphatic ones. That observation could be rationalized by the tendency of bromine compounds to lose Br_2 to HBr under the influence of heat and the resistance of aromatic compounds to lose halogen, as reflected by the reported bond energies (Table 16). In addition to bond energies, other (secondary) forces such as hydrogen-bonding interactions and cohesive energies of functional groups affect flammabilities of materials.[55]

Some support for these hypotheses comes from our observation that tetrabromophthalic anhydride was a more effective flame retardant for wool than the corresponding tetrachloro analog.

Some of the functional groups present in wool could also participate in blocking the propagation of the free-radical chain during burning or could combine with reactive species, as illustrated in Eqs. (6)–(8). Since disulfide bonds have a low bond energy (about 60 kcal/mol), they may be cleaved

TABLE 16
Bond Dissociation Energies[a]

Bond	Dissociation energy (kcal/mol)
CH₃-Cl (aliphatic)	80
C₆H₅-Cl (aromatic)	88
CH₃-Br (aliphatic)	67
C₆H₅-Br (aromatic)	71
CH₃-I (aliphatic)	54
C₆H₅-I (aromatic)	57
S-S	54–60
H-I	71.4
H-Br	87.5
H-Cl	103.2
I-I	36.1
Br-Br	46.1
Cl-Cl	58.0

[a]Data from M. Kotake, Constants of Organic Compounds, Asakura Publishing Co., Tokyo, Japan, 1963, pp. 537; and J. D. Roberts and M. C. Caserio, Basic Principles of Organic Chemistry, Benjamin, New York, 1964, p. 77; and Einhorn[62]

heterolytically under the influence of heat to give thiyl radicals [Eq. (6)], which can then combine with R—CH₂· to terminate free-radical chain propagation [Eq. (7)]. Amino groups in wool may react with aldehydes, phosgene, and other volatile compounds formed during the combustion process and thus retard flame propagation. Although such reactions have not been demonstrated, they are plausible and could account for the known greater flame resistance of wool compared to other polyamide-type fibers, such as nylon.

8. Safety of Flame-Resistant Fabrics

The safety of flame-resistant fabrics, with regard to factors other than flammability, seems to have been almost completely neglected in most studies. No matter how effective a particular fire-retardant treatment may be, it should not be adopted if the treated fabric elicits skin irritation or an allergic response. For this reason, we evaluate our most promising treated fabrics for dermal toxicity.[101] In this test, a swatch of the fabric is kept in contact with the ear of a rabbit for 24 hr; the ear is observed for redness or swelling. If no primary irritation occurs, the test is continued for about a week. In addition, the treated cloth should be tested to see whether it triggers formation of antibodies or elicits an allergic response, as it may do, even though the reagents and the native wool itself are not antigenic. Thus, although wool itself (W) and the flameproofing compound (hapten, H) may both be nonantigenic, the combination (WH) of the two may be antigenic.[57,221,256] The combination should be tested for antibody formation. In one procedure, the modified wool is ground and dispersed in Freund's adjuvant, and the dispersion is injected into a rabbit. A similar procedure should be followed with H. The serum from the rabbit, which may contain antibodies against WH and H, is then injected into the skin of a guinea pig. If antibodies are present, a blue spot will appear on the skin at the site of injection. Another technique is a test for insoluble antigens with fluorescent antibodies. Safety of the treated fabric should finally be tested on human subjects under proper medical supervision.

Procedures and protocols for evaluating the safety of potential flameproofing and other compounds with potential dermal toxicity compounds were described recently in a book edited by Maibach.[172] However, as noted earlier, evaluation of the flame retardants themselves may not always be valid because they may react differently in combination with the fabric. Zirconium compounds that are used in cosmetics, for example, appear to cause skin and lung irritations when applied, in aerosol or spray forms, to the skin,[143] but they are, as far as I know, innocuous when present in wool treated for flame resistance. A potential flameproofing compound is different chemically when it reacts covalently or noncovalently with wool; i.e., it is changed by covalent or noncovalent reaction with wool fibers. Furthermore, because wool is a protein, it may, in contrast to other textiles, successfully compete with another keratin-like material, the skin, for any compound and thus minimize or retard any leaching of nonreactive flameproofing compounds from the wool onto the skin. Because of the increasing worldwide use of flameproofed textiles, problems of dermal safety and toxicity of flame-resistant fabrics will be paramount in future studies.

The physiological and toxicological effects of smoke produced during combustion are also important in flammability research.[62,63,117,118,157] Einhorn[62] offers an excellent discussion of this problem. He points out that many victims of fires die without apparent skin injury. Some of these deaths are presumably due to the toxic effect of smoke inhalation and asphyxiations. Thus research is urgently needed to establish the pyrolytic decomposition products and the nature of the smoke produced by native and flame-resist-treated fabrics as well as its chemical composition, density, and toxicity.[52,201,202,252]

The combined technique of gas–liquid chromatography (GLC) and mass spectroscopy offers a way to study the volatile products of the combustion of flameproofing compounds and textiles.[62] The sample is volatilized in a gas chromatographic chamber and separated into its components, which are then individually injected into a mass spectrometer connected to the GLC apparatus. Since mass spectroscopy involves fragmentation of organic compounds under the impact of high-energy, analysis of the mass spectroscopic data frequently permits identification of combustion products.

Mass spectroscopy is also a useful technique to study the mechanism of fragmentation of compounds used as flame retardants, as illustrated with 2,4,6-tribromo-3-aminobenzenesulfonic acid (tribromometanilic acid). The analysis was carried out in a 79-eV electron impact DuPont (CEC) 21-110 mass spectrometer by use of the direct insertion probe.[86] The ion source temperature was 250°C. The mass spectrum indicated a monoisotopic molecular weight of 407 and an isotopic cluster characteristic of three bromine atoms. This molecular weight corresponds to $C_6H_4NSO_3Br_3$. The most intense peak in the high-mass portion of the spectrum corresponds to $C_6H_5NSO_3Br_2$ (monoisotopic molecular weight of 329) and thus results from a second compound of this composition either in the original mixture or as a product of thermal decomposition. Hydrobromic acid (HBr) evolves concurrently with the two components having higher molecular weights to yield the base peaks of the spectrum m/e 80 and 92, which further suggests that thermal decomposition accounts for the anomalous $C_6H_5NSO_3Br_2$ compound. These results show that mass spectroscopy is useful for identification but might also indicate thermal decomposition pathways of flameproofing compounds and treated textiles. Those pathways are undoubtedly related to similar fragmentations and degradations that take place during combustion, since, as mentioned earlier, there may be some analogy between mass spectroscopy and the combustion process.

In fires, particulate content of the smoke might be as important as the density of smoke in impairing the vision of victims so that they would be unable to escape during a fire. The introduction of fire retardants into textiles, however, might decrease, rather than increase, smoke density and toxicity.

Possible effects of flame retardants on the environment[128] and the carcinogenicity and mutagenicity of flame retardants also need to be tested[1,38,179,180,234] in the light of the recent discovery that tris-(2,3-dibromopropyl) phosphate has been found to be positive in the Ames Salmonella Test.[38,206] We have tested DMSO solutions of tetrabromophthalic anhydride at concentrations of 10, 100, 1000, and 10,000 μg/plate in the Ames *Salmonella typhimurium* tester strains TA100, TA1535, TA98, and TA1537 both with and without an *in vitro* rat liver metabolizing system.[180]

The results are given in Table 17. Neither tetrabromophthalic anhydride nor tetrabromophthalic acid exhibited mutagenic activity in any of the four tester strains either with or without the *in vitro* metabolizing system. At the 10,000 μg/plate level, both compounds inhibited the growth of the tester strain. The growth inhibition was much less marked in the presence of the

TABLE 17

Reversion of *Salmonella typhimurium* strains TA 1535, TA 100, TA 98, and TA 1537 Following Exposure to Tetrabromophthalic Anhydride, Tetrabromophthalic Acid, or Tris(2,3-dibromopropyl) Phosphate[181]

Test material	Micrograms of test material per plate	Revertants per plate[a]							
		TA 1535		TA 100		TA 98		TA 1537	
		S-9[b]	No. S-9	S-9	No S-9	S-9	No S-9	S-9	No S-9
Dimethylsulfoxide only (solvent control, mean ±S.D.)	—	12 ± 5 (8)[c]	19 ± 6 (8)	147 ± 19 (8)	162 ± 24 (8)	33 ± 6 (4)	34 ± 4 (4)	9 ± 2 (4)	13 ± 6 (4)
Tetrabromophthalic anhydride	10	9	21	124	132	31	30	14	13
	100	8	19	129	133	32	28	12	14
	1,000	13	17	136	130	34	32	15	13
	10,000	7*	*	100*	*	29*	*	11	*
Tetrabromophthalic acid	10	13	25	136	178	38	30	10	10
	100	19	25	138	146	44	34	11	7
	1,000	20	29	161	62*	34	34	14	7
	10,000	12*	*	125*	*	32*	*	10*	*
Tris(2,3-dibromopropyl) phosphate	10	61	16	264	144				
	100	604	19	1000	151				
	1,000	1000	37						

[a] Values are means of replicate determinations, except control values, which are reported as the mean and standard of all control plates run concurrently with the test materials reported. Underlined values exceed the mean control value plus 3 S.D. The symbol * indicates growth inhibition of the background lawn. Positive controls to test the response of each strain and the activity of the metabolizing system were included with each determination: TA 1535, 5 μg of N-methyl-N'-nitro-N-nitrosoguanidine, > 1000 revertants; TA 100 and TA 98, 1 μg of aflatoxin B_1 (+S−9), > 1000 revertants; TA 1537, 100 μg of 9-aminoacridine, > 1000 revertants.
[b] S−9 denotes the metabolizing mixture according to Ames et al.,[1] 50 μl of liver supernatant per plate.
[c] Number of determinations.

metabolizing system. Tris(2,3-dibromopropyl) phosphate was a highly effective mutagen in strains TA 100 and TA 1535 in the presence of the *in vitro* metabolizing system, as previously shown.[38,206]

The mutagenicity of tris(2,3-dibromopropyl) phosphate may be due to alkylation of sensitive sites in DNA by carbonium ions derived from cleavage of aliphatic carbon–bromine bonds. The aromatic carbon–bromine bonds present in TBPA are relatively inert, so that this compound and its corresponding diacid are less likely to act as biological alkylating agents. However, tetrabromophthalic anhydride is known to acylate protein amino groups in dimethylformamide solutions,[88,89,253,254] and it could, in principle, acylate amino and other functional groups in DNA.[68] The absence of a mutagenic effect in our tests with TBPA implies that acylation of nucleic acids does not occur under the conditions employed.

Although no single mutagenicity test is sufficient to detect all possible adverse genetic effects, these results provide no indication of a mutagenic effect of tetrabromophthalic anhydride and tetrabromophthalic acid.

ACKNOWLEDGMENTS

I take great pleasure in thanking E. Menefee for constructive contributions to the discussion on wool structure, P. G. Gordon, B. N. Hoschke, and W. H. Ward for constructive reading of the manuscript, and my colleagues whose names appear in the References for excellent scientific collaboration.

Reference to a company or product name does not imply approval or recommendation of that product by the U.S. Department of Agriculture to the exclusion of others may be suitable.

This chapter constitutes Paper XV in a series on "Flame-Resistant Wool."

9. References

1. Ames, B. N., McCann, J., and Yamasaki, E., Methods for detecting carcinogens and mutagens with *Salmonella*/mammalian-microsome mutagenicity test, *Mutat. Res.* **31**, 347–364 (1975).
2. Amyantov, N. I., and Titov, V. A., A method for brominating metanilic acid (in Russian), *Anilinokras. Prom.* **2** (8–9), 24–26 (1932).
3. Anabi, S., Isaacs, P., and Lewin, M., Flame-retardancy of wool cotton blends by chlorinated polymer deposition, in *IUPAC International Symposium on Macromolecules, Jerusalem, Israel, Abstracts*, p. 453, International Union of Pure and Applied Chemistry (July 1975).
4. Anon., New FR tests and standards, *Am. Text. Rep.* **4**, 13 (Dec. 1975).
5. Anon., Plastic fires create new hazards for both firemen and public, *J. Am. Med. Assoc.* **234**, 211–213 (1975).
6. Anon., Wool triumphs over flames, *Agric. Res.* **23**, 3–4 (1974).

7. Anon., Technical Information About Silicone (Superwash) Textile Finishes, Dow Corning Bulletin 22-151A-01 (1974).

8. Anon., American Burning, The Report of the National Commission on Fire Prevention and Control, Washington, D.C. (1972).

9. Anon., Fires cost billions yearly, Chem. Eng. News, 16 (Oct. 18, 1971).

10. Anon., Flammability of fabrics, Ciba Rev. (4) (1969). (I thank Dr. E. P. Martin of Zurich, Switzerland for sending me a copy of this publication.)

11. Anon., Test Procedure Specifications for Flame-Retardant Chemicals, State of California, Office of the Fire Marshall, 107 South Broadway, Room 9035, Los Angeles, Calif. (1974).

12. Asquith, R. S., and Otteburn, M. S., Cystine–alkali reactions in relation to protein cross-linking, in Protein Crosslinking: Nutritional and Medical Consequences (M. Friedman, ed.), Plenum Press, New York (1977), pp. 93–122.

13. Asquith, R. S., and Otteburn, M. S., Self-crosslinking in keratin under the influence of dry heat, J. Appl. Polym. Sci. Appl. Polym. Symp. 18, 277–287 (1971).

14. Asquith, R. S., and Otteburn, M. S., Basic amino acids in heated keratin, J. Text. Inst. 60, 208–210 (1969).

15. Bajpai, L. O., Whewell, C. S., and Woodhouse, J. M., Action of tetrakis(hydroxymethyl) phosphonium chloride on wool, Nature London 187, 602–603 (1960).

16. Barnako, D., Flammable fabrics, J. Am. Med. Assoc. 221, 189 (1972).

17. Beck, P. J., Gordon, P. G., and Ingham, P. E., Thermogravimetric analysis of flame-retardant-treated wools, Text. Res. J. 46, 478–483 (1976).

18. Beck, P. J., Gordon, P. G., and Stephens, L. J., Mechanism of flame-retardation of wool, in Proceedings of the Fifth International Wool Textile Research Conference, Aachen, Germany (1975) (K. Ziegler, ed.), German Wool Research Institute, Aachen, Vol. II, pp. 549–558 (1976).

19. Bell, J. W., Clegg, D., and Whewell, C. S., The action of heat on wool, J. Text. Inst. 51, T1173–T1182 (1960).

20. Bell, J. W., Hutchinson, C. R., and Whewell, C. S., The effect of metal salts on the thermal degradation of wool, J. Text. Inst. 57, T43–T44 (1966).

21. Beninate, J. V., Boylston, E. K., Drake, G. L., and Reeves, W. A., Imparting flame resistance to fibrous textiles from alkaline medium, U.S. Pat. 3,607,356; Chem. Abs. 76, 15735p (1972).

22. Benisek, L., Improvement in the natural flame resistance of wool, Part III. Vanadium, molybdenum, and tungsten complexes, in Proceedings of the Fifth International Wool Textile Research Conference, Aachen, Germany (1975) (K. Ziegler, ed.), German Wool Research Institute, Aachen, Vol. V, pp. 31–46 (1976).

23. Benisek, L., Burning behavior of carpets: The advantages of wool and flame-resistant wool, Text. Res. J. 45, 373–382 (1975).

24. Benisek, L., Flammability. Position of wool, Prog. Fire Retardant Ser. 5, 57–93 (1975).

25. Benisek, L., Textilveredlung 10 (12), 482–489 (1975); Chem. Abs. 84 (32), 414 (1975).

26. Benisek, L., Current flammability methods and specifications and the position of wool. Part 2. Specialized test methods; Carpet tests, Wool Sci. Rev. 51, 29–42 (1975).

27. Benisek, L., The reaction of wool with potassium hexafluorozirconate, Text. Res. J. 45, 351–353 (1975).

28. Benisek, L., Flame retardance of protein fibers, in Flame-Retardant Polymeric Materials (M. Lewin, S. M. Atlas, and E. M. Pearce, eds.), Vol. 1, pp. 137–191, Plenum Press, New York (1975).

29. Benisek, L., Current flammability methods and specifications and the position of wool. Part 1. General tests for flammability, Wool Sci. Rev. 50, 40–54 (1974).

30. Benisek, L., Improvement of the natural flame-resistance of wool. Part I: Metal-complex applications, J. Text. Inst. 65, 102–108 (1974).

31. Benisek, L., Improvement of the natural flame-resistance of wool. Part II: Multi-purpose finishes, *J. Text. Inst.* **65**, 140–145 (1974);

32. Benisek, L., Improvement of the natural flame-resistance of wool. Part III. Vanadium, molybdenum, and tungsten complexes, *J. Text. Inst.* **67**, 226–228 (1976);

33. Benisek, L., Improvement of the natural flame-resistance of wool. Part IV: Wool rich blends with polyester fibre and rayon, *J. Text. Inst.* **67**, 261–265 (1976).

34. Benisek, L., Edmondson, G. K., and Greenwood, B. D., Washable, permanent-press, and flame-resistant wool. Part I. Aqueous emulsions of a polyorganosiloxane, *Text. Inst. Ind.* **14**, 343–347 (1976).

35. Benisek, L., Die Entflammabarkeit von Fasern un die Flamm-resistenz der Wolle, *Textilveredlung* **8**, 318–326 (1973).

36. Benisek, L., New aspects of flame protection using wool, *Text. Mfr.* **99**, 36–39 (1972).

37. Benisek, L., Use of titanium complexes to improve the natural flame retardancy of wool, *J. Soc. Dyers Colour.* **87**, 277–278 (1971).

38. Blum, A., and Ames, B. N., Flame retardant additives as possible cancer hazards, *Science* **195**, 17–23 (1977).

39. Blum, D. E., quoted in Ref. 10, p. 2.

40. Bradbury, J. H., The morphology and chemical structure of wool, in *IUPAC International Symposium on Macromolecules, Jerusalem, Israel, Abstracts*, pp. 49–50, International Union of Pure and Applied Chemistry (July 1975).

41. Bradbury, J. H., Structure and chemistry of keratin fibers, *Adv. Protein Chem.* **27**, 111–211 (1973).

42. Carroll-Porczynski, C. Z., Fabric flammability: New testing methods and equipment, *Text. Inst. Ind.* **10**, 188–194 (1971).

43. Cavins, J. F., and Friedman, M., Automatic integration and computation of amino acid analyses, *Cereal Chem.* **45**, 172–176 (1968).

44. Chauffe, L., Windle, J. J., and Friedman, M., An electron spin resonance study of melanin treated with reducing agents, *Biophys. J.* **16**, 565–572 (1975).

45. Chaigneau, M., and Le Moan, G., Analysis and evolution of compounds formed by the pyrolysis and combustion of wool, *Analysis* 4 (1), 28–33 (1976); *Chem. Abs.* **84**, 154, 813 (1976).

46. Cole, R., Flameproofing of keratin fibers, *Ger. Offen.* 2,009,121 (Jan. 21 (1971); *Chem. Abs.* **74**, 113,168 (1971).

47. Cook, G. A., Meierer, R. E., and Shields, B. M., Combustibility tests of several flame-resistant fabrics in compressed air, oxygen-enriched air, and pure oxygen, *Text. Res. J.* **37**, 591–599 (1967).

48. Crawshaw, G. G., Duffield, P. A., and Mehta, P. N., Flammability and flameproofing of wool fabrics, *J. Appl. Polym. Sci.* Appl. Polym. Symp. **18**, 1183–1197 (1971).

49. Crighton, J. S., and Happey, F., Differential thermal analysis of keratin and related protein fibers, in *Symposium on Fibrous Proteins*, 1967, New York and Sydney, (W. G. Crewther, ed.), pp. 409–420, Plenum Press, New York (1968).

50. Date, M., and Fukuoka, S., Fireproofed textiles, *Jpn. Kokai* **75**, 116, 800 (Sept. 12, 1975); *Chem. Abs.* **84**, 61101q (1976).

51. Davis, C. A., Relationship of flammability measurements, *Text. Chem. Color.* **1**, 540–548 (1969).

52. Davis, M. H., and Scroggie, J. G., Reaction of a basic chromium (III)–zirconium (IV) sulphate complex with amino acids, *Aust. J. Chem.* **27**, 279–286 (1974).

53. Dean, P. A. W., and Evans, D. F., Spectroscopic studies of inorganic fluoro-complexes. Part I. The ^{19}F nuclear magnetic resonance vibrational spectra of hexafluorometallates of groups IVA and IVB, *J. Chem. Soc. (A)* 698–701 (1967).

54. De Boos, A. G., Chemical testing and analysis, *Text. Prog.* **6**, 1–33 (1974).

55. Delman, A. D., Recent advances in the development of flame-retardant polymers, *J. Macromol. Sci. Rev. Macromol. Chem.* **3**, 281–311 (1969).
56. Diamond, J., and Friedman, M., Multifunctional derivatives from polyhalogenated cyclic anhydrides as flameproofing agents for wool, in *172nd Meeting of the American Chemical Society, San Francisco, Calif. Abstracts*, CELL, 076 Washington, D.C. (1976).
57. Draize, J. H., Dermal Toxicity in Appraisal of the Safety of Chemicals in Foods, Drugs, and Cosmetics, The Association of Food and Drug Officials of the United States, Texas State Dept. of Health, Austin, Texas (1959), p. 46.
58. Drake, L. G., Flammability: Yesterday, today, tomorrow, *Text. Chem. Color.* **8** (12), 184–190 (1976).
59. Drake, G. L., Jr., and Chance, L. H., Flame-retardant fabrics safeguard your life, in *Yearbook of Agriculture*, USDA, Washington, D.C. (1968), p. 179.
60. Dupraz, C. A., How to apply a flame resistant finish to wool fabric, *Am. Dyest. Rep.* **60**, 54–56 (1971).
61. Edmondson, G. K., and Benisek, L., Solvent-applied flame-resistant treatments for wool, cotton, and wool–cotton blends, *J. Text. Inst.*, **68**, 230–239 (1977).
62. Einhorn, I. N., Physiological and toxicological aspects of smoke produced during combustion of polymeric materials, *Environ. Health Perspectives* **11**, 163–189 (1975).
63. Einsele, U., and Tarakcioglu, I., Investigation into the combustion gases from different textile fibres, *Melliand Textilber.* **58**, 52–59 (1977).
64. Eisenberg, B. J., and Weil, E. D., A new durable flame retardant for cellulosics, *Text. Chem. Color.* **6**, 180–182 (1974).
65. Fincher, K. W., Guise, G. B., and White, M. A., Machine-washable flame-resistant wool, *Text. Res. J.* **43**, 623–625 (1973).
66. Fong, W., Whitfield, R. E., Miller, L. A., and Brown, A. H., Wool fabric stabilization by interfacial polymerization. II. Developmental studies of process variables, Am. Dyest. Rep. **51**, 325–334 (1962).
67. Friedman, M., unpublished results.
68. Friedman, M., Effect of lysine modification on chemical, physical, nutritive, and functional properties of proteins, in *Food Proteins* (J. R. Whitaker and S. Tannenbaum, eds.), Avi, Westport, Conn. pp. 446–483 (1977).
69. Friedman, M., ed., *Protein Crosslinking: Biochemical and Molecular Aspects*, Plenum Press, New York (1977).
70. Friedman, M., ed., *Protein Crosslinking: Nutritional and Medical Consequences*, Plenum Press, New York (1977).
71. Friedman, M., Crosslinking amino acids—stereochemistry and nomenclature, in *Protein Crosslinking: Nutritional and Medical Consequences* (M. Friedman, ed.), Plenum Press, New York (1977), pp. 1–28.
72. Friedman, M., Research strategies for flame-resistant wool and wool blends, in *IUPAC International Symposium on Macromolecules, Jerusalem, Israel, Abstracts*, p. 451, International Union of Pure and Applied Chemistry (1975).
73. Friedman, M., ed., *Protein-Metal Interactions*, Plenum Press, New York (1974), p. 513.
74. Friedman, M., *The Chemistry and Biochemistry of the Sulfhydryl Group in Amino Acids, Peptides, and Proteins*, Permagon Press, Elmsford, N.Y. (1973).
75. Friedman, M., Removal of mercury from liquids using keratin derivatives, U.S. Pat. 3,725,261 (1973).
76. Friedman, M., Flame-resistant wool, U.S. Pat. 3,669,610 (1972).
77. Friedman, M., Effect of enzyme-containing detergent on strength of untreated woolen fabrics, *Text. Res. J.* **41**, 315–318 (1971).
78. Friedman, M., Solvent effects in reactions of amino groups in amino acids, peptides, and proteins with α,β-unsaturated compounds, *J. Am. Chem. Soc.* **89**, 4709–4713 (1967).

79. Friedman, M., Ash, J. F., Bry, R. E., and Simonaitis, R. A., Multipurpose finishes for wool, in *174th Meeting of the American Chemical Society, Chicago, Ill., Abstracts*, CELL 35, Washington, D.C.) (1977).
80. Friedman, M., Ash, J. F., Bry, R. E., and Simonaitis, R. A., Moth-resistant wool and wool blends, submitted for publication (1978).
81. Friedman, M., and Ash, J. F., Production of flame-resistant wool and wool blends during dyeing, *Proceedings of the Fifth International Wool Textile Research Conference, Aachen, Germany (1975)* (K. Ziegler, ed.), German Wool Research Institute, Aachen, Vol. V, pp. 47–54 (1976).
82. Friedman, M., Ash, J. F., and Fong, W., Dyebath application of flame-retardants for flame-resistant wool, *Text. Res. J.* 45, 994–996 (1975).
83. Friedman, M., Ash, J., and Fong, W., Non-discoloring flame-resistant wool, U.S. Pat. 3,927,962 (1975).
84. Friedman, M., and Fong, W., Flame-resistant wool, U.S. Pat. 3,950,129 (1976).
85. Friedman, M., Ash, J. F., and Fong, W., Dyebath application of chlorendic acid for flame-resistant wool, *Text. Res. J.* 44, 555–556 (1974).
86. Friedman, M., and Haddon, W. F., unpublished results.
87. Friedman, M., Harrison, C. S., Ward, W. H., and Lundgren, H. P., Sorption behavior of mercuric and methylmercuric salts on wool, *J. Appl. Polym. Sci.* 17, 377–390 (1973).
88. Friedman, M., and Koenig, N. H., Two-solvent process for flame-resistant wool with tetrabromophthalic anhydride, *Proceedings of the Fifth International Wool Textile Research Conference, Aachen, Germany (1975)* (K. Ziegler, ed.), German Wool Research Institute, Aachen, Vol. V, pp. 67–72 (1976).
89. Friedman, M., and Koenig, N. H., Modification of wool, U.S. Pat. 4,007,006 (1977).
90. Friedman, M., and Lundin, R. E., unpublished results.
91. Friedman, M., and Masri, M. S., Interaction of mercury compounds with wool and related biopolymers, in *Protein–Metal Interactions* (M. Friedman, ed.), Plenum Press, New York (1974), pp. 505–550.
91a. Friedman, M., and Masri, M. S., Sorption behaviour of mercuric salts on modified wool and polyamino acids, *J. Appl. Polym. Sci.* 17, 2179–2186 (1973).
92. Friedman, M., and Millard, M. M., Radiation effects in protein analysis by X-ray photo-electron spectroscopy, *Fed. Proc.* 35, 1457 (1976).
93. Friedman, M., and Noma, A. T., Methods and problems in chromatographic analysis of sulfur amino acids, in *Protein Nutritional Quality of Foods and Feeds*, Part 1 (M. Friedman, ed.), Dekker, New York (1975), pp. 521–548.
94. Friedman, M., and Noma, A. T., Cystine content of wool, *Text. Res. J.* 40, 1073–1078 (1970).
95. Friedman, M., and Romersberger, J. A., Relative influences of electron-withdrawing functional groups on basicities of amino acid derivatives, *J. Org. Chem.* 33, 154–157 (1968).
96. Friedman, M., and Tillin, S., Reactions of amino acid, peptides, and proteins with α,β-unsaturated compounds. XXX. Partly-reduced-alkylated wool, *Text. Res. J.* 44, 578–580 (1974).
97. Friedman, M., and Tillin, S., Flame-resistant wool, *Text. Res. J.* 40, 1045–1047 (1970).
98. Friedman, M., and Thorsen, W. J., Flame-resistant wool–cotton and wool–cotton-polyester blends, *Text. Res. J.* 46, 70–72 (1976).
99. Friedman, M., and Waiss, A. C., Jr., Mercury uptake by agricultural products and by-products, *Environ. Sci. Technol.* 6, 457–458 (1972).
100. Friedman, M., and Wall, J. S. Additive linear free energy relationships in reaction kinetics of amino groups with α,β-unsaturated compounds. *J. Org. Chem.* 31, 2888–2894 (1966).

101. Friedman, M., Whitfield, R. E. and Tillin, S., Flame-resistant wool. V. Enhancement of the natural flame-resistance of wool, *Text. Res. J.* **43**, 212–217 (1973).
102. Friedman, M., and Williams, L. D., The ninhydrin reaction. VII. Stoichiometry of formation of Ruhemann's Purple in the ninhydrin reaction, *Bioorg. Chem.* **3**, 267–280 (1974).
103. Friedman, M., and Williams, L. D., The ninhydrin reaction. VI. The reaction of ninhydrin with keratin proteins, *Anal. Biochem.* **54**, 333–345 (1973).
104. Galil, F., New tests for ranking safer FR garments, *Text. Chem. Color.* **8** (3), 38–43 (1976).
105. Gilbert, S., and Liepins, R., Treatment for improving flame retardancy of wool and minimizing toxic gas evolution in burning, *J. Appl. Polym. Sci.* **16**, 1009–1016 (1972).
106. Golborn, P., and Duffy, J. J., Dialkyl alkyl and aromatic sulfonamidoethyl phosphonates, U.S. Pat. 3,959,551 (1976).
107. Gordon, P. G., Private communication (1977).
108. Gordon, P. G., The composition of fluorozirconate and fluorotitanate on wool, *J. Text. Inst.* **66**, 97–102 (1975).
109. Gordon, P. G., and Roberts, G. A., Response, *Text. Res. J.* **45**, 353–354 (1975).
110. Gordon, P. G., and Roberts, G. A., The reaction of wool with potassium hexafluorozirconate, *Text. Res. J.* **44**, 414–421 (1974).
111. Gordon, P. G., and Stephen, L. J., The treatment of wool with hexafluorotitanate, *J. Soc. Dyers Colour.* **90**, 239–245 (1974).
112. Gordon, P. G., McMahon, D. T. W., and Stephens, L. J., Investigation into the mechanisms of flame retardation on wool, *Text. Res. J.*, **47**, 699–711 (1977).
113. Gottlieb, I. M., A theory of flame-retardant finishes, *Text. Res. J.* **26**, 156–157 (1956).
114. Gregorski, K. S., An X-ray diffraction study of thermally-induced structural changes in β-keratin, in *Protein Crosslinking: Biochemical and Molecular Aspects*, (M. Friedman, ed.), Plenum Press, New York (1977), pp. 329–344.
115. Grossman, V. B., and Inglis, A. S., Determination of zirconium in biological materials by atomic absorption spectrophotometry, *Anal. Chem.* **43**, 1903–1905 (1971).
116. Hack, R., Müller-Schulte, D., Meltzow, W., and Zahn, H., Thermal behaviour of untreated flame-resistant wool, in *Proceedings of the Fifth International Wool Textile Research Conference, Aachen, Germany (1975)* (K. Ziegler, ed.), German Wool Research Institute, Aachen, Vol. II, 538–548 (1976).
116a. Harlan, J. W., and Feairheller, S. H., Chemistry of the crosslinking of collagen during tanning, in *Protein Crosslinking: Biochemical and Molecular Aspects* (M. Friedman, ed.), Plenum Press, New York (1977), pp. 425–440.
117. Hilado, C. J., Marcussen, W. H., Furst, A., Kouritides, D. A., Parker, J. A., and Fewell, L. F., Relative toxicity of pyrolysis products of some aircraft seat materials, *J. Combust. Toxicol.* **4** (1), 16–20 (1977).
118. Hilado, C. J., Relative toxicity of pyrolysis products of polymeric materials using various test conditions and ranking systems, *J. Consumer Prod. Flammability* **3** (4), 288–297 (1976); **4**, 40–59 (1977).
119. Haly, A. R., and Snaith, J. W., Differential thermal analysis of wool—the phase transition endotherm under various conditions, *Text. Res. J.* **37**, 898–907 (1967); **40**, 142–149 (1970).
120. Hendrix, J. E., Anderson, T. K., Clayton, T. J., Olson, E. S., and Barker, R. H., Flammability measurements on thermal decomposition of textiles, *J. Fire Flammability* **1**, 107–139 (1970).
121. Hindson, W. R., Flame resistance in textiles, *Text. J. Aust.* **49**, 16 (1974).
122. Hoschke, B. N., Personal communication (1977).
123. Hoschke, B. N., Flame resistance of pure wool for racing drivers, *Text. J. Aust.* **48**, 21–23 (1973).
124. Hoschke, B. N., Flame-resistant nomex blends, *Text. Res. J.* **44**, 956–958 (1974).
125. Hoschke, B. N., An evaluation of resin-shrinkproofed wool in lightweight wool polyester

blend fabrics, *Proceedings of the Fifth International Wool Textile Research Conference, Aachen, Germany (1975)*. (K. Ziegler, ed.), German Wool Research Institute, Aachen, Vol. IV, pp. 581–592 (1976).

126. Humfeld, H., Elmquist, R. E., and Kettering, J. H., The Sterilization of Wool and Its Effect on Physical and Chemical Properties of Wool Fabric, Technical Bulletin No. 588, USDA, Washington, D.C. (1937), pp. 1–27.

127. Husler, J. W., *At. Absorp. News.* **8**, 1–2 (1969).

128. Hutzinger, O., Sundstrom, G., and Safe, S., Environmental chemistry of flame retardants. Part I. Introduction and principles, *Chemosphere*, **5**, No. 1, 3–10 (1976).

129. Ingham, P. E., Tin-based flame retardants for wool, *Tin Its Uses* **105**, 5–7 (1975); *Chem. Abs.* **195**, 258j (1975).

130. Ingham, P. E., Flame Resist Treatments for Woolly Sheepskins, Communication No. 30, Wool Research Organization of New Zealand, Christ Church (1974).

131. Ingham, P. E., private communications.

132. Ingham, P. E., The pyrolysis of wool and the action of flame retardants, *J. Appl. Polym. Sci.* **15**, 3025–3041 (1971).

133. Ingham, P. E., and Haden, D. D., *Fifth International Wool Textile Research Conference, Aachen, Germany, Abstracts* (K. Ziegler, ed.), German Wool Research Institute, Aachen, 297–298 (1975).

134. Ingham, P. E., and Haden, D. D., Treatments for wooly sheepskins to prevent staining during tanning and to confer flame-resist properties, in *Proceedings of the Fifth International Wool Textile Research Conference, Aachen, Germany (1975)* (K. Ziegler, ed.), German Wool Research Institute, Aachen, Vol. V, pp. 95–105 (1976).

135. Isaacs, J. L., The oxygen index flammability test, *Mod. Plast.* **47**, 124–130 (1970).

136. Jenkins, A. D., and Wolfram, L. J., The chemistry of the reaction between tetrakis (hydroxymethyl) phosphonium chloride and keratin, *J. Soc. Dyers Color.* **79**, 55–61 (1963).

137. Johnson, A., The development of flame-resistant fabrics containing wool, *J. Text. Inst.* **39**, 561–577 (1948).

138. Johnson, P. R., A general correlation of the flammability of natural and synthetic polymers, *J. Appl. Polym. Sci.* **18**, 491–504 (1974).

139. Jones, F. W., Shrink-resisting wool with aqueous emulsions of a polyorganosiloxane, *J. Text. Inst.* **67**, 32–33 (1976).

140. Kappas, A., and Maines, M. D., Tin: A potent inducer of heme oxygenase in kidney, *Science* **192**, 60–62 (1977).

141. Kasarda, D. D., and Black, D. R., Thermal degradation of proteins studied by mass spectrometry, *Biopolymers* **6**, 1001–1004 (1968).

142. Kaswell, E. R., Some thoughts and information on nonflammable products, *Text. Chem. Coror.* **4**, 33–40 (1972).

142a. Kirkpatrick, A., and Maclaren, J. A., Further studies of sulfated wool, *Text. Res. J.*, **44**, 753–755 (1974).

143. Klimmer, O. R., and Doll, W., The problem of toxicity and tissue storage of soluble zirconium compounds in short- and long-term feeding to warm-blooded animals, *Arzneim. Forsch.* 1286–1290 (1964); cf. also *Drug Cosmet. Ind.* **115**, 20 (Aug. 1974); *Federal Register*, **40** (No 109), pp. 24328–24344 (1975); and Smith, I. C. and Carson, B. L. Zirconium, *Trace Metals in the Environment*, Vol. 3, Ann Arbor Science Publishers, Ann Arbor, Michigan (1977).

144. Koenig, N. H., and Friedman, M., Comparison of wool reactions with selected mono- and bifunctional reagents, in *Protein Crosslinking: Biochemical and Molecular Aspects* (M. Friedman, ed.), Plenum Press, New York (1977), pp. 355–382.

145. Koenig, N. H., and Friedman, M., Properties of wool treated with sulfonyl chlorides, *172nd Meeting of the American Chemical Society, San Francisco, Calif., Abstracts* CELL 077, Washington, D.C. (1976).

146. Koenig, N. H., Chemical modification of wool in aprotic swelling media, *J. Appl. Polym. Sci.* **21**, 455–465 (1977).
147. Koenig, N. H., and Friedman, M., Combined application of reactive compounds in nonaqueous swelling solvents for flame- and shrink-resistant wool, *Text. Res. J.* **47**, 139–141 (1977).
148. Koenig, N. H., and Friedman, M., Process for simultaneously flameproofing and shrink-proofing wool, U.S. Pat. 4,029,471 (1977).
149. Koenig, N. H., and Friedman, M., Insect-proofing wool with zinc acetate, U.S. Pat. 3,927,969 (1975).
150. Koenig, N. H., and Friedman, M., Shrink-proofing of wool with cyclic acid anhydrides and zinc acetate, U.S. Pat. 3,867,095 (1975).
151. Koenig, N. H., and Friedman, M., Zinc–wool keratin reactions in nonaqueous solvents, in *Protein–Metal Interactions* (M. Friedman, ed.), Plenum Press, New York (1974), pp. 81–95.
152. Koenig, N. H., and Friedman, M., Modification of wool with vinylpyridine and zinc chloride, U.S. Pat. 3,749,553 (1973).
153. Koenig, N. H., and Friedman, M., Surface modification of wool and other fibrous materials by 4-vinylpyridine and zinc chloride, *Text. Res. J.* **42**, 319–320 (1972).
154. Koenig, N. H., Muir, M. W., and Friedman, M., Reaction of wool with zinc acetate in dimethylformamide, *Text. Res. J.* **44**, 717–719 (1974).
155. Koenig, N. H., Muir, M. W., and Friedman, M., Properties of wool modified with activated vinyl compounds, *Text. Res. J.* **43**, 682–688 (1973).
156. Kop'ev, M. A., Tyuganova, M. A., Rubtsova, I. K., Shner, S. M., and Bocharova, L. P., Fire-resistant woolen materials, *Tekst. Promst. Moscow* (10), 69–70 (1975); *Chem. Abs.* **84**, 45862h (1976).
157. Krahne, B., Decomposition of combustion gases and fumes from untreated textile materials and their flame-retardant versions, *Melliand Textilber.* **58**, 64–70 (1977).
158. Kulkarni, V. G., The separation of cortical cells and the pyrolysis of wool keratin, *Text. Res. J.* **45**, 89–90 (1975).
159. Kyryla, W. C., and Papa, A. J., eds., *Flame Retardancy of Polymeric Materials*, Vols. 1 and 2, Dekker, New York (1973).
160. La France, N., Ziegler, K. L., and Zahn, H., Formation of sulphur-containing compounds in wool during carbonization, *J. Text. Inst.* **51**, T1168–T1172 (1960).
161. Le Blanc, R. B., What's available for flame retardant textiles, *Text. Ind.* **28**, 28–41 (1976).
162. Le Blanc, R. B., Apparel flammability standards, *Am. Dyest. Rep.* **65**, (1), 19 (1976).
163. Lee, K. S., X-ray studies of heated keratins, *Text. Res. J.* **46**, 779–785.
164. Leveau, M., Comportement thermique des fractions corticales de la Laine, *Bull. Inst. Text. France* **80**, 57–63 (1959).
165. Lewin, M., Flame-retarding of wool cotton blends. *171st Meeting of the American Chemical Society, New York, April 1976, Abstracts*, CELL 67, Washington, D.C. (1976).
166. Lewin, M., Isaacs, P. K., and Shaf, B., Flame retardance of wool by sulphamates, in *Proceedings of the Fifth International Wool Textile Research Conference, Aachen, Germany (1975)* (K. Ziegler, ed.), German Wool Research Institute, Aachen, Vol. V, pp. 73–84 (1976).
167. Lundgren, H. P., and Ward, W. H., Levels of molecular organization in α-keratins, *Arch. Biochem. Biophys. Suppl.* **1**, 78–111 (1962).
168. Lundgren, H. P., and Ward, W. H., The keratins, in *Ultra Structure of Protein Fibers* (R. Borasky, ed.), Academic Press, New York (1963), pp. 39–122.
169. Litchfield, E. L., and Kubala, T. A., Flammability of fabrics and other materials in oxygen-enriched atmospheres. II. Minimum ignition energies, *Fire Technol.* **5**, 341–345 (1969).
170. Lyons, J. W., *The Chemistry and Uses of Fire Retardants*, Wiley-Interscience, New York (1970).

171. Lyons, D. W., and Robbins, H. G., The CPSC (Consumer Product Safety Commission) and textile flammability, *Mod. Text.* **LVII**, 7 (April 1976).

172. Maibach, H. I., *Animal Models in Dermatology*, Churchill Livingston, Edinburgh (1975).

173. Masri, M. S., and Friedman, M., Effects of chemical modification of wool on metal ion binding, *J. Appl. Polym. Sci.* **18**, 2367–2377 (1974).

174. Masri, M. S., and Friedman, M., Interactions of keratins with metal ions: Uptake profiles, mode of binding, and effects on properties of wool, in *Protein–Metal Interactions* (M. Friedman, ed.), Plenum Press, New York, (1974), pp. 551–587.

175. Masri, M. S., and Friedman, M., Competitive binding of mercuric chloride by wool and container from dilute solutions, *Environ. Sci. Technol.* **7**, 951–953 (1973).

176. Masri, M. S., and Friedman, M., Mercury uptake by polyamine carbohydrates, *Environ. Sci. Technol.* **6**, 745–746 (1972).

177. Masri, M. S., Reuter, F. W., and Friedman, M., Binding of metal cations by natural substances, *J. Appl. Polym. Sci.* **18**, 675–681 (1974).

178. Masri, M. S., Reuter, F. W., and Friedman, M., Interaction of wool with metal cations, *Text. Res. J.* **44**, 298–300 (1974).

179. McAnn, J., Choi, E., Yamasaki, E., and Ames, B. N., Detection of carcinogens and mutagens in the *Salmonella*/microsome test: Assay of 300 chemicals, *Proc. Nat. Acad. Sci. USA*, **72**, 5135–5139 (1972); **73**, 950–954 (1973).

180. MacGregor, J. T., and Friedman, M., Nonmutagenicity of tetrabromophthalic anhydride and tetrabromophthalic acid in the Ames *Salmonella*/microsome mutagenicity test, *Mutat. Res.* **56**, 81–84 (1977).

181. Mehta, R. D., The effect of pH of the THPC formulation on the flame-resistance and dimensional stability of cotton/wool blend fabrics, *174th American Chemical Society Meeting, Chicago, Ill. Aug. 28–Sept. 2, 1977, Abstracts*, CELL 34, Washington, D.C. (1977).

182. Menefee, E., Physical and chemical consequences of keratin crosslinking, with application to the determination of crosslink density, in *Protein Crosslinking: Biochemical and Molecular Aspects* (M. Friedman, ed.), Plenum Press, New York (1977), pp. 307–328.

183. Menefee, E., Charge separation associated with dipole disordering in proteins, *Ann. N.Y. Acad. Sci.* **238**, 53–67 (1974).

184. Menefee, E., Relation of keratin structure to its mechanical behavior, *J. Appl. Polym. Sci. Appl. Polym. Symp.* **18**, 809–821 (1971).

185. Millard, M. M., and Friedman, M., X-Ray photoelectron spectroscopy of BSA and ethyl vinyl sulfone modified BSA, *Biochem. Biophys. Res. Commun.* **70**, 445–451 (1976).

186. Millard, M. M., Masri, M. S., Friedman, M., and Pavlath, A., Fiber and textile surface analysis by X-ray photoelectron spectroscopy, *Proceedings of the Technical Conference, American Association Textile Chemists and Colorists, New Orleans, La.*, (1974), pp. 246–257.

187. Miller, B. and Martin, J. R., The flammability behaviours of mixtures containing wool and synthetic fibers, in *Proceedings of the Fifth International Wool Textile Research Conference, Aachen, Germany (1975)* (K. Ziegler, ed.), German Wool Research Institute, Aachen, Vol. V, pp. 55–64 (1976).

188. Miller, B., and Martin, J. R., A methodology for the interpretation of thermal and flammability behaviour of multicomponent fibrous polymer systems, *J. Fire Flammability* **6**, 105–118 (1975).

189. Mitchell, T. W., and Feughelman, M., The mechanical properties of wool fibers in water at temperatures above 100° C, *Text. Res. J.* **37**, 660–666 (1967).

190. Nader, R., Ralph Nader reports, *Ladies Home Journal*, 58 (Feb. 1976).

191. Needles, H. L., Fixation of flame retardants in wool, in *Fourth Semiannual Report GI 43105*, National Science Foundation, Washington, D.C. (1976), pp. 209–227; cf. also *J. Consumer Product Flammability*, **4** (2), 156–159 (1977).

192. Nelson, K. H., Brown, W. D., and Strauch, S. J., Rapid determination of bromine-containing flame retardants on fabrics, *Text. Res. J.* **43**, 357–361 (1973).
193. Norton, G. P., and Nicholls, C. H., The effect of heating wool containing alkali, *J. Text. Inst.* **51**, T1183–T1192 (1960).
194. O'Brien, S. J., and Weyker, R. G., The application of Pyroset CP Flame retardant to wool, *Text. Chem. Color.* **3**, 185–188 (1971).
195. Ohe, H., and Matsumara, K., The relation between the heat of combustion and the oxygen index for high-polymeric materials, *Text. Res. J.* **45**, 778–784 (1975).
196. Otterburn, M. S., Healy, M., and Sinclair, W., The formation, isolation, and importance of isopeptides in heated proteins, in *Protein Crosslinking: Nutritional and Medical Consequences* (M. Friedman, ed.), Plenum Press, New York (1977), pp. 239–262.
197. van Overbeke, M., Mazingue, G., and Desputes, P., Thermal degradation of wool, *Bull. Inst. Text. Fr.* **30**, 273–288 (1952).
198. Pailthorpe, M. T., and Nicholls, C. H., Unpaired electrons in 365 nm irradiated fibrous proteins, *Radiation Res.* **49**, 112–122 (1972); cf. also Ref 74, Chap. 10.
199. Pardo, C. E., Jr., Fong, W., O'Connell, R. A., and Simpson, J. E., Du Pre, A. M., LeVeen, E. P., and Murphy, E., Wurset wool shrink resistant process proves feasible under plant conditions, *Am. Dyest. Rep.* **64**, 36–38 (1975).
200. Pavlant, A. E., and Lee, K. S., Wool modification induced by low temperature glow discharge, *Proceedings of the Fifth International Wool Textile Research Conference, Aachen, Germany (1975)* (K. Ziegler, ed.), German Wool Research Institute, Aachen, Vol. III, pp. 263–274 (1976).
201. Petajan, J. H., Voorphee, K. J., Packham, S. C., Baldwin, R. C., Einhorn, I. N., Grunnet, M. L., Dinger, B. G., and Birky, M. M., Extreme toxicity from combustion products of a fire-retarded polyurethane foam, *Science* **187**, 742–744 (1974).
202. Pearce, E. M., and Liepins, R., Flame retardants, *Environ. Health Perspectives* **11**, 59–69 (1975).
203. Pittman, A. G., Ward, W. H., and Wasley, W. L., Rapid dyeing and flame-resist treatments for wool in hot ethylene glycol, *Text. Res. J.* **46**, 921–924 (1976).
204. Pleasance, H. D., Wool's protective, safety applications, *Text. J. Aust.* **50**, 20 (Oct. 1975).
205. Porter, Sylvia, When the fabrics around us burn, *San Francisco Chronicle* (June 19, 1975). (She cites a survey of deaths, injuries, and material losses caused by fires conducted by the Consumer Product Safety Commission's National Electronic Injury Surveillance System and its U.S. Household Fire Experience Survey. Information was gathered from hospital emergency rooms; does not include minor burns treated in physicians' offices.)
206. Prival, M. J., McCoy, E. C., Gutter, B., and Rosenkranz, H. S., Tris(2,3-dibromopropyl) phosphate: Mutagenicity of a widely used flame retardant, *Science* **195**, 76–78 (1977).
207. Reuter, F. W., Secor, G. E., and Friedman, M., A method for bromine determination in wool fabric by X-ray fluorescence spectrometry, *Text. Res. J.* **46**, 463–465 (1976).
208. Reuter, F. W., Hautala, E., Randall, J. M., Masri, M. S., and Friedman, M., Energy dispersive X-ray fluorescence spectroscopy applied to metallic ion scavenging properties of agricultural waste products, *Proceedings of the Second International Conference on Nuclear Methods in Environmental Research, Columbia, Mo., July 1974* (J. R. Vogt and W. Meyer, eds.), University of Missouri, Columbia, pp. 168–186 (1975).
209. Richards, H. R., Mesure quantitative de la visibilité dans des atmosphères contenant de la fumée, *Bull Inst. Text. Fr.* **25** (156), 675–687 (1971).
210. Richards, H. R., and Patel, A. B., A proposed method for making wool shrink-resistant and stretchable without damaging fiber structure, *Text. Chem. Color.* **1**, 54–62 (1969).
211. Roberts, G. A., Wool in miscellaneous commercial applications, *Text. Inst. Ind. Aust.* **3**, 26 (Dec. 1975).
212. Rosenthal, A. J., Flame-retardant fiber blend, U.S. Pat. 4,035,542 (1977). (Brom*bis*-phenol, Chloro*bis*phenol).

213. Sanuki, H., Effects of porosity and humidity on flammability of wool assemblies, *J. Jpn. Res. Assoc. Text. End Uses* **16** (5), 151–155 (1975).
214. Sanuki, H. Flammability and flameproofing of wool, *Sen'i Gakkaishi* **31** (6), 161, 169 (1975); *Chem. Abs.* **83**, 98594 (1975).
215. Sawamura, K., Fireproofing of wool, *Jpn. Kokai* **75**, 46, 997 (April 26, 1975); *Chem. Abs.* **83**, 116, 863b (1975).
216. Schmitt, J., and Dardis, R., Cost–benefit analysis of flammability standards, *Text. Chem. Color.* **8**, 56–59 (1976).
217. Schmitt, R. H., Grove, E. L., and Brown, R. D., The equivalent conductance of hexafluorocomplexes of group IV (Si, Ge, Sn, Ti, Zr, Hf), *J. Am. Chem. Soc.* **82**, 5292 (1960).
218. Schwenker, R. F., Jr., and Dusenbury, J. H., Differential thermal analysis of protein fibers, *Text. Res. J.* **30**, 800–801 (1960).
219. Schwimmer, S., and Friedman, M., Genesis of volatile sulphur-containing food flavors, *Flavour Ind.* 137–145 (1972).
220. Segall, W. M., A fire in our bosom, *Text. Chem. Color.* **1**, 67–68 (1969).
221. Sela, M., Antigens and Antigenicity, *Naturwissenschaften* **56**, 206–211 (1969).
222. Sello, S. B., and Stevens, C. V., Flame resistant washable wool, in *Proceedings of the Fifth International Wool Textile Research Conference, Aachen, Germany (1975)* (K. Ziegler, ed.), German Wool Research Institute, Aachen, Vol. V, pp. 85–94 (1976).
223. Shirakawa, T., Hiraoka, S., and Mizuno, K., Finishing of textiles, *Jpn. Kokai* **75**, 111, 399 (1975).
224. Simpson, W. S., Efficient flameproofing agents for wool, *J. Appl. Polym. Sci. Appl. Polym. Symp.* **18**, 1177–1182 (1971).
225. Simpson, W. S., The Flame-Proofing of Wool, Communication No. 8, Wool Research Organization of New Zealand (Inc.), Christchurch (Aug. 1970).
226. Smith, G. W., High resolution proton magnetic resonance studies of para-disubstituted benzenes, *J. Mol. Spectrosc.* **12**, 146–170 (1964).
227. Snow, J. T., Finley, J. W., and Friedman, M., Oxidation of sulfhydryl groups to disulfides by sulfoxides, *Biochem. Biophys. Res. Commun.* **64**, 441–447 (1975).
228. Stamm, G., Feuer, Schutz oder Gefährdung durch Textilien?, *Textilveredlung* **8**, 320–326 (1977).
229. Stamm, G., Heutiger Stand der Brenn- und Entflammbarkeitsprüfung, *Textileveredlung* **8**, 341–349 (1977).
230. Suchecki, S. M., The fire storm has passed, *Text. Ind.* **140**, 22–27 (Feb. 1976).
231. Suchecki, S. M., Is FR legislation succeeding?, *Text. Ind.* **140**, 47–49 (Feb. 1976).
232. Sugiura, T., Mieno, K., and Yonemoto, K., Fireproofing of textiles, *Jpn. Kokai* **75**, 65, 697 (1975); *Chem. Abs.* **84**, 3252c (1976).
233. Sugiura, T., Mieno, K., and Yonemoto, K., Fireproofing of textiles, *Jpn. Kokai* **75**, 65, 698 (1975); *Chem. Abs.* **84**, 3252a (1976).
234. Sundstrom, G., Hutzinger, O., and Safe, S., Identification of hexabromobiphenyl as a major component of flame retardant Fire-Master RBP-6, *Chemosphere* **1**, 11–14 (1976).
235. Sweetman, B. J., and Maclaren, J. A., The reduction of wool keratin by tertiary phosphines, *Aust. J. Chem.* **19**, 2347–2354 (1966).
236. Symm, R. H., Reactive finishes applied from chlorinated solvents, *Tex. Chem. Color.* **1**, 161–164 (1969).
237. Tesoro, G. A., and Rivlin, J., Flammability behavior of experimental blends, *Text. Chem. Color.* **3**, 156–160 (1971).
238. Tesoro, G. A., Flame-retardant fabrics: Are researchers on the right track, *Text. Chem. Color.* **1**, 307–310 (1969).
239. Tillin, S., Pardo, C. E., Fong, W., and Friedman, M., Flameproof and shrinkproof wool, *Text. Res. J.* **42**, 135–136 (1972).

240. Thelen, J., and Knott, J. Essai d'interprétation des traitements ignifuges de la laine a base de sels de titane et de zirconium, *Proceedings of the Fifth International Wool Textile Research Conference, Aachen, Germany (1975)* (K. Ziegler, ed.), German Wool Research Institute, Aachen, Vol. V, pp. 107-118 (1976).

241. Thorsen, W. J., Flame-resisting wool with halo-organic compounds. Part I. Water-soluble or emulsifiable bromine derivatives incorporating one or more reactive centers. *Text. Res. J.* **46**, 100-103 (1976).

242. Tovey, H., and Vickers, A., Hazard analysis of fires involving blankets, *Text. Chem. Color.* **8**, 23-26 (1976).

243. Tyuganova, M. A., Kop'ev, M. A., Kozhanova, T. Ya., and Muromtseva, G., Production of fireproof wool (USSR), *Temat. Sb. Nauk. Tr. Mosk. Tekstiln. int.* **1** (4), 47-51 (1974); from *Ref. Zh. Khim. Abs.* 445895 (1975).

244. Uejima, H., Kudo, K., Nanba, T., Ida T. Yamamura, H., and Echigo, Y., Fireproofing of carpets, *Jpn. Kokai* **75**, 69, 358 (1973); *Chem. Abs.* **84**, 1233, 351 (1976).

245. Van Rensburg, N. J. J., SAWTRI Simultaneous Shrink-Resist and Flame-Retardant Treatment for All-Wool Fabrics. I. Preliminary Trials with Chlorine/Hercosett/THPOH and Chlorine THPOH, SAWTRI Technical Report No. 332, South African Wool Textile Research Institute, Port Elizabeth, South Africa (1976).

246. Van Rensburg, N. J. J., and Barkhuysen, F. A., Treatment of all-cotton and 67/33 cotton/wool blends with THPOH and liquid ammonia, a preliminary report, SAWTRI Bull. **8**, 26-33 (Sept. 10, 1976).

247. Van Rensburg, N. J. J., and White, J. E., A study of the effect of different softening agents on the limiting oxygen index of flame-retardant cotton fabrics, *SAWTRI Bull.* **9**, 25-28 (1975).

248. Walker, F. K., and Harrison, W. J., The exothermic gaseous oxidation of scoured wool, *N. Z. J. Sci.* **8**, 106-121 (1965).

249. Walker, F. K., Harrison, W. J., and Read, A. J., Ignition of wool in air. Part I—Ignition temperature of dry wool, *N. Z. J. Sci.* **10**, 32-51 (1967).

250. Wasley, W. L., and Pittman, A. G., Flameproofing wool textiles, U.S. Pat. 3,936,267 (1976).

251. Watt, I. C., Properties of wool fibers heated to temperatures above 100° C, *Text. Res. J.* **45**, 728-735 (1975).

252. Weil, E. D., and Aaronson, A. M., Phosphorus flame retardants—some effects on smoke and combustion products, presented at the University of Detroit Polymer Conference on Recent Advances in Combustion and Smoke-Retardants of Polymers, May 25-27, 1976. (I thank Dr. Weil for a preprint of this paper.)

253. Whitfield, R. E., and Friedman, M., Flame-resistant wool. III. Chemical modification of wool with chlorendic anhydride and related haloorganic acid anhydrides, *Text. Chem. Color.* **5**, 76-78 (1973).

254. Whitfield, R. E., and Friedman, M., Flame-resistant wool. IV. Chemical modification of wool with haloorganic acid halides, *Text. Res. J.* **42**, 533-535 (1972).

255. Williams, M. J., THPC-treated wool: Amino acid analysis, *Text. Chem. Color.* **2**, 41-44 (1970).

256. Wilson, R. H., A note on skin tests of flame-retardant materials, *Text. Res. J.* **32**, 424-425 (1962).

257. Young, C., Burn-resistant fabrics: There's hope ahead, *FDA Papers* **5** (6), 14-18 (1971).

258. Ziegler, K., Schmitz, I., and Zahn, H., Introduction of new crosslinks into protein, in *Protein Crosslinking: Biochemical and Molecular Aspects* (M. Friedman, ed.), Plenum Press, New York (1977), pp. 345-354.

9

Smoke and Tenability: A Perspective on the Materials Approach to the Fire Problem

Jack Kracklauer

1. Introduction: Combustion versus Fire

Combustion is a chemical phenomenon in which an oxidant is reacted with a fuel to produce energy and numerous combustion products. Thermodynamics and combustion chemistry are two highly developed and precise scientific disciplines. The combustion engineering discipline has evolved to the point that combustion efficiency in commercial diesel engines which transport more than 75% of the output of the American economy is greater than 99%. Unfortunately, there is another aspect to this phenomenon. Uncontrolled or accidental combustion is called fire. Available estimates from the National Fire Protection Association[1] indicate that in 1975 fire losses occurred at a rate of approximately 6% of the value of all new construction[2] in the United States. More importantly, the United States has the highest fire death rate of any industrialized nation.[3] A shocking comparison recently presented in *Fire Journal*[4] serves to illustrate the magnitude of this problem. Between 1961 and 1972, the Department of Defense records indicate that 46,000 Americans were killed in the Vietnam war. "In the same time period 144,000 people—three times as many—died because of fires in this country,

Jack Kracklauer • Market Development, Arapahoe Chemicals, Boulder, Colorado.

with hardly a whimper of protest being raised!" For 1975 these fire deaths were estimated[5] at 11,800 or 55.4 per million inhabitants. This problem is of sufficient magnitude that a blue ribbon panel at the federal level studied the problem and issued a report called *America Burning* in 1973.[6] This report states that "in the modern environment of synthetic materials, smoke and toxic gases have become increasingly important hazards" and indicates that in one study 53% of the victims succumbing at a fire die of smoke and toxic gas involvement, not from direct fire or burns.[7] The same report recommends that a national goal of a 50% reduction of fire losses be established. Recent data[8] indicate that 72% of fire deaths occur in residential occupancies. The foregoing would seem to position the problem of smoke generation from polymeric materials in accidental fires in the forefront of concern for the scientific and regulatory communities. While the problem is evident and the concern is real, there is a critical difference between the environments that must be dealt with by combustion engineering and fire safety engineering. Combustion engineering specifies its environment by design of the equipment. Fire safety engineering, on the other hand, must deal with a totally uncontrolled environment and a combustion occurrence which is a result of an abberation of normal performance. A poignant example of this complex problem is the result[9,10] of two recent attempts to characterize the critical physical and thermal occurrences during fire growth in the most common scenario[11] of residential fire fatalities, the bedding fire. On two separate occasions, a full-scale (2.4 × 3.7 × 2.4 m) mock-up of a residential bedroom was constructed and completely furnished with a bed, desk, bookshelf, television, dresser, bedclothes, and curtains. One of the purposes of the first two fires in this mock-up facility was to evaluate alternative instrumentation locations and types. Independent of the details of this instrumentation, the time to flashover was observed in both tests. In the authors words,[12] "although an attempt was made to duplicate for the second test the structure, furnishings and ignition of the first bedroom test, a different fire history resulted (full-room involvement in 7 minutes as opposed to 17½ minutes for the first test)." Again, in the words of the author after the first test,[13] "full understanding of fire will be possible only when small changes in circumstances do not have unpredictable, drastic effects on the results." While a full understanding may not be possible, an analysis of the real-world fire problem can be fruitful in indicating the nature of the phenomenon, the role of materials in the problem, and hopefully directions the technical community can pursue in an attempt to make progress toward the goal of a 50% reduction in fire loss.

2. Fire: Ignition and Materials

As mentioned in the introduction, fire in the built environment is the result of an aberration of normal performance. A recent report[7] has attempted to collate the available statistics to determine the most likely scenarios for fatal fires from the perspective of ignition source and items ignited. It is valuable to review this information since ignition is the first critical link in the sequence of events leading to the statistics of fire fatalities. Four sources were used to generate the analysis listed in Table 1.

This analysis suggests that 10% of the scenarios detailed above are not, in fact, subject to solution by materials modification. These are the flammable liquid fires where materials are not likely to make any significant contribution during the ignition phase of the fatal fire. In addition to flammable liquids, apparel fires constitute 12% of the fatal fire scenarios. These apparel fires are largely independent of occupancy, and since extensive legislation already exists in the flammable fabric area and 72% of the fatal fires occur in residential occupancies,[15] we shall not address this area further.

The remaining 44% of the detailed scenarios will now be analyzed. Clearly, the most hazardous ignition risk (29%) is smoking materials. While it would seem simple from a risk management viewpoint to ban smoking as a substantive means to reduce our national fire death toll, this would be an unacceptable invasion of personal freedom of choice. This is clearly indicated by the air bag controversy relative to auto safety. Even more directly, the

TABLE 1
Top Fire Death Scenarios[14]

Occupancy	Ignition source	Item ignited	% of U.S fire deaths
Residential	Smoking	Furnishings	27
Residential	Open flame	Furnishings	5
Transportation	Several (accidents)	Flammable liquids	4
Usually residential	Heating and cooking equip.	Apparel	4
Residential	Heating and cooking equip.	Furnishings	4
Independent	Several	Apparel–flammable liq.	3
Residential	Heating and cooking equip.	Flammable liquids	3
Residential	Open flame	Flammable liquids	3
Independent	Open flame	Apparel	3
Residential	Heating and cooking equip.	Interior finish	2
Residential	Electrical equip.	Interior finish	2
Independent	Smoking	Apparel	2
Residential	Electrical equip.	Structural	2
Residential	Smoking	Trash	2
			66
		Others, all less than 2% of total	34
			100

proven link between smoking and cancer risk has failed to significantly reduce the rate of consumption of cigarettes in the United States. Consequently, it is unlikely that the risk of fire death from smoking, which on a nationwide annual basis is approximately 1 in 64,000, will have any significant impact on either legislation or public acceptance of smoking as a fire risk.

The other three most likely ignition sources from this scenario ranking are heating and cooking equipment at 6%, open flames at 5%, and electrical sources at 4%. In most of these occurrences, malfunction of an appliance, furnace, heater, or electrical system is the source of ignition. These are real-world events, and, again, materials science or modification is not likely to have any significant impact on their occurrence. One scenario which is currently under study is the open flame ignition caused by children playing with matches. Numerous groups are involved in an effort to develop a child-resistant matchbook, and recent efforts[16] appear promising.

In the larger perspective, however, it would appear that ignition sources are attributable to the failure—either personal in the case of smoking materials or mechanical—of the accoutrements of our twentieth-century lifestyle. Since these accouterments are not, in fact, highly hazardous or failure prone with the possible exception of smoking materials, it would appear that we must move further along the causative sequence in fire to find an appropriate point of attack for hazard reduction from the fire problem. The next step, assuming that sources of ignition do and will exist in the built environment, is the first item ignited. Table 1 also contains this information. Furnishings represent the largest hazard, with 36% of the fatal fires attributed to this source. This is consistent with another review[17] which also concluded that combustible contents were the most probable first item to ignite in fatal dwelling fires. This review proceeded further in its analysis and concluded that surface finish involvement was responsible for the spread of smoke and fire in 55% of these fatal fires. This fact, coupled with the additional information in Table 1 that surface finish and structural members are the first item to ignite in 6% of the scenarios studied, would indicate that the investigation of the role of materials in surface finish applications is the most likely to have a major impact on fire fatalities. An additional incentive to studying this area is that development of criteria for improved fire performance of surface finish materials can be enforced through the existing mechanism of the National Model Building Codes. Furnishings, on the other hand, would require development of a new regulatory mechanism. The only current legislation on furnishings is a California law[18] regulating upholstered furniture and a federal law regulating mattresses.[19] Both of these combustible contents regulations are currently enforced only at the manufacturer level, whereas the building codes through the inspection mechanism maintain control when substantial renovation or alteration of existing structures[18] is completed. Consequently, the materials aspect of smoke control from accidental fires involving surface finish materials will serve as the focus for the following sections of this chapter.

3. Smoke: A Hazard Analysis

Analysis of fire fatalities will typically report smoke inhalation as the most common cause of death. In the limited data[20] available on actual autopsies of fire fatalities carbon monoxide is the primary cause of death in 50–80% of the cases studied. While carbon monoxide is the primary cause of death in accidental fires, the first response of an individual in a fire environment should be to try to escape. It should be recognized that without proper protection and equipment such as that used by the fire services (helmets, fire coats, and self-contained breathing apparatus), the only safe and logical alternative is to leave the scene of a fire and call for professional help. This is the true meaning of the concept of tenability. The detailed analysis of fire fatalities in Maryland from which the autopsy results above were obtained also indicates that[21] "victim location shows that 80% of all victims (in this study) physically able to escape were alerted and attempted to escape." This information clearly focuses the question of tenability on which property of a fire is the first to limit an individual's ability to egress. Table 2 presents the various criteria that have been proposed relative to visibility limitation of egress potential.

These criteria are based on the visibility loss caused by the liquid aerosol and carbon particulate components in smoke. This aspect of tenability criteria was identified as the key aspect to be considered relative to survivability in fire exposure and was well summarized in the 1961 Los Angeles Fire Department Report[29] on operation school burning as follows:

> The smoke itself does not contain a high enough concentration of dangerous gases to be lethal in the early stages of a fire. However, the untenable smoke, by its irritant properties and its obscuration of normal visibility, does immobilize the occupants within the area of the building where they happen to be located. They are then "trapped" within the building, and, unless rescued promptly, may be killed by lethal heat and gases which follow shortly.

The criteria presented in Table 2 are based on the visibility aspect of smoke since the human response factors of eye irritation and panic are difficult to generalize and impossible to model with instruments. Optical density is defined by the following equation:

$$\text{optical density} = \log_{10} \frac{100}{T}$$

where T = light transmittance in percent.

TABLE 2
Proposed Visibility Criteria for Tenability in Fire Environment

Criterion (optical density/m)	Basis	Reference
0.051	20% transmission across 45-ft distance	22
0.27	Visibility of 40-W exit sign across 12 ft	23
0.26	16% transmission across 10 ft	24
0.043	Considered human eye response factors	25
0.02	Based on tenability considerations	26, 27
0.062	Aircraft mock-up, 20-ft distance, visual acuity < 20:100	28

The criteria are expressed as optical density per meter since the distance to the exit is an obviously important factor in egress potential and will vary from scenario to scenario. Inspection of the numerical values in Table 2 reveals two rather obvious groupings, one at OD/m = 0.02 and the other at 0.26. The 0.051-OD/m criterion of the Los Angeles tests becomes 0.23 if the "distance required for visibility" drops from the 13.7 m (45 ft) used in that reference to the 3 m (10 ft) used by Gross to develop his 0.26 criterion. The other three references have used the lower criterion in an attempt to recognize the significance of panic and eye irritation. The Lockheed study[28] of a mock-up of an aircraft interior provides a graphic indication of the irritant aspect of smoke and visibility. The procedure used in this study was to place four observers outside of the aircraft cabin mock-up. Two of those observers were directly exposed to the fire environment, while the other two viewed the same tests through windows. In two of the tests[30] smoke buildup was adequate to cause the observers using windows to report eventual degradation of visibility to the 20:200 level (they could only see the large E on a standard eye test chart at 6.3 m or 20 ft). The unshielded observers withdrew from the test chamber at approximately 5 min from the start of the test when the photocells were measuring an OD/m of 0.04–0.07. The protected observers, on the other hand, reported visibility at the 20:110 level at 8–9.5 min when the photocells were off scale (OD/m > 1.0) in one case and recording OD/m of 0.07 in the second. Since the time required to egress after a person becomes aware of a fire has been estimated to be 2 min,[26,27] the difference of 3–4.5 min between the shielded and unshielded observers is quite significant. This information adds perspective to the divergence of values in Table 2. A simple way to address these criteria values may be to consider the low criterion (i.e., OD/m = 0.02) as conservative and the higher value (i.e., OD/m = 0.26) as optimistic. To assess the time rate of development of smoke and the other tenability limiting factors of carbon monoxide and temperature, criteria are needed for these two factors also. Again, the literature provides conservative and optimistic criteria, as presented in Table 3.

The primary reason for the dramatic difference between the conservative and optimistic criteria of Table 3 is that the optimistic criteria are based on conditions considered untenable for brief periods of 5–15 min, while the lower criteria are considered incapacitating in the longer time frame of 60 min[32] Data to be presented subsequently will demonstrate that the criteria selected will depend on the type of fire hazard. For example, smoldering ignition results in slow buildup of smoke and combustion products, making the conservative criteria seem most reasonable. Flaming ignition, on the other hand, can result in untenable conditions developing very rapidly (less than 5 min) so the optimistic criteria seem most realistic in that scenario.

Having found criteria for tenability limits in accidental fires, it now is a simple matter to compare these tenability limits to actual fire data records to determine the relative importance of these three tenability limiting factors. Obviously, a complete record of temperature, smoke opacity, and carbon monoxide is not available for accidental fires in the built environment which result in fatalities. However, there have been four test programs recently which have consisted of full-scale fire tests which produced complete data records which can be used in conjunction with the criteria of Table 3 to determine the real time required to reach the escape limiting conditions attributable to each of the fire hazard factors. The results of this analysis are presented in Table 4.

These data clearly indicate that smoke is the first of the fire hazard conditions to limit egress in these test fires. Since the first five tests involved fully furnished occupancies typical of residential accommodations, it can be inferred that there is a good basis for the conclusion previously presented from the Los Angeles tests, namely that smoke is the primary egress problem in spite of the fact that heat or carbon monoxide are the true lethal components in the fire environment. It is interesting to note that the first and second bedroom test fires are those previously used to illustrate the variability of real fires. The time to flashover in these two fires varied by 10 min. In spite of that variability, the time to the criteria level of both visibility loss and

TABLE 3
Fire Conditions Limiting Escape

	Conservative[27]	Optimistic[31]
Heat (°F)	150	300
Carbon monoxide (%)	0.04	1.0
Visibility (OD/m)	0.02	0.26

TABLE 4
Time to Reach Tenability Criteria in Large-Scale Fire Tests

Test	Facility	Item ignited	Criteria[a] applied	Time to reach tenability criteria			Flashover	Reference
				Smoke (min)	Heat (min)	Monoxide (min)		
Furnished	Bedroom I	Mattress	Optimistic	$3\frac{1}{3}$	$5\frac{1}{2}$	$18\frac{1}{3}$	$17\frac{1}{2}$	9
Furnished	Bedroom II	Mattress	Optimistic	4	6	$6\frac{1}{3}$	7	10
Furnished	Apartment	Sofa	Conservative[b]	30	$41\frac{1}{3}$	After flashover	$41\frac{3}{4}$	26
Furnished	Apartment	Pillow	Conservative[b]	$6\frac{1}{3}$	28	After flashover	30	26
Furnished	Apartment	Toaster	Conservative[b]	$11\frac{1}{3}$	24	After flashover	26	26
Unfurnished	Room	Wood crib	Optimistic	$3\frac{5}{6}$[c]	ND	ND	$7\frac{1}{3}$	33

[a]From Table 3.
[b]Smoke criterion = 0.02 OD/m; temperature criterion = 300°F.
[c]Average of three fire tests.

temperature varied by less than 1 min. This would suggest that the state of the art may not be so confused as a superficial analysis would indicate. Further support for that conclusion is found in the data record supporting the third test in Table 4. The apartment fire using sofa ignition was run twice because of an instrument failure during the first test. The author concludes[34] that the "development of the fire through the smoldering stage was nearly a complete parallel..." to the first fire test, with flaming combustion occurring at 40½ and 39½ min in the two tests. An additional feature of this test series was the exposure of rats in the fire test environments to determine whether any unusually toxic combustion products would cause mortality sooner than the measured heat and carbon monoxide levels would lead one to expect. There was no indication of unusually toxic materials causing mortality prior to flashover. It was observed, however, that the rats exposed outside the room of fire origin in the bedroom test where the fire was not allowed to progress through flashover were "...strongly affected by the fire environment."[35] This abnormal behavior was first observed at 26 min—a full 20 min *after* the visibility (egress) criteria had been exceeded.

Finally, the last result listed in Table 4 provides an interesting comparison and contrast to the other data. These data are from a sterile test with no furnishings included in the test room. A wood crib was used for ignition, and surface finish materials were used as the combustibles studied. The criteria application in this test[36] was also different from that in previous tests. Smoke was measured as it exited from the test enclosure. The rate of smoke effluent was then integrated over time, and the criterion of 0.26 OD/m was applied to a normal-sized residential occupancy of dimensions $7.6 \times 12.2 \times 2.4$ m ($25 \times 40 \times 8$ ft). The concept behind this treatment is that a wood crib simulates a rapidly developing fire and it is unlikely that a resident would be unaware of such a fire if he were in the room of fire origin. The hazard represented by this scenario is that of the living room fire at night when the occupants are asleep in bedrooms remote from the living room. The life hazard posed by this scenario is for the occupants to be able to escape prior to the heat and other fire gases producing life-threatening conditions which the previous apartment test fires had indicated occur at or after flashover. Flashover is the point in fire development when the fire itself goes beyond the room of origin and can be characterized as a period of explosive fire growth. The reality of this scenario is illustrated by a recent residential fire analysis reported in *Fire Journal*.[37] Figure 1 presents the layout of this residence and represents the extent of the fire which was extinguished within 5 min after the fire service arrived. The neighbors who were the first to arrive reported that the first sound they heard was the living room windows breaking out and that subsequently they heard the heat detectors sound. The analysis indicated that this was apparently a long smoldering fire which delayed activation of the six heat detectors set for 136° F. After becoming aware of the fire,

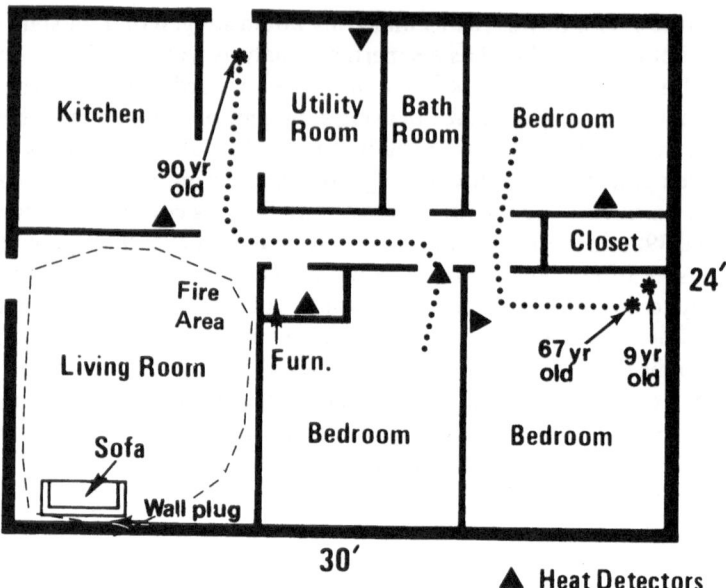

the 67-year-old grandfather went into the southeast bedroom and succumbed while shielding the 9-year-old boy with his body. The boy was rescued but succumbed 37 hr later to "smoke inhalation." The NFPA analysis of this fatal fire points out that "results of full scale experiments in typical dwelling fires indicate that detectable quantitatives of smoke precede detectable degrees of heat in nearly all incidents. . ."[38] and emphasizes that the six heat detectors installed in this residence do not conform to the minimum level of protection for residential fire warning equipment.[39] The other fatality was found 10 ft from the kitchen door—less than 15 sec from safety. The sole survivor of this fire reported being awakened by the alarm from the heat detector in the southeast bedroom. She went to another bedroom to awaken her husband, and they returned to rescue the 9-year-old boy. She was rescued from the southeast bedroom by neighbors, but the smoke and heat were so dense that the fire service personnel had to use breathing apparatus to find and evacuate the boy.

This unfortunate example of a real fire is intended to underscore a very important point which is addressed in the unfurnished room test in Table 4. Namely, prior to flashover, fire itself is a highly localized phenomenon, while the smoke produced by the fire can rapidly build up untenable, i.e., life-threatening, conditions throughout the occupancy. A dramatic example of the speed of transition from a smoldering sofa fire to flashover is provided by the duplicate apartment fires with sofa ignition previously described. The first fire was extinguished (with sprinklers) 55 sec after flaming combustion began (at 39½ min from initiation of the test) and immediately prior to flashover. The actual fire damage was limited to the center section of the sofa and the wall behind it. The second test fire, which was an almost exact duplicate of the first with transition to flaming combustion occurring only 50 sec later, was extinguished 85 sec after flaming began, and the living room and kitchen both demonstrated the effects of a postflashover fire, i.e., serious heat damage not observed in the first fire extinguished before flashover. This leads one to conclude that flashover required less than 30 sec to occur. It is further worth noting that in these fires the visibility criterion for escape was exceeded 11–13 min prior to flashover. The preceding analysis of the findings of the three furnished occupancy tests and the fire analysis clearly indicate the necessity of expansion of fire hazard focus from the room of fire origin to the full dimensions of the occupancy. To illustrate this concept, if one assumed that the door to the room of fire origin was initially closed and fire buildup occurred at the same rate as in the open-door tests with the door being blown open when the flashover temperature of 540° C was reached (approximately 8 min), extrapolation of the average smoke accumulation into the full occupany provides an OD/m of 1.1, more than fourfold higher than the optimistic criteria for visibility. Similar treatment for the flashover temperature yields a well-mixed average temperature of only 53°C or one third of the temperature criterion for tenability. The preceding analyses lead us to the following conclusions, which will serve as the basis for investigating the materials response aspect of fire hazard:

1. Smoke is the primary tenability hazard in accidental fire in the built environment.
2. Prior to flashover the fire itself is confined to the room of origin.
3. Smoke is a total occupancy hazard in the preflashover fire environment, not a localized problem.
4. Recent large-scale enclosure and occupancy tests indicate that heat, not carbon monoxide or other toxicants, is the most likely second tenability criterion to be reached, but usually 10 min after smoke prohibits escape.

4. Smoke: The Measurement Problem

The test results reviewed in the previous section have used optical measurement of visible light obscuration. This is appropriate because of the human hazard posed by smoke in accidental fires. The appropriate definition of smoke from this perspective was proposed in 1971 in an ASTM Review[40] as follows: "particulate matter and dispersed liquids suspended in air—principally unburned carbon resulting from incomplete combustion." This definition highlights the fact that optical measurement of smoke is in fact a secondary measurement of this aspect of the fire environment and that mass measurement would be more direct. The vast majority of smoke assessment techniques use optical measurement in an attempt to allow correlation with the human tenability concern in hazard assessment from building fires. This question of correlation of small-scale fire hazard tests with "actual fire conditions" was emphasized dramatically in the 1974 Federal Trade Commission action[41] against foam plastics manufacturers. Since there is much confusion in the area of small-scale smoke measurement caused by the divergent indications of smoke generation given by the various methods, in this section we shall address the question of correlation for the three most reasonable small-scale tests.

Before proceeding with this correlation analysis, it is necessary to address two questions of considerable technical merit. The first is the technical basis for optical assessment of smoke which is a complex mixture of particulate and aerosol components. This is based on the assumption that Lambert's law pertains where optical density is described[42] by the following equation:

$$D = \log \frac{I_0}{I} = \sigma \frac{L}{2.303}$$

where I_0 and I are the measured intensities of the incident and obscured light, L is the path length, and σ is the attenuation coefficient. Foster[43] has described optical density as a function of mass concentration of particulates, particulate density, radius, effective cross-sectional scattering and absorption areas; and light path length. Numerous groups[44-47] have studied the relationship between optical density and mass concentration of smoke particulates and found a linear relationship. The University of Utah group lead by Seader[48] has reported a broad correlation, while the other groups report differences in the linear relationships from material to material. Table 5 represents King's results indicating[49] wood smoke differs in its correlation between gravimetric concentration and optical density from four synthetic materials. Pyrolysis studies of cellulose have shown significant amounts of aerosol and organic tars in the vapor phase products which may have been lost during the drying procedure King used prior to weighing the collected smoke. King also reports that ". . .smoke generated under nonflaming conditions results in less obscuration per unit mass of smoke than under flaming conditions." The magnitude of this difference is 50% and is probably attributable to the difference in heat intensity between smoldering (2.5 W/cm^2) and flaming (approximately 5 W/cm^2) conditions used by King since Bankston *et al.*[45] have shown a difference in mean particle diameter of 0.5 μm at

TABLE 5
Correlation of Optical and Gravimetric
Smoke Measurements[a]

	Slope of correlation line	
Material	Flaming	Nonflaming
ABS plastic	0.27	0.33–0.56
Polystyrene	0.27	0.38–0.52
Rigid polyvinyl chloride	0.28	0.52
Flexible polyvinyl chloride	0.30	0.46
Red oak	0.40	0.83–1.21
Average	0.30	0.60

[a]Data from Ref. 49.

3.2 W/cm^2 versus ~1 μm at 6.2 W/cm^2. In summary, then, the data would indicate that there is a correlation between the direct or gravimetric smoke density and the optically measured smoke density in laboratory test methods.

The second point to be considered is the more general definition of smoke recently proposed to ASTM[50]:

> Smoke: A resulting characteristic of fire, is defined as the airborne products evolved when a material (system, component or assembly) decomposes by heat or burning. Smoke may contain liquid or solid particulates (principally unburned matter resulting from incomplete combustion), or gases or any combination thereof.

This definition is unquestionably correct, but its breadth introduces a serious complication by encompassing both toxic gas components along with the visibility-reducing aerosols and particulates. The analysis of the preceding section has shown that large-scale fire tests indicate these to be separate problems which occur at significantly different times during fire growth. It would not seem reasonable, therefore, to consider the development of a test method to measure both simultaneously when conceptually and practically they are separate problems.

Most smoke test methods have been developed as adjuncts to flammability test methods. This resulted from the misconception that flame and heat were the primary life safety concerns in accidental fires. A broad variety of smoke measurement techniques has been proposed or developed in the last 30 years, and the two ASTM reviews cited earlier[40,50] provide some perspective. The purpose of this review is to restrict its focus to those tests where correlation to large-scale fires has been studied. This effectively limits the field of investigation to three techniques. The first is the ASTM E-84,[51] also known as the Steiner tunnel. Physically this is a 7.6 × 0.6 m (25 × 2 ft) tunnel constructed of fire brick (see Fig. 2). The sample to be tested is mounted on the

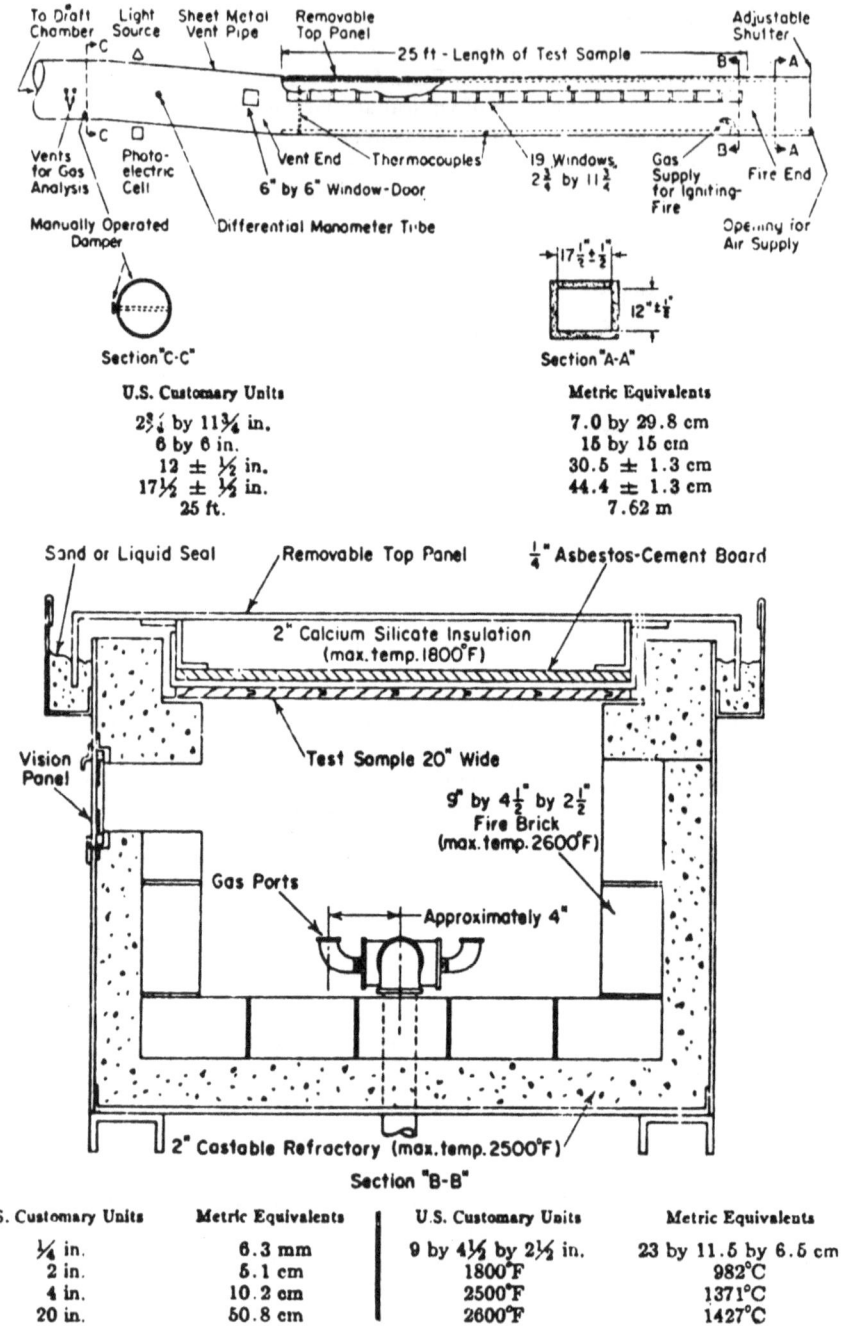

FIGURE 2. Configuration of the E-84 tunnel equipment.[51a]

ceiling of the tunnel furnace and exposed to an igniting methane gas flame of 5000 Btu/min which is impinged on one end of the sample for a distance of 1.4 m (4.5 ft). An air flow of 10.2 m^3/min (240 ft/min velocity) is maintained through the tunnel for the 10-min duration of the test. The method is primarily designed to measure flame spread, but a photocell and light source mounted on the exit stack from the tunnel continuously record transmittance versus time for the 10-min test period. The smoke results are usually reported as a smoke density classification, where the integrated area under the percent transmission time curve is expressed as a ratio to the same integrated area determined for red oak—the reference standard for this test method. It is possible, however, to obtain the actual plots from which the rate of smoke evolution, the time to maximum smoke generation, and the actual integrated (total) amount of smoke generated during the test can be derived. The E-84 test method is widely used in the model building codes as a criteria test for surface finish materials. An acceptance level for smoke density classification of 450* has been adopted, based on some early work at Underwriters Laboratory, which led to the criteria level of OD/m = 0.051 presented in Table 2 and discussed in Sec. 2. Table 6 presents data from Underwriters laboratories[52,53] on the effect of air velocity in the tunnel test on smoke density classification. These data indicate that in this test method both synthetic and natural products show a pronounced decrease in the amount of smoke produced as the air velocity is increased. It should be recognized that a minor part of this effect is probably attributable to cleaner burning at higher ventilation rates. The more significant effect, however, is the reduction in heat input to the sample caused by the increasing air flow rates. One can calculate, based on the 730°C temperature measured[54] 0.6 m from the burner under standard air flow conditions (note: heat flux at 0.9 m measured at 6.5 W/cm^2), that the temperature increases to 1000°C at the low air flow and decreases to 560°C at the high air flow. It is interesting to compare these results to the results obtained by Bankson *et al.*[55] and Tsuchiya and Sumi[42] in their studies of the effect of radiant flux (Bankston) and increasing pyrolysis temperature (Tsuchiya) of the smoke generation properties of some similar materials. Both groups used flow systems to characterize smoke generation, but both used radiant pyrolysis as contrasted with the flaming combustion in the tunnel test. Tsuchiya's work indicated[56] that increasing temperature in the range of 500–700°C decreased the smoke-generating characteristics of wood—the opposite of the indications from the tunnel results in Table 6. He did observe, however, an increase in smoke generation with temperature for rigid urethane foam, polyvinyl chloride and polyisocyanurate foam which correlates with the results in Table 6. Figures 3 and 4 present the results of Bankston's studies

---

*Numerical SDC indices do not define the hazard presented by this material (or products made from this material) under actual fire conditions.

TABLE 6
E-84 Air Velocity Effect on Smoke Developed Classification[a]

	Air velocity (ft/min)					
	177		240[b]		312	
Material	SDC[c]	%[d]	SDC[c]	%[d]	SDC[c]	%[d]
Red oak	271	271	100	100	36	36
⅜-in. plywood	325	232	140	100	40	28
Rigid polyvinyl chloride	1490	130	1147	100	658	57
1.9-pcf isocyanurate foam (A)	236	236	100	100	50	50
1.9-pcf urethane foam (B)	657	110	595	100	27	5
1.9-pcf urethane foam (C)	390	122	319	100	21	7
Average		185		100		31

[a]Data from Refs. 52 and 53.
[b]Standard condition.
[c]Numerical SDC indices do not define the hazard presented by this material (or products made from this material) under actual fire conditions.
[d]% = percent relative to 240 ft/min standard condition.

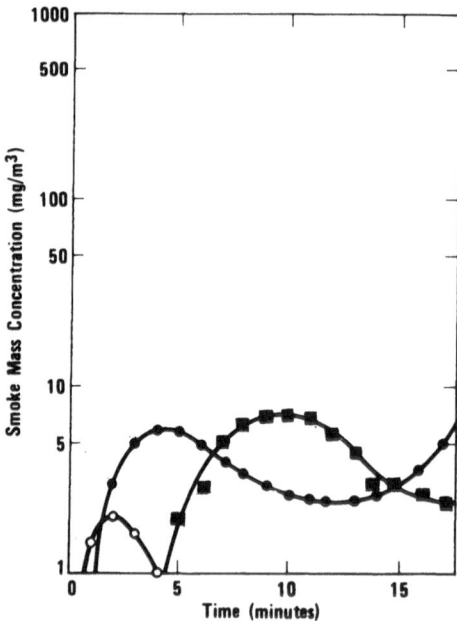

FIGURE 3. Mass rate of smoke generated from three materials.[55] Exposure conditions: radiant = 3.2 W/cm² air. ○ urethane, ● wood, ■ PVC.

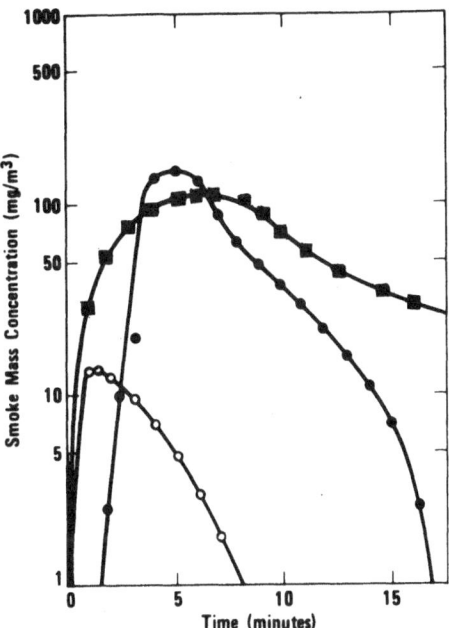

FIGURE 4. Mass rate of smoke generation from three materials.[55] Exposure conditions: radiant = 6.2 W/cm² air. ○ urethane, ● wood, ■ PVC.

and clearly demonstrate a strong increase in smoke generation rate and amount of smoke when the intensity of the pyrolyzing radiation is increased from 3.2 W/cm² (similar to the intensity of exposure at the highest air flow rate) to 6.2 W/cm² (similar to the standard air flow rate). It is interesting to note that the smoke maxima at 3.2 W/cm² expressed as a percentage of the maxima at 6.2 W/cm² for the three materials are wood, 4%; urethane, 6.5%; and PVC, 57%; which agree reasonably well with the percentage results in Table 6 for the similar materials under comparable exposure conditions. On the other hand, it appears discomforting that two superficially similar laboratory tests would give opposite conclusions concerning the effect of pyrolysis intensity on the smoke generation of wood. Tsuchiya's work was based on experiments in nitrogen, air, and 50:50 mixture, and his conclusions were derived by averaging these results. Bankston's results were obtained in air. Oxygen measurements during the tunnel test[57] (at standard air flow conditions) indicated minimum oxygen concentrations in the range of 13–20% with similar materials. It is worth observing that oxygen concentrations in the large-scale tests reported earlier ranged between 14 and 21% at the point the smoke criteria levels were exceeded. In addition, heat flux measurements in a full-scale corner test using a 9-kg (20-lb) wood crib were found to be

$6.4 \ W/cm^2$ at a height of 1 m above the crib at the threshold of flashover and 8.0 after flashover.[58] Consequently, it appears that the tunnel test may represent an accurate simulation of material smoke generation characteristics based on oxygen concentration and pyrolysis intensity considerations.

One final aspect of the E-84 method is the progressive involvement of the sample related to its flammability (rate of flame spread). This is the basic design feature of the E-84 tunnel and is integrally involved with the measurement of smoke in the method. This is similar to the way surface finish involvement proceeds in large-scale tests, since the three dominant factors governing smoke generation in large-scale fires are the rate of combustion involvement of the material, the heat exposure of the material, and the inherent smoke generation properties of the material under those conditions. The design of the tunnel test would seem to take the first two factors into account in assessing the third factor. It is worth noting that the heat flux exposure measured in large-scale crib fire tests is of the same order of magnitude as that measured in the tunnel. The average effluent air flow from the doorway in the unfurnished room test of Table 6 at flashover conditions[59] was 2.9 m/sec through an approximate area of 1.1 m^2. Inspection of the visual record would indicate the crib fire plume area to be approximately 2.9 m^2, which indicates a velocity across the surface finish material under test of 1.1 m/sec (213 ft/min), which is of the same magnitude as the standard conditions in the tunnel test when the limited accuracy of the approximations used are considered. Consequently one would assume that the E-84 test method should show good correlation to large-scale tests of surface finish materials.

The second test of merit is the smoke density chamber (more commonly known as NBS smoke chamber) developed by the National Bureau of Standards[60] (see Fig. 5). The instrument is an 0.51-m^3 (18-ft^3) box which contains a vertically oriented light path of 914-mm (36 in.) length from a bottom-mounted light source (17.5-W optical lamp) to a top-mounted photometer.[61] The optical system is 20.3 cm (8 in.) from the two corner walls diametrically opposite the sample mounting bracket. The vertically exposed sample is normally 65.1 mm ($2\frac{9}{16}$ in.) square and is posi-

---

FIGURE 5. Configuration of NBS smoke chamber.[61] A, Phototube Enclosure; B, Chamber; C, Blowout Panel; D, Hinged Door with Window; E, Exhaust Vent Control; F, Radiometer Output Jack; G, Temperature (Wall) Indicator; H, Temperature Indicator Switch; I, Autotransformers; J, Voltmeter (furnace); K, Fuse Holders; L, Furnace Heater Switch; M, Gas & Air Flowmeters; N, Gas & Air Shutoff Valves; O, Light Intensity Controls; P, Light Voltage Measuring Jack; Q, Light Source Switch; R, Line Switch; S, Support Frame; T, Indicating Lamps; U, Photometer Readout; V, Rods; W, Glass Window; X, Exhaust Vent; Y, Inlet Vent; Z, Access Ports.

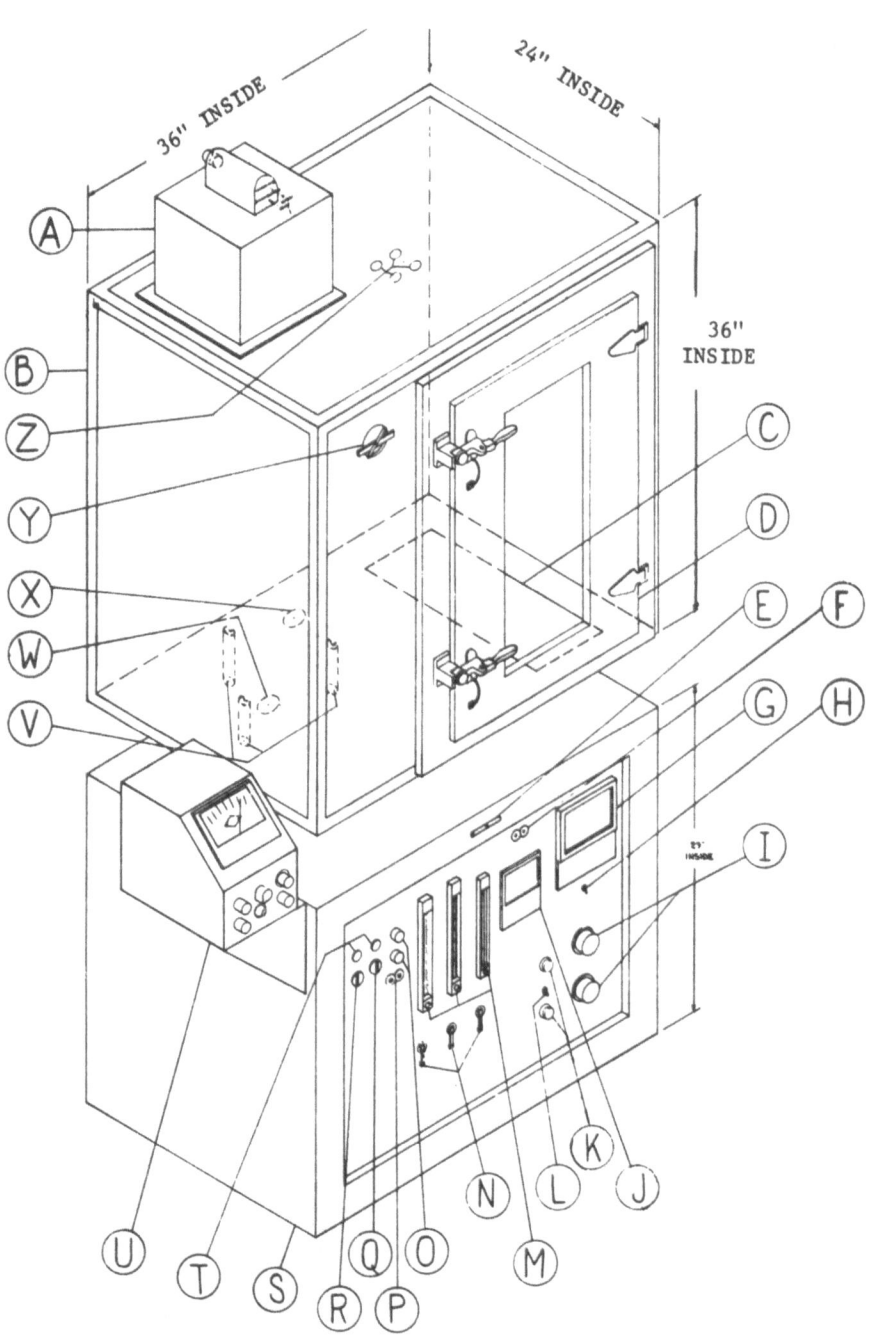

tioned 38.1 mm (1½ in.) away from a circular radiant furnace of 76.2 mm (3 in.) inside diameter. This radiant furnace is adjusted to 2.5 W/cm^2 output as measured at the center of the exposed sample surface. Two modes of operation are possible with this test method. Nonflaming conditions expose the sample to the radiant furnace only. Flaming exposure supplements the radiant furnace with a series of six gas jets impinging at various angles which are positioned in front of the sample and fired with 50 $cm^3/$ min of propane premixed with air at a 10:1 ratio. While the flame exposure has not normally been considered anything more than a pilot flame to ignite the combustion products liberated during pyrolysis by the radiant panel, the 50 $cm^3/$min propane flow, in fact, contributes an additional heat input to the sample of 2.4 W/cm^2. Conceptually, the essential features of this test method are the following: It is a closed system with no ventilation; the heat from the radiant heater does cause some thermal convection, which is the only source of mixing in the chamber; the vertical light path was chosen to avoid stratification problems which might be anticipated in the absence of an internal mixing device. The results obtained with this method are a continuous recording of light transmission on a 5-decade scale. These light transmission versus time outputs are normally expressed as specific optical density by correcting the measured percent transmission for sample size and chamber dimensions using the following formula[62]:

$$\text{specific optical density} = D_s = \frac{V}{AL}\left(\log_{10}\frac{100}{T}\right)$$

where

V = volume of the chamber
A = surface area of sample exposed
L = length of light path

The geometrical factor for calculating D_s in the NBS smoke chamber is 132. The reason for including these geometrical factors was the concept that D_s values obtained in the smoke density chamber could then be used to predict smoke density in ". . . other enclosure volumes, on the basis of several assumptions, such as the assumption of similar smoke–air mixing and complete surface involvement of the material under consideration."[62] An example of this proposed extension of small-scale results is presented in the following calculation for two surface finish materials studied in the unfurnished room test analyzed earlier.[33] The report includes the measured D_m values, which are the maximum D_s observed in the standard smoke density chamber test as well as the minimum OD/m values recorded in the test structure. The calculation is conducted as follows:

V = volume of theoretical house = 222.5 m^2

A = estimated area producing smoke for gypsum board = 2.9 m^2

L = light path = 3 m

geometrical factor = $\dfrac{V}{AL}$ = 25.6

measured D_m for gypsum board = 33

by definition $\dfrac{D_m}{V/AL}$ = $\log_{10} \dfrac{100}{T}$

Therefore, the NBS chamber predicts 5.0% transmission at minimum. Actual measured minimum transmission in room = 0.5% transmission. Repeating these calculations for a totally involved material, we obtain the following values:

A = area involved (total at flashover) = 8.7 m^2

$\dfrac{V}{AL}$ = 8.52

for particle board, measured D_m = 398

as before, predicted percent transmission = 1.9×10^{-45}%

measured minimum percent transmission in room = 0.15%

It should be immediately recognized that one obvious problem with this example is the fact that the occupancy in the room test was hypothetical and the smoke exiting the room through the open door could not build up but was exhausted. As pointed out earlier, one of the conceptual problems in applying this extrapolation is that the actual area producing smoke is unknown in most real fires since flame spread is a progressive phenomenon. There are two more basic problems with this method, however. The first is a direct result of the fact that the design of the instrument, as an accumulation chamber, forces smoke—a very strong absorber of radiant energy—to build up between the radiant energy source and the sample. The report summarizing the observations after the second bedroom fire (Table 2) draws the following conclusions concerning the role of the smoke layer[63]:

Smoke:
1. Blocked radiation from the hot upper walls and ceiling.
2. Constituted the major source for radiant energy transferred within the layer.
3. Provided a significant portion of the energy fed back to the burning bed.[63]

These conclusions are fully supported by the results of a recent study of larger fires in a free-burning configuration. Palmer states[64] the following: "this study confirms that the thermal radiation of a fire is strongly absorbed by the smoke." Both of these recent findings add a new dimension to the significance of smoke in fire growth as well as tenability. As it relates to the smoke density chamber, however, it seems clear that buildup of smoke between the radiant source and the sample might be expected to significantly reduce the actual exposure received by the sample. This may account for the discrepancy found between a ventilated and nonventilated evaluation of smoke generation from wood in a horizontal rather than vertical configuration. The radiant furnace is modified for this modification, resulting in a somewhat longer radiation path. Cassanova *et al.*[65] report that ventilation of the enclosure results in the horizontal mounting producing a considerably higher smoke generation than when the sample is mounted vertically. Breden, on the other hand, found no significant difference between horizontal and vertical mounting in a nonventilated chamber. Gaskill[66] has studied the effect of ventilation directly in the NBS smoke chamber. His work indicates that a change from unventilated conditions to a ventilation rate of 20 volume changes/hr (an average velocity of only 0.3 m/min or 1 ft/min) drops the maximum specific optical density in the NBS chamber for red oak from 395 to 125 under nonflaming exposure and from 75 to 5 under flaming conditions. Similar testing[67] of rigid PVC showed a decrease from 490 to 120 nonflaming and 530 at both conditions with flaming conditions. These results are in glaring contrast to the results obtained on similar materials by Bankston and presented in Figs. 3 and 4. Part of the reason for these differences is the design of Bankston's apparatus where smoke is measured in the effluent air stream while Gaskill measured the "residual" smoke opacity in the smoke density chamber itself. The more substantive difference is the fact that Bankston's apparatus does not permit any buildup of smoke between the radiant source and the sample while the standard NBS smoke chamber does. The most dramatic example of this problem is an unpublished comparison of seven rigid urethane foams in the NBS smoke chamber under nonflaming exposure at 2.5 W/cm^2 and 5.0 W/cm^2, a comparison similar to the 3.2 and 6.2 W/cm^2 comparison with urethane in Figs. 3 and 4. Without smoke buildup at the panel, an increase in smoke concentration of more than 500% was observed going from 3.2 to 6.2 W/cm^2. In terms of optical density per meter the slope of King's work[47] allows estimates at 3.2 W/cm^2, OD/m $= 1.4 \times 10^{-3}$, and at 6.2 W/cm^2, OD/m $= 9 \times 10^{-3}$. This is in glaring contrast to the results of a similar comparison conducted with the NBS smoke chamber which indicates only a 75% increase from OD/m $= 1.0$ at 2.5 W/cm^2 to OD/m $= 1.75$ at 5.0 W/cm^2. This difference is unacceptably large and most probably attributable to the reduction in energy transfer to the sample caused by smoke buildup in the chamber. It should be observed that the 2.5 W/cm^2 exposure level is well below the exposure intensity measured[58] in a crib fire of 6.4 W/cm^2. Unfortunately, it does not appear that simply increasing the output of the radiant panel can correct this deficiency in the nonventilated NBS chamber.

The second deficiency is common to both the E-84 tunnel and the NBS smoke chamber. Neither can effectively measure smoke generation from materials that melt and flow under high-temperature exposure. This problem has been well documented by Jacobs for the NBS Chamber, and his results are presented in Table 7.[68] The first two materials demonstrate that the addition of silica filler does not significantly affect smoke in the absence of melt flow (dripping). The other entries demonstrate that for smoke systems, i.e., polyethylene and nylon, the effect can be severe. This is a problem that must be recognized in testing thermoplastic systems in both smoke and flammability testing. In summary, then, there are some large differences between the standard conditions used in the NBS smoke chamber and large-scale fire conditions. The two most serious are the low heat flux and the accumulation feature of this method.

The final test method of significance will be mentioned briefly because it is a laboratory screening tool only. The technique was described in 1974[70] and consists of a 14-cm-diameter cylindrical combustion chamber with a 7.6-cm chimney and an air flow control system. A 3.8 × 1.2 × 0.3 cm sample is combusted for 30 sec with a micro burner flame fueled at 90 cc/min with propane. The smoke is entrained in a 55 m/min air flow (measured at the

TABLE 7
Effect of Dripping on Smoke Measurement[69]

Polymer	Unfilled		Filled with 10 pph of SiO$_2$	
	D_m flaming	Extent of dripping	D_m flaming	Extent of dripping
ABS	1060	None	1070	None
Aromatic polymer^a	735	None	650	None
Polypropylene	150	Severe	425	None
Polyethylene	65	Severe	235	None
Nylon 6	20	Severe	170	None
Polycarbonate	215	Severe	400	None
Polysulfone	235	Severe	200	None
Polystyrene	700	Severe	1030	None
EPDM rubber	130	Severe	360	None
Olefin rubber (thermoplastic)	100	Severe	125	Severe
Polyurethane (thermoplastic)	320	Severe	500	None

^{a}Uniroyal product trademarked Arylon.

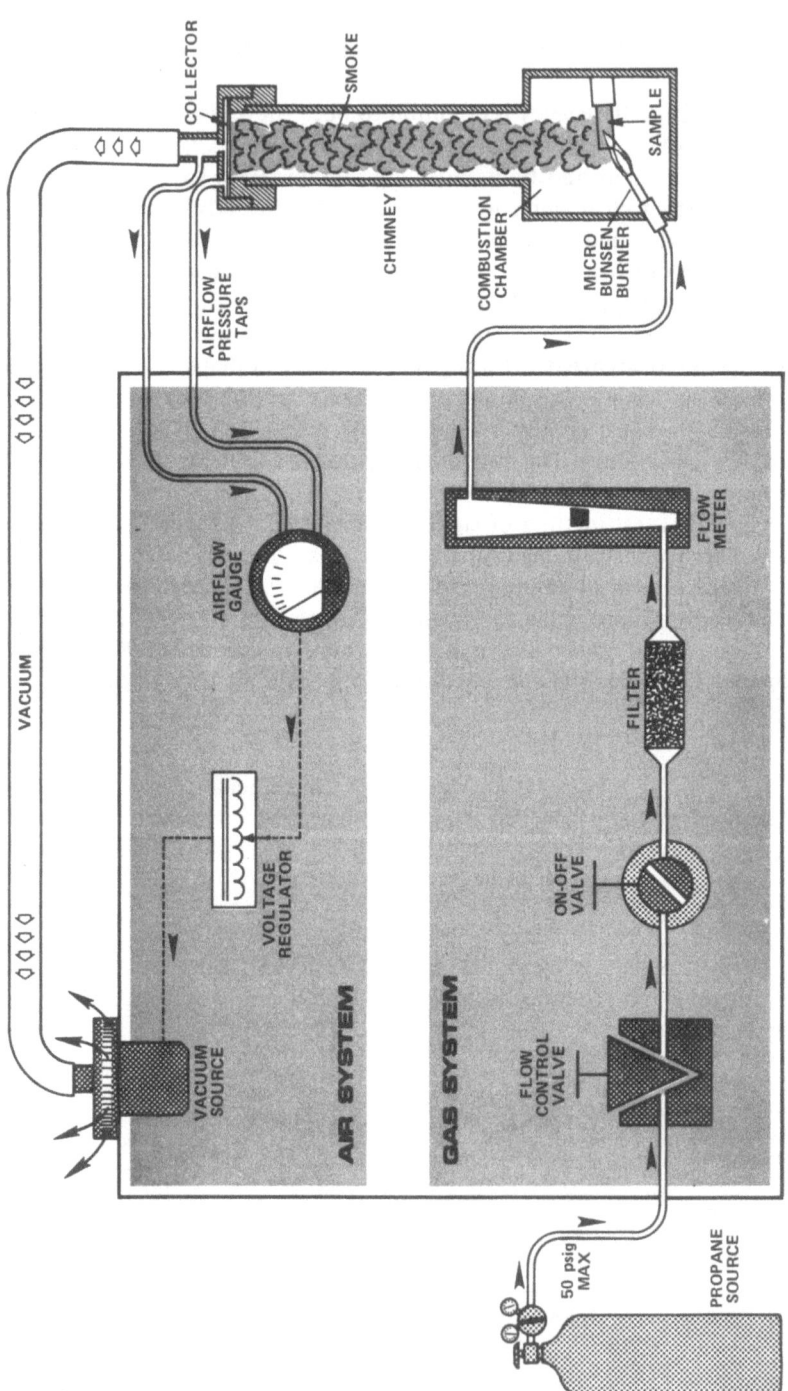

FIGURE 6. Conceptual schematic of Arapahoe smoke chamber.

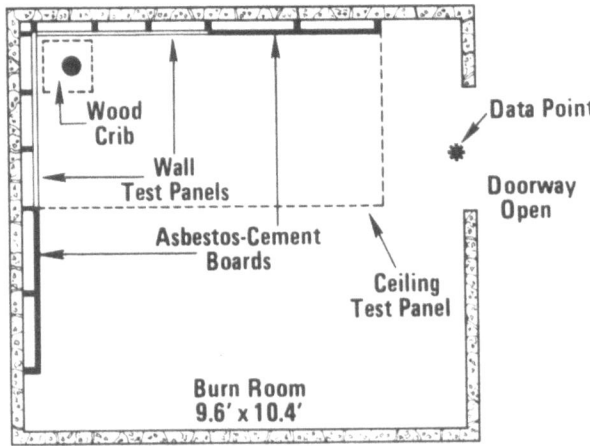

FIGURE 7. NBC compartment fire test.[33]

entrance to the combustion chamber) and collected on a fiberglass filter pad. The smoke collected during the 30-sec combustion is weighed and expressed as a percent of the material actually burned. An additional feature of this method, known as the Arapahoe smoke chamber (ASC), is the separate estimate of char (see Fig. 6). The primary reasons for including this instrument in this review are that it requires a small sample (less than 2 g), it is a flow method, it provides a gravimetric estimate of smoke generation, and it can be very simply adjusted to study the effects of air velocity and heat flux. It is essentially a microanalytical technique not suitable or recommended for measuring the performance of composite materials. It can, however, be a potent research tool in developing materials of reduced smoke generation. Two recent reports[71,72] have indicated the possibility of correlating the gravimetric results from the Arapahoe smoke chamber with both NBS smoke chamber and E-84 tunnel smoke results. Since neither of these correlations has been verified at this time, they will not be reviewed here. It is promising, however, that a small-scale technique based on gravimetric smoke measurement has shown some degree of correlation with the larger-scale techniques reviewed earlier.

5. Correlation: Studies of Small- and Large-Scale Smoke Tests

Historically, the first study pertinent to this subject was more of a fact-finding investigation than a correlation study. It was an extensive program conducted at Underwriters Laboratories[23] in the mid-1960's. The purpose of this study was to attempt to determine if there was a relationship between the smoke-developed rating from the E-84 tunnel test and the loss of visibility of exit signs in a large room. The test setup consisted of a 0.5 × 3.5 × 4.3 m (31⅓ × 22⅓ × 14 ft) room into which the effluent from the E-84 tunnel test was dispersed. The design of the facility forced mixing in the upper 30% of the room so that laminar flow of the smoke layer in a downward fashion was achieved. A series of six exit signs with 40-W bulbs were placed on the wall 3.5 m from viewing ports. The visibility criterion reported in Table 2 of 0.27 OD/m is based on three of the six exit signs still being visible with a measured transmission of 10%. A total of 39 materials was investigated in this study that covered a broad range of material types, flammability, and smoke ratings. The conclusions drawn in the report indicate that the 21 materials with smoke ratings below 200* provided good visibility for 6 min. The six materials in the 200* to 325* classification range provided good to marginal visibility for periods up to 6 min, and the 12 materials with smoke ratings above 325* produced ratings of fair to full obscuration. Inspection of the raw data reveals that of the 25 materials with smoke ratings of < 300* only 1 provided less than 8 min of time before reaching a 6% transmission (OD/m level of 0.33). These results suggest it may be reasonable to use smoke ratings from the E-84 tunnel as acceptance criteria for surface finish materials.

The first large-scale test to study the correlation between the small-scale test methods of the previous section and large-scale fires was a program conducted by IITRI in 1969 which studied the fire response of seven interior finish materials. The correlation study was reported in 1971.[73] The test facility was a 3 × 4.6 × 2.4 m (10 × 15 × 8 ft) room connected to a 1.8 × 2.4 × 15.2 m (6 × 8 × 50 ft) corridor. Three levels of heat input were used to simulate high intensity, low intensity, and closet fires. The finish materials (8.9 m² for each test) were separately exposed on the walls and ceilings of both the room and the corridor. The results of the smoke correlation study are presented in Table 8. This analysis indicates that the E-84 tunnel provided good correlation at 7 of the 12 conditions studied and fair correlation (which the author indicated would be usable) in 1 more. The NBS maximum optical density under flaming conditions provided good correlations in only 1 of the 12 conditions with 2 additional fair ratings. The nonflaming results provided only 1 fair correlation, while the average of flaming and nonflaming values showed 2 fair correlations. While this series of tests does not indicate acceptable breadth of correlation for either small-scale test, the E-84 tunnel clearly provides better correlation than the NBS smoke chamber with this extensive series of large-scale results. A second

---

*Numerical smoke index does not define the hazard presented by this material (or products made from this material) under actual fire conditions.

TABLE 8

Degree of Correlation of Small-Scale Smoke Tests with Large-Scale Results[a]

Material location/fire intensity	Time of comparison (t, min)	E-84 tunnel[b]	NBS chamber, D_m	
			Flaming	Nonflaming
Room ceiling/high intensity	4	Good	Fair	Poor
	10	Poor	(Average = fair)	
Room wall/high intensity	4	Poor	Poor	Poor
	10	Poor	Poor	Fair
Room wall/low intensity	4	Poor	(Average = fair)	
	10	Fair	Poor	Poor
Corridor ceiling/closet intensity	4	Good	Good	Poor
	10	Good	Poor	Poor
Corridor wall/high intensity	4	Good	Poor	Poor
	10	Good	Poor	Poor
Corridor wall/closet intensity	4	Good	Fair	Poor
	10	Good	Poor	Poor

[a]Data from Ref. 74.
[b]For E-84, the integrated optical density, not the smoke density classification, was used.

more limited program conducted by NBS[33] also afforded the opportunity to compare correlation of these two small-scale tests with large-scale fire tests.[75] The facility used in this series was a 2.9 × 3.2 × 2.4 m (9.5 × 10.5 × 7.9 ft) room with an open doorway measuring 0.9 × 2 m (35 × 80 in.)—see Fig. 7. Three 1.2 × 2.4 m (4 × 8 ft) panels of the materials to be tested were used to cover two walls at a corner and the ceiling immediately above the corner. Two sizes of wood cribs—33 and 6.3 kg (72 and 14 lb)—were used to initiate the tests. Previous tests in the same enclosure[76] had demonstrated that crib fires duplicated the time history and intensity of upholstered furniture fires. One unique feature of this more recent test series was a factorial design with duplication which allowed estimation of the variability of smoke measurement in this full-scale test. The standard deviation was found to be 0.08 OD/m[77] or 12%. In addition it was found that increasing the size of the crib by a factor of 5 (from 6.3 to 33 kg) only increased the smoke generated by 57% (from OD/m = 0.56 to 0.88).[77] Finally, a series of eight tests were conducted in this facility using surface finish materials which had been evaluated in both the E-84 tunnel and the NBS smoke chamber. Since both of these test methods report a smoke quantity, the smoke generated in the room test was measured at the doorway and summed based on the simultaneously measured effluent air flow rate. The correla-

TABLE 9
Smoke Test Correlation Analysis:
Laboratory Versus Full Scale[78]

Laboratory method	Correlation coefficient[a]
E-84 smoke index	0.81
NBS D_m flaming	0.50
NBS D_m nonflaming	0.66

[a]Correlation coefficient used is R^2.

tions between the two small-scale tests and the results of the large-scale
tests are presented in Table 9.

This analysis clearly demonstrates that the E-84 smoke index correlates to a much higher degree with the data from these seven materials in
room fire exposure. In perspective, then, it would appear that the sim-
ilarities to actual fire parameters in heat flux, air flow rate, and progressive
surface involvement which are design features of the E-84 tunnel are responsible for reasonable correlation with large-scale fire tests. It should
be observed that numerous individuals and groups have pointed out the
inadequacies of the 2-decade optical system and the area ratio approach
used for calculating E-84 smoke indices. In light of the basic design strengths
of the instrument, it seems worthwhile to improve the optical system and
calculation protocol. Conceivably, these modifications could improve the
already good correlation between the E-84 smoke index and large-scale fire
tests.

While considering the question of correlations and large-scale tests,
there is one final point which needs to be made. In the currently evolving
art or science of large-scale fire testing, a considerable amount of effort is
being invested in a test method known as corner testing. Corner tests have
the questionable advantage of being open (or two sides), which makes
dramatic photography and visual observation easier. The geometry is less
complex, which also simplifies analysis. One of the more cogent analyses
of this method compared the extensive corner tests conducted at Underwriters Laboratories[53] and the room corner program of Fang,[33] just reviewed above. This recent analysis[79] concludes that an open corner test
is a superior flammability test to an enclosed room test. This conclusion is
based on extrapolation of the E-84 FSC versus temperature correlation
which Fang based on 7 materials not including any foam plastics. The corner test analysis was based on 28 materials including 16 cellular plastics.
While Fang did not study cellular plastics in the room test, the UL program
did and found that all gave flashover when a 9.1-kg (20-lb) wood crib was

used for ignition.[80] Six materials were evaluated in both the corner using a 9.1-kg crib and the room using the same ignition source. Five of the 6 materials clearly failed the room exposure, with the sixth material, a thermoplastic foam, giving marginal performance in spite of the fact that 91% of the surface of the test room was affected during the test. These same 6 materials gave passing performance in the corner test, with a wood sample being the only failure (failure defined as full ceiling involvement and ceiling temperature greater than 500°C). The key deficiency in open corner tests—both configurations 2.4 and 7.6 m (8 and 25 ft) high—is the absence of a lintel to constrain the free convective egress of the smoke formed by the material under test. In addition to the primary importance of smoke in tenability concerns, smoke has been implicated as a major contributing factor in fire growth by two recent studies,[63,64] previously discussed. Consequently, it would appear that the serious researcher in the area of fire hazard analysis and control should restrict his focus to room enclosure tests or, at the very least, use a corner test with a substantial lintel on the open sides (2.4-m walls with a 0.91-m canopy on open sides), as has been recently studied in Canada.[82] In conclusion, then, the limited studies which allow comparison of small-scale tests with large-scale fire programs demonstrate that the E-84 tunnel test smoke index should be preferred over the NBS smoke chamber as a tool which has shown correlation with large-scale results.

6. Smoke Hazard Control: The Fire Detector Aspect

Early detection and warning are the most effective means of reducing the death toll in accidental fires. The strategy of requiring smoke detectors in residential occupancies is being widely studied and incorporated into codes and regulations. While this strategy is necessary and very worthwhile, it should not be considered a panacea which by itself will solve the problem. An interesting contrast of the behavior of smoke detectors in occupancy tests has appeared in two recent reports.[26,27] In the Factory Mutual Study, four fully furnished occupancy tenability fires were conducted with a total of eight detector–location combinations studied. All of these fires proceeded to a flaming combustion and flashover in less than 60 min and only three detector–location combinations (38%) resulted in adequate warning (2 min or more) before untenable conditions developed. The NBS tests were conducted with only 1 combustible item present for each test. A total of 40 tests was conducted, with 24 being smoldering ignitions. It was observed that most of the smoldering ignitions required more than 60 min to begin flaming and that some never began flaming.

Possibly as a result of these milder fire conditions, a much higher success ratio was found (in the 75 to 85% range).[82] One of the primary problems with smoke or fire detectors is the high ratio of false alarms to true alarms. The significance of this problem is highlighted by a report of an inspection of an apartment complex by a fire marshal 18 months after installation. A total of 186 detectors were tested, and 34 (18%) were found inoperative.[83] Subsequent inspection revealed that 31 of the 34 inoperative units had been deactivated by the tenants to silence the warning signal. This author of this article—a fire marshal—made the following statement:

> ... the Fire Marshals Association of North America ... voted to oppose the passage of the tentative amendment requiring smoke detectors in living units. I am certain that there is not a fire marshal in the United States who opposes the concept of providing "life safety" for people. What they are opposed to is a mandatory requirement for installation of a device that has no reliability data, provides no control over its installation, and has no means provided by which inspection or enforcement can be achieved.[84]

This suggests that while smoke detectors should form an integral part of the systems approach to the fire problem, at the current development of this art they cannot be relied upon as the role alternative for reducing fire losses. An example from an actual fire record provides the final emphasis on this point. A fire in a nursing home in 1976 resulted in eight fatalities— none in the room of origin—which were attributable to the heavy smoke conditions.[85] The occupancy was protected by four ionization-type smoke detectors on each floor, and one was located less than 4.5 m (15 ft) from the room of fire origin. The staff was alerted to the fire at approximately 6:30 a.m. by screams of an occupant of the fire room. The smoke detector sounded an alarm at 6:44 a.m., but an employee responding to that alarm found the fire so fully developed he could not extinguish it. It should be emphasized that there are numerous other examples in this same journal which demonstrate successful avoidance of a disastrous fire attributable to early detection due to fire detectors. This example is only intended to underscore the fact that detection does not represent the total solution to the fire loss problem.

7. Smoke Hazard: The Materials Perspective

There are two elements to the materials perspective on smoke generation during accidental combustion exposure. The first and most obvious is materials selection; that is, choose materials which provide equivalent performance in their intended application but generate significantly less smoke. This can best be illustrated by a specific example taken from one of the flammability studies[53] previously reviewed. The data in Table 10

TABLE 10
Materials Selection Based on Flammability *and* Smoke [a]

Sample code	Materials	E-84[b] ratings	
		Flame spread	Smoke developed
Z	Spray urethane	18	860
X	Spray urethane	21	1017
R	Foil-faced urethane boardstock	27	169
S	Same as R without foil	22	98
A	Isocyanurate boardstock	23	100
P	Urethane boardstock	23	343
B	Urethane boardstock	26	595
Y	Spray urethane	28	189
C	Urethane boardstock	28	318

[a]Data from Ref. 53.
[b]Numerical flame-spread and smoke-developed indices do not define the hazard presented by this material (or products made from this material) under actual fire conditions.

list numerous low-density rigid foams which have similar flammability characteristics as measured by the E-84 tunnel. All of these low-density foams are within 20% of the same density, which is in the range used for thermal insulation.[86] All passed the corner test. Four of the nine examples listed have smoke ratings well below the 300* maximum smoke developed which previous work (see Sec. 5) had indicated would provide 8 min of acceptable visibility, while the other five foams could be expected to exceed visibility limits in less than 6 min. While a 2- to 4-min time advantage may not sound dramatic, it becomes very significant when compared to the estimated 2-min egress time estimate. Consequently, this example has shown the possible significance of proper selection from among alternative materials which seem to offer the same functional value (insulation) in their intended application.

The second element to the materials perspective is probably more substantive but definitely more complex. It is the approach of material design or modification to achieve smoke reduction. The substance of this opportunity can be realized by consideration of the breadth of application and excellent blend of cost–performance provided by a common polymeric material such as polyvinyl chloride. This polymer enjoys a market in excess of 1×10^9 metric tons annually. It does generate large quantities of smoke when exposed to accidental fire, as was indicated in Table 6. Poly-

*Numerical smoke index does not define the hazard presented by this material (or products made from this material) under actual fire conditions.

TABLE 11
Smoke-Suppressant Efficacy in Vinyl Floor Tile[88]

Product	NBS smoke chamber			E-84 tunnel[a]	
	D_m flaming	D_m nonflaming	Avg. % change	Smoke index	% change
Vinyl floor tile					
Standard	260	320		611	
Smoke suppressed[b]	274	400	+16	214	−65
Conductive floor tile					
Standard	356	467		396	
Smoke suppressed[b]	267	480	−9	202	−49

[a] Numerical smoke index does not define the hazard presented by this material (or products made from this material) under actual fire conditions.
[b] Smoke suppressed with FE 55®, a trademark of Arapahoe Chemicals.

vinyl chloride is a polymer which responds very well to the highly developed art of plastics additive technology. One area of this technology which has been developing in recent years is that of smoke-suppressant additives.[87] Table 11 contains an example of the type of improvement which can be achieved in a typical rigid PVC surface finish material through the use of smoke-suppressant additives. As in the case of the data in the preceding table, a reduction in smoke index as measured by the E-84 tunnel to below 300* can be expected to have a significant impact on time available for egress. Unfortunately, however, similar improvement is not demonstrated in the results obtained with the NBS smoke chamber. This discrepancy between these two test methods is not surprising in light of the difference in heat flux between the two techniques mentioned in Sec. 4. Figures 3 and 4 clearly illustrate that smoke generation increases dramatically with heat flux, and the technology illustrated in Table 11 functions partially by an endothermic char catalytic mechanism. This energy flux sensitivity of smoke-suppressant efficacy was dramatically demonstrated in the NBS smoke chamber with a PVC compound. Two rigid PVC formulations, identical except for 0.2% FE 55 added to the smoke-suppressed formulation, were compared under standard flaming conditions, and the smoke-suppressed formulation provided a 20% reduction in D_m. The radiant furnace was then turned off and the gas flow to the burners tripled. The comparison of the two formulations then demonstrated a 66% reduction in D_m caused by the smoke suppressant. These examples are intended only as

*Numerical smoke index does not define the hazard presented by this material (or products made from this material) under actual fire conditions.

illustrative examples since the recent patent and technical literature contains many examples of smoke-suppressant additives proposed for use with many different polymer systems. Unfortunately, most of the data is obtained in small-scale laboratory tests which have not been shown to correlate with large-scale test results.

A second area which has shown some promise for the improvement of smoke generation characteristics of materials is the use of fillers.[65] Table 12 contains some illustrative data in unsaturated polyester resin systems.[89] This filler technology has been widely applied in both unsaturated polyester resins and styrene butadiene rubber used for carpet backing. It offers cost advantages over the smoke-suppressant additive and materials selection alternatives, but the high loadings required ($>$ 50%) limit its applicability and can have a dramatic negative impact on physical properties.

One final point needs to be made in the consideration of the materials aspect of smoke hazard control. Table 13 contains the results of a carpet flammability evaluation[90] comparing two alternative flame-retardant formulations.

These carpets were identical except for the flame-retardant formulations included in the polymeric backing applied. Small-scale laboratory results had indicated that the organic flame retardant would produce 50% less smoke than the inorganic system when evaluated as films of the backing alone. Fabrication of the final composite system and evaluation in the E-84 tunnel revealed only a 30% smoke reduction for the organic system. While selection of a low-smoke-producing flame-retardant system is a key materials

TABLE 12
Comparison of Flammability and Smoke from Unsaturated Polyester Systems[89]

System	Weight %	E-84 tunnel[a]		NBS smoke chamber, D_m flaming
		Flame spread	Smoke developed	
Polyester resin	75			
1-in. chopped fiberglass	25	280	1000	—
Polyester resin	40			
Hydrated alumina, unground	40	100	635	247
1-in. chopped fiberglass	20			
Polyester resin	32			
Hydrated alumina[b]	48	55	350	190
1-in. chopped fiberglass	20			

[a]Numerical flame-spread and smoke density indices do not define the hazard presented by this material (or products made from this material) under actual fire conditions.
[b]Treated hydrated alumina, Hyflex, trademark of Solem Industries.

TABLE 13
Comparison of the Flame-Retarded Carpet Systems

	E-84 tunnel[a] results	
System	Flame spread	Smoke index
Nylon carpet with inorganic flame retardant	67	1037
Nylon carpet with organic flame retardant	67	711

[a]Numerical flame-spread and smoke indices do not define the hazard presented by this material (or products made from this material) under actual fire conditions.

modification approach, it is extremely important to recognize that composite materials perform quite differently from noncomposite components. The unfortunate aspect of this problem is that small-scale laboratory tests such as the NBS smoke chamber and the Arapahoe smoke chamber are totally unsuitable for evaluation of composites. It is possible, of course, to run samples of composites and obtain numbers. The point is that the small size of the samples and relatively small heat source used are incapable of indicating the interactions of components in a composite which will occur in large-scale exposure. Consequently, the results obtained in small-scale tests will be at best meaningless and possibly misleading. An example of this problem is the performance of samples R and S in Table 10 when evaluated in a full-scale room test. Small-scale laboratory tests including the E-84 tunnel had indicated that the foil-covered sample (R) was slightly inferior in performance to the same identical material with the foil removed (S). Evaluation of both in a room with a 9.1-kg crib ignition revealed that 100% of the exposed foam (S) was damaged, while the foil-covered sample showed only 18% surface involvement. Since one of the dominant factors in smoke generation is the area of the sample involved in the fire, the 100% higher smoke produced by sample R in the E-84 test would be reduced by a factor of 5 when the area involved in large-scale exposure is taken into account. While it is easy to see how a superficial review of this complex area could lead one to the conclusion that smoke is not a material property but an artifact of combustion conditions,[91] in fact, smoke generation from surface finish materials in fire environments is a reproducible phenomenon capable of characterization and, to a significant extent, amenable to modification by a properly directed formulation program.

8. Smoke Hazard Assessment: Summary

Smoke has been shown to be one of the primary factors affecting tenability, which is the key life safety concern during accidental fires in the built environment. Ignition, the first step in the causative sequence, is seen as an aberration of normal performance in a usually reliable routine functioning of man and his environment. Since all organic materials will burn under sufficiently severe exposure and our twentieth-century life-style will not readily accept concrete chairs and steel homes, an alternative approach is necessary. Hazard analysis demonstrates that smoke is in fact the first critical and limiting condition reached in most full-scale experiments specifically designed to model occupancy fires. It has been found to precede both heat and toxicants by a substantial margin in the dynamic environment of a developing occupancy fire. Subsequent analysis compared small-scale tests to large-scale tests to assess the key factors affecting smoke generation from materials. These factors can be summarized as follows:

- Smoke generation by materials is a strong function of exposure intensity.
- The smoke generated during fire exposure is itself a strong contributor to exposure intensity.
- Materials orientation, configuration, and position relative to other combustibles are significant factors affecting smoke generation.
- Ventilation, forced by heating, ventilating, and air-conditioning systems or natural by convection, is a major factor to be considered.
- The smoke generated during an accidental fire creates a problem in tenability and egress at great distances from the room of fire origin, while the fire itself remains a localized problem until flashover.

Analysis of two popular small-scale tests relative to these factors led to the hypothesis that the E-84 tunnel test smoke index should correlate with large-scale test results. Subsequent review of the limited data available from smoke analysis of large-scale tests suggested the tunnel test is the appropriate method to use in selecting materials and improving performance in fire tests.

A brief review of smoke and fire detector performance led to the conclusion that these devices are a necessary adjunct to the systems approach to the smoke hazard problem, but they cannot and should not be considered the total solution to this problem at this time. Finally, consideration of the materials aspect of the smoke and tenability problem led to the following three conclusions:

- Selection of materials based on a reasonably supported criterion of smoke generation can improve the time to escape potential in accidental fire based on the best data available at this time.
- Modification of the formulation of polymeric materials as part of the product design effort and evaluated against criteria based on large-scale testing can result in significant improvements in test

response. The significance of these improvements in actual fires or large-scale tests has not yet been verified.

- Composites of two or more materials should be expected to perform differently from the individual components, and, at the current time, this cannot be predicted by laboratory tests but must be evaluated in full-size modeling experiments (room tests).

In conclusion, we can state that smoke and tenability do not comprise a precise science at this time. They are, however, quantifiable phenomena which are the subject of a rapidly developing area of technology. This subject cannot be simply characterized due to its complexity, but the current state of the art as presented does seem to hold promise for significant advances in the near future.

9. References

1. Fires and Fire Losses Classified 1975, *Fire J.* **70** (6), 17 (Nov. 1976).
2. Business Roundup, *Fortune*, **XCIV** (6), 16 (Dec. 1976).
3. D. Harlow, *Fire J.* **69** (6), 43 (Nov. 1974).
4. A. Phillips, *Fire J.* **70** (2), 11 (March 1976).
5. Fire and Fire Losses Classified 1975, *Fire J.* **70** (6), 19 (Nov. 1976).
6. America Burning, The Report of the National Commission on Fire Prevention and Control, U.S. Government Printing Office, Washington, D.C. (1973).
7. *Chem. Eng. News*, **51**, 9 (July 16, 1973).
8. F. Clarke and J. Ottoson, *Fire J.* **70** (3), 22 (1976).
9. P. Croce and H. Emmons, The Large-Scale Bedroom Fire Test, July 11, 1973, Factory Mutual Serial No. 21011.4, RC74-T-31, Factory Mutual Research Corporation, Norwood, Mass. (July 1974).
10. P. Croce, A Study of Room Fire Development: The Second Full-Scale Bedroom Fire Test of the Home Fire Project, July 24, 1974, Factory Mutual Serial No. 21011.4, RC74-T-31, Factory Mutual Research Corporation, Norwood, Mass. (June 1975).
11. See Ref. 8, p. 117.
12. See Ref. 10, p. 50.
13. See Ref. 9, p. 49.
14. See Ref. 8, p. 117.
15. See Ref. 8, p. 22.
16. D. Byrd and J. Burgess, *Fire J.* **70** (3), 11 (May 1976).
17. W. Christian, *Fire J.* **68** (1), 22 (Jan. 1974).
18. California Assembly Bill No. 1522, Chapter 749, 1972, effective Oct. 1, 1975.
19. N. Kroepfler, *Fire J.* **70** (5), 53 (Sept. 1976).
20. B. Halpin, R. Fisher, and Y. Caplan, Fire Fatality Study, paper presented at International Symposium on Toxicity and Physiology of Combustion Products, University of Utah, Salt Lake City, March 22–26, 1976, p. 11.
21. See Ref. 20, p. 10.
22. Los Angeles Fire Department, Operation School Burning, No. 2, National Fire Protection Association Boston (1961).
23. J. Bono and B. Breed, *Fire Technol.* **2**, 146 (1966).

24. D. Gross, J. Loftus, and A. Robertson, Method for Measuring Smoke from Burning Materials, ASTM Std. No. 422 (1967), pp. 166–204.
25. T. Wakamatsu, Calculation of Smoke Movement in Buildings, Research Paper No. 34, Japan Building Research Institute (Aug. 1968).
26. G. Heskestad, Escape Potentials from Apartments Protected by Fire Detectors in High Rise Buildings, NTIS No. PB 234014, Factory Mutual Research Corporation, Norwood, Mass. (June 1974), p. 117.
27. R. Bukowski, T. Waterman, and W. Christian, Detector Sensitivity and Siting Requirements for Dwellings, National Bureau of Standards Report NBB-GCR-75-51, NTIS No. PB 247483.
28. E. Lopez, Smoke Emission from Burning Cabin Materials and the Effect on Visibility in Wide Bodied Jet Transports, Report No FAA-RD-73-127, NTIS No. AD 776963, Lockheed Aircraft Corporation for FAA, Atlantic City, New Jersey (March 1974).
29. See Ref. 22, p. 34.
30. See Ref. 28, App. IV, Figs. 3 and 6.
31. J. Kracklauer and C. Sparkes, A Critical Review of Smoke Hazard Criteria for Construction Products, paper presented at Plastics in Building Conference, SPE RETEC, Boston, Mass. (Nov. 9, 1976).
32. R. Montgomery, C. Reinhart, and J. Terrill, Comments on Fire Toxicity, paper presented at the Polymer Conference Series at the University of Utah, Salt Lake City, (July 11, 1974), Table VIII, p. 41.
33. J. Fang, Fire Buildup in a Room and the Role of Interior Finish Materials, National Bureau of Standards Technical Note 879, National Bureau of Standards, Washington, D.C. (June 1975).
34. See Ref. 26, p. 94.
35. See Ref. 26, p. 108.
36. See Ref. 26, p. 36.
37. NFPA Fire Analysis Department, *Fire J.* **70** (1), 43 (Jan. 1976).
38. See Ref. 37, p. 45.
39. National Fire Protection Association, Boston, 74-1974, Standard for the Installation, Maintenance and Use of Household Fire Warning Equipment (1974), Secs. 2-4.3.2–2-4.3.5, p. 74-11.
40. ASTM Committee E5 on Fire Tests of Materials and Construction Task Group, Task Group of Subcommittee 4, *Mater. Res. Stand.* **11** (4), 16 (April 1971).
41. Federal Trade Commission Consent Order Docket C 2596, 84FTC1253 (Nov. 4, 1974), p. 1272.
42. Y. Tsuchiya and K. Sumi, *J. Fire Flammability* **5**, 68 (Jan. 1974).
43. W. Foster, *Br. J. Appl. Phys.* **10**, 416 (Sept. 1959).
44. J. Kracklauer, C. Sparkes, and R. Legg, *Plast. Technol.* 46 (March 1976).
45. C. Bankston, R. Cassanova, E. Powell, and B. Zinn, *J. Fire Flammability* **7** (April 1976).
46. I. Einhorn, M. Birky, M. Grunnet, S. Packham, J. Petajar, and J. Seader, The Physiological and Toxicological Aspects of Smoke Produced During the Combustion of Polymer Materials, Rann Program GI 33650 Summary Report, NTIS, Springfield, Va. (1975).
47. T. King, *J. Fire Flammability* **6**, 222 (April 1975).
48. J. Seader and S. Ou, *Fire Res.* **1**, 3 (1977).
49. See Ref. 47, p. 226.
50. ASTM E5-02 Task Group on Smoke Test Methods, *ASTM Standardization News* 18 (Aug. 1976).
51. ASTM E-84-75, Test for Surface Burning Characteristics of Building Materials, *Annual Book of ASTM Standards*, Part 18, American Society for Testing and Materials, Philadelphia (1975).

51a. T. Lee and C. Huggett, Interlaboratory Evaluation of the Tunnel Test (ASTM E-84) Applied to Floor Coverings, NBSIR 73-125, National Bureau of Standards, Washington, D.C. (March 1973), p. 48.

52. J. Bono and B. Breed, Study of Smoke Ratings Developed in Standard Fire Tests in Relation to Visual Observations, Bulletin of Research Number 56, Underwriters Laboratories, Inc., Northbrook, Illinois (April 1965), p. 27.

53. G. Castino, J. Beyreis, and W. Metes, Flammability Studies of Cellular Plastics and Other Building Materials Used for Interior Finishes, Subject 723, Underwriters Laboratories, Inc., Northbrook, Illinois (June 13, 1975), p. 46.

54. See Ref. 50, p. 23.

55. C. Bankston, R. Cassanova, E. Powell, and B. Zinn, Properties of Smoke Produced by Burning Wood, Urethane and PVC Samples under Different Conditions, Research Report of NSF/Rann Grant No. ERT73-03168A01, School of Aerospace Engineering, Georgia Institute of Technology, Atlanta (Dec. 1975).

56. See Ref. 42, p. 71.

57. See Ref. 53, p. 47.

58. See Ref. 53, p. 48.

59. See Ref. 33, p. 18.

60. ASTM STP 422, Symposium on File Test Methods, American Society for Testing and Materials, Philadelphia (1966).

61. T. G. Lee, Interlaboratory Evaluation of Smoke Density Chamber, NBS Technical Note 708, National Bureau of Standards, Washington, D.C. (Dec. 1971).

62. See Ref. 50, p. 24.

63. See Ref. 10, p. ii.

64. T. Palmer, *J. Fire Flammability* 7, 469 (Oct. 1976).

65. R. Cassanova, E. Powell, C. Bankston, and B. Zinn, The Effects of PVC Additives on the Properties of Smoke Produced During Nonflaming Combustion, paper no. 76-43 presented at the Western States Section of the Combustion Institute Meeting October 18–19, 1976, San Diego, p. 11.

66. J. Gaskill, *J. Fire Flammability* 1, 191 (July 1970).

67. See Ref. 66, p. 196.

68. M. Jacobs, *J. Fire Flammability* 6, 347 (July 1975).

69. See Ref. 68, p. 349.

70. J. Kracklauer, C. Sparkes, and R. Legg, A New Research Tool for Smoke Characterization, paper presented at the 32nd Annual Technical Conference of the Society of Plastics Engineers, San Francisco, Calif. (May 13–16, 1974).

71. J. Seader and S. Ou, *Fire Res.* 3, in press (1977).

72. Arapahoe Smoke Chamber/E-84 Tunnel Correlation for Rigid Urethane Foams, Arapahoe Smoke Chamber Technical Service Report TSR 101 V12, Boulder, Colorado (Aug. 1976).

73. W. Christian and T. Waterman, *Fire Technol.* 7 (4), 332 (Nov. 1971).

74. See Ref. 73, p. 339.

75. See Ref. 33, p. 29.

76. J. Fang, Measurements of the Behavior of Incidental Fires in Compartments, NBSIR 75-679, National Bureau of Standards, Washington, D.C. (March 1975).

77. See Ref. 33, p. 42.

78. See Ref. 31, p. 7.

79. W. Christian, G. Castino, and J. Beyreis, *J. Fire Flammability* 8, 41 (Jan. 1977).

80. See Ref. 53, p. 57.

81. M. D'Souza and J. McGuire, ASTM E-84 and the Flammability Thermosetting Plastics, *Fire Technol.* 13 (2), 85 (May 1977).

82. E. Gallagher, *Fire J.* **71** (2), 39 (March 1977).
83. H. Boyd, *Fire J.* **71** (1), 83 (Jan. 1977).
84. See Ref. 83, p. 84.
85. The Cermak House Fire, *Fire J.* **70** (5), 18 (Sept. 1976).
86. See Ref. 53, pp. 8 and 9.
87. J. Kracklauer and C. Sparkes, *Plast. Eng.* **30** (6), 57 (June 1974).
88. See Ref. 31, p. 5.
89. J. Keating, Flame and Smoke Management in Polyester Resin Systems, paper presented at Technical Proceedings, 32nd Annual Conference SPI Reinforced Plastics/Composites Institute, February 8–11, 1977, Washington D.C., Sec. 13F, p. 3.
90. Arapahoe Chemicals, unpublished data (1976).
91. E. Weil and A. Aaronson, Phosphorus Flame Retardants—Some Effects on Smoke and Combustion Products, paper presented at SPI Cellular Plastics Division Meeting, Montreal, Canada (Nov. 18, 1976), p. 14.

Index